Study Guide

BIOLOGY
Life on Earth
THIRD EDITION

Gerald Audesirk

Teresa Audesirk

Prepared by
Joseph P. Chinnici
Virginia Commonwealth University

MACMILLAN PUBLISHING COMPANY
New York

MAXWELL MACMILLAN CANADA, INC.
Toronto

MAXWELL MACMILLAN INTERNATIONAL
New York   Oxford   Singapore   Sydney

Macmillan Publishing Company
866 Third Avenue, New York, New York 10022

Maxwell Macmillan Canada, Inc.
1200 Eglinton Avenue East
Suite 200
Don Mills, Ontario M3C 3N1

Printing: 2 3 4 5 6 7   Year: 3 4 5 6 7 8

ISBN   0-02-304831-X

ACKNOWLEDGEMENTS

My thanks go to the following individuals who were instrumental in bringing this project to reality. Bob Rogers and Kristin Watts, from Macmillan Publishing Company, offered me the chance to review early drafts of the new Audesirk and Auderisk text and then encouraged me to gather the team to write the Study Guide, Instructors Manual, Test Bank, and Laboratory Guide. Don Fritsch, Margaret May, J. Lewis Payne, and Gail Turner, from the Biology Department of Virginia Commonwealth University, each contributed to one or more of these books, and the five of us worked closely together to create these mutually compatible texts. Most importantly, Dianna Terlizzi Chinnici spent many hours reviewing the contents and accuracy of this text. It is a much better companion to the Audesirks' text due to her efforts. Of course, any remaining errors are solely my responsibility.

Also, I wish to acknowledge David J. Cotter, of Georgia College. Many of the Unit Examination questions and the Chapter 11 genetics problems, as well as several figures, are adapted from the Study Guide for the first edition of the Audesirks' text, of which he is the author.

TO THE STUDENT - PLEASE READ

This study guide should be used with the textbook <u>Biology</u>: <u>Life</u> <u>on</u> <u>Earth</u>, third edition, by G. and T. Audesirk. To learn the most from your general biology course (and to earn the highest possible grade), you need to take advantage of the three interrelated elements of the course: the lectures, the textbook, and the study guide.

1. **Attend all lectures**, be attentive in class, and take good and complete notes. On the evening after each lecture, recopy the notes into a more readable and organized form, relying on the textbook and study guide for help in understanding confusing and difficult topics and for filling in "gaps" in the notes. You'll be pleasantly surprised how much you will learn and remember by simply rewriting and augmenting classroom notes.

2. **Read the textbook chapters <u>before</u> they are covered in lecture**. This will make the lecture more comprehensible. If you understand the lecture better, it will be less frustrating and more fun to go to class and your confidence level about biology will rise.

3. After reading the textbook and attending lecture, **use this study guide to prepare for the inevitable exams** that go with college life. This study guide has a detailed outline of each chapter in the textbook, a set of review questions with answers (or directions to find the answers), and unit exams.

FORMAT OF THE STUDY GUIDE.

This guide is divided into 46 chapters and six unit exams, corresponding to the organization of the textbook. Each chapter has a similar format, consisting of:

1. A detailed **outline** of the material in the textbook, organized to allow you to easily find answers to the review questions. All important concepts from the textbook are included in the outline, along with some helpful hints for remembering the material.

2. A set of **review questions** in a variety of formats. After reading the textbook, attending the lecture, rewriting your lecture notes, reviewing the study guide outline, and studying, you should attempt to answer these review questions.

3. **Answers to the review questions**. Refer to these <u>after</u> answering the review questions. Looking at the answers before trying to answer them yourself can mislead you into thinking you know the answers when you may not. Use the answers provided in this guide to <u>check</u> your answers for accuracy, not to generate your answers.

After each unit of material, take the **unit exam**. It consists of about 10 dichotomous (either/or) questions from each chapter in the unit. After completing each unit exam, check your answers against the **answer key**. For incorrect answers, review your notes and the study guide and study the material again until you really understand it.

HINTS FOR DOING WELL ON OBJECTIVE-TYPE LECTURE EXAMS.

Nowadays, most colleges and universities have large sections of general biology, and professors determine grades by giving computer-graded "objective" examinations consisting of multiple choice, matching, and true-false questions. If this describes your course, the following hints may help improve your grades, even if you don't "test well" on objective exams.

1. **Learn the meaning of biological terms**. A glossary of key words appears after each chapter in the textbook. Learn the meaning of these words - they are often asked on objective exams.

2. **Answer questions you know first**. If you don't know the answer to a question, skip over it and come back to it later. Sometimes you will get a clue to the correct answer from other questions on the exam.

3. **Read all the choices** for multiple choice questions. Then, **try to eliminate** as many obviously incorrect choices as possible, and pick the choice that appears most correct from those remaining. If your professor says there is no penalty for guessing, **answer every question**.

4. Most biology professors have had no training in constructing objective exams, and often fall into subconscious patterns of asking questions. Psychologists have noted the following trends on multiple choice exams, some of which may apply to your situation.

   A. A choice that is much longer or much shorter than the others tends to be the correct answer. Examples (c. is the correct answer in each case):

      America was discovered by:

      a.   the Indians
      b.   either an Italian sailing for Spain or the Vikings
      c.   the Ukrainians
      d.   the Chinese

Biology is the study of:

a.   plants, chemicals, rocks, and stars
b.   cells and protoplasm
c.   life
d.   the creation of life by God as told in the Book of Genesis

B.   If two choices are exactly the same except for one word, one of these choices is usually correct. Example (c. is the correct choice):

An acidic solution:

a.   is pure water
b.   has a pH greater than 7
c.   has a pH less than 7
d.   cannot be produced in the human body

C.   If you are sure that two choices are correct and another choice is "all of the above", that is usually the correct answer.

D.   If part of a statement is false, the entire statement is false. Example (False is correct):

True or False: Humans, dogs, cats, mice, and lizards are mammals.

E.   Statements with absolutes such as always or never are usually false, and those with usually, or seldom are more likely to be true.

ESSAY EXAMS.

On **essay exams,** the more time you spend forming an outline, the more precise your answer will be and the less likely it is that you will leave out important points. Begin with a clear, concise introductory statement summarizing your answer. The rest of the answer should take the "bikini approach": big enough to cover the material but small enough to keep it interesting.

Table of Contents

CHAPTER 1: AN INTRODUCTION TO LIFE ON EARTH.

CHAPTER OUTLINE.

This chapter discusses the characteristics of life and gives a brief overview of the vast diversity of living organisms. Basic scientific principles are introduced as well as the scientific method. Finally, a brief introduction to the mechanism and evidence for evolution is presented.

1.   The earth contains an astonishing array of diverse living organisms.

2.   **Characteristics** of living things:

      A.   They are <u>complex</u> and <u>highly</u> <u>organized</u>, being made of:
          1)   Atoms organized into
          2)   Molecules, which are organized into
          3)   Organelles (subcellular structures), which are organized into
          4)   Cells, tissues, organs, and systems, which are organized into
          5)   Organisms, which are organized into
          6)   Populations and communities.

      B.   They actively maintain their complex structure (<u>homeostasis</u> = "to stay the same"): energy is constantly needed to maintain complex, highly organized structures.

      C.   They <u>grow</u>, increasing in size and complexity due to conversion of molecules acquired from the environment into specific molecules of the organism's body.

      D.   They <u>acquire</u> and <u>convert</u> <u>nutrient</u> <u>molecules</u> using energy which originates from the sun.

      E.   They <u>perceive</u> <u>and</u> <u>respond</u> <u>to</u> <u>environmental</u> <u>stimuli</u>: All organisms, from bacteria to plants to humans, modify their behavior in response to changes in the outside world.

    F.   They <u>reproduce</u>: before an organism dies, it can perpetuate its genes (hereditary molecules of DNA) by producing offspring. Rarely, the DNA mutates (changes due to replication errors) causing changes in the offspring receiving it.

    G.   Their populations can <u>evolve</u>: due to slight variations among the offspring produced by an organism, evolution occurs as organisms with useful variations (called <u>adaptations</u>) produce more offspring than organisms with less useful variations.

3.   A brief introduction to the vast **diversity** of living organisms.

    A.   Biologists have grouped organisms into five categories called <u>Kingdoms</u>, based on three characteristics:
          1)   Types of cells in the organism.
          2)   Number of cells in the organism.
          3)   How the organisms acquire energy.

    B.   <u>Kingdom Monera</u> (bacteria and cyanobacteria).
          1)   The most primitive Kingdom since the organisms are:
               a.   <u>Unicellular</u> (entire body is one cell).
               b.   <u>Prokaryotic</u> (cells lack a membrane-bound or true nucleus, are small in size, and lack most cellular organelles).
          2)   Some monerans are <u>autotrophs</u> ("self-feeders"), capturing energy from the sun and using it to make food, a process called <u>photosynthesis</u> (using photons of light energy to make food molecules).
          3)   Some monerans are <u>heterotrophs</u> ("other-feeders"), incapable of photosynthesis. These get food by absorbing the dead remains of other organisms.

    C.   <u>Kingdom Protista</u>.
          1)   These organisms are:
               a.   Unicellular.
               b.   <u>Eukaryotic</u> (cells have a membrane-bound or true nucleus, are larger in size, and have cellular organelles).
          2)   Some are autotrophic (<u>Unicellular Algae</u>).
          3)   Others are heterotrophic, either:
               a.   Absorbing decaying food molecules (<u>absorption</u>), or
               b.   Ingesting chunks of undigested food (<u>ingestion</u>).

    D.   <u>Kingdom Fungi</u>.
          1)   These organisms are mostly:
               a.   <u>Multicellular</u> (body having many cells).
               b.   Eukaryotic.
           2)   Fungi are heterotrophic and absorptive, secreting digestive enzymes out of their bodies to decompose dead plant and animal bodies.

    E.   <u>Kingdom Plantae</u>.
          1)   Plants are multicellular and eukaryotic.
          2)   Plants are autotrophic.
          3)   Some types (or Divisions) of plants:
               a.   <u>Multicellular Algae</u>: primitive plants that live in water.

      b.   <u>Mosses</u>: most primitive land plants, having neither true roots nor <u>vascular</u> <u>tissue</u> (cellular tubes transporting water and nutrients throughout the body).

      c.   <u>Vascular</u> <u>Plants</u>: the <u>ferns</u>, <u>conifers</u> (cone-bearing plants), and <u>flowering plants</u>. They possess vascular tissue for transport of molecules between roots, stems, and leaves.

F.   <u>Kingdom Animalia</u>.

1)   Animals are multicellular and eukaryotic.

2)   Animals are heterotrophic and mostly ingestive, eating chunks of undigested food.

3)   Some types (or Phyla) of animals.

      a.   <u>Sponges</u>: simplest animals; aquatic; pass water through pores in their body walls to filter out minute plants and animals for food.

      b.   <u>Coelenterates</u>: aquatic; includes the jellyfish, corals and sea anemones; catch food in their stinging tentacles; have a common opening for both the mouth and anus.

      c.   <u>Annelids</u>: segmented worms (earthworms, leeches and tube worms); live in moist or wet environments; digestive tract has separate mouth and anal openings for more efficient digestion.

      d.   <u>Molluscs</u>: includes the snails, clams, scallops, and octopuses; must have a damp or wet environment.

      e.   <u>Arthropods</u>: most diverse group of animals, including insects, spiders, centipedes, millipedes, crabs, shrimp, and lobsters; have waterproof external skeleton ("<u>exoskeleton</u>") with jointed appendages; most successful of the animals, living on land and in both fresh and salt water habitats.

      f.   <u>Echinoderms</u>: "spiny skinned" marine (salt water) animals including starfish, sand dollars, sea urchins, and sea cucumbers; have a spiny internal skeleton ("<u>endoskeleton</u>").

      g.   <u>Chordates</u>: includes the vertebrates (fish, amphibians, reptiles, birds, mammals); fresh and salt water, and terrestrial; most have a bony endoskeleton covered with muscle and skin.

4.   **Scientific principles**: a set of unproven assumptions essential to biology. These are:

A.   <u>Natural</u> <u>Causality</u>: all earthly events can be traced to preceding natural causes. Supernatural intervention has no place in science.

B.   <u>Uniformity</u> <u>in</u> <u>Time</u> <u>and</u> <u>Space</u>: natural laws do not change with time or distance. Example: scientists assume that gravity always has worked as it does today and works the same way everywhere in the universe.

C.   <u>Common</u> <u>Perception</u>: assumes all humans individually perceive normal and aesthetic events through their senses in fundamentally the same way. However, our interpretation of such events may differ. Example: appreciation of rock music or the morality of abortion.

5.   **Scientific method**: how scientists study the workings of life. It consists of four interrelated

operations.

A.   <u>Observation</u>: the beginnings of scientific inquiry. Example: Maggots appear on fresh meat left uncovered.

B.   <u>Hypothesis</u>: a tentative testable explanation of an observed event based on an educated guess about its cause. Example: Maggots appear on fresh meat left uncovered because flies land on the meat and lay eggs.

C.   <u>Experiment</u>: a study done under rigidly controlled conditions based on a prediction stemming from the hypothesis. Example: If the fresh meat is covered with a fine gauze to keep the flies away, no maggots should appear in the meat.

D.   <u>Conclusion</u>: a judgement about the validity of the hypothesis, based on the results of the experiment.
1)   Example: Maggots did not appear in the meat covered with gauze but did appear on fresh meat left uncovered in the same place at the same time. Therefore, the hypothesis is supported by the results of the experiment.
2)   When a hypothesis is supported by the results of many different kinds of experiments, scientists are confident enough about its validity to call it a theory or a law.
3)   Scientific conclusions must always remain tentative and be subject to revision if new observations or experiments demand it.

6.   **Real science**.

A.   Besides the Scientific Method, real scientific advances often involve accidents and acumen, lucky guesses, controversies between scientists and the unusual insights of brilliant scientists. Two examples of "real science:"
1)   The discovery of penicillin by Alexander Fleming.
2)   The study of color vision in bees by Karl von Frisch.

B.   Often, scientific advances occur by a process of <u>successive disproofs</u>.
1)   Various alternative explanations are disproven in a series of experiments, leaving only one possible correct hypothesis.
2)   But, unless scientists are absolutely sure <u>all</u> possible alternatives have been disproven, they cannot be certain that the remaining hypothesis is the only correct explanation.

7.   **Evolution**: the theory that present day organisms descended, with modification, from pre-existing forms; the unifying concept in biology.

A.   <u>Mechanisms</u>. As proposed by the English naturalists Charles Darwin and Alfred Russell Wallace in the 1850s, evolution occurs due to three natural processes:
1)   <u>Genetic</u> <u>variation</u> among members of a population: these ultimately arise from <u>gene</u> <u>mutation</u> (alterations in the structure of a gene).
2)   <u>Inheritance</u> of variations from parents to offspring.
3)   <u>Natural</u> <u>Selection</u>: survival and enhanced reproduction of organisms with favorable variations (adaptations) that best meet the challenges of the

environment.
- a. Over time, organisms with the best adaptations (variations in structure, physiology, or behavior that aid survival and reproduction) will replace those with less favorable variations.
- b. Natural selection preserves genes that help organisms flourish and discards the rest.
- c. Ultimately, natural selection has unpredictable results because environments tend to change dramatically (e.g., ice ages). What helps organisms survive today may be a liability tomorrow.

4) Thus, over thousands of generations, the interplay of environment, variation, and natural selection result in evolution (modification of the genetic makeup of a species). Species that do not adapt to changing environments become extinct.

B. Evidence for evolution is threefold:
1) Fossil records show gradual changes in organisms over millions of years. Example: fossil horses.
2) Related modern organisms show remarkable similarities in physiology and structure. Example: limbs of birds, bats, dogs, sheep, seals, and porpoises indicate recent common ancestry.
3) Effects of natural selection can be observed at the present time. Example: evolution of pesticide resistance among insects.

****

REVIEW QUESTIONS:

1.-5.   TRUE or FALSE?

1.   The earth contains only a few diverse types of organisms.
2.   Since all living organisms are made of cells, they all have a simple body structure.
3.   Molecules are organized into atoms.
4.   Cells exhibit homeostasis since they actively maintain their structure.
5.   Cells give off energy as they acquire and convert nutrient molecules.
6.   Which of the following are not characteristics used to categorize organisms into Kingdoms?
- A.   Types of cells present.
- B.   Numbers of cells present.
- C.   Presence or absence of cell walls.
- D.   How the organisms acquire energy.
- E.   How the organisms move.

7.-16.   **MATCHING TEST:** the five Kingdoms

Choices: A.   Monera   D.   Plantae
           B.   Protista   E.   Animalia
           C.   Fungi

7.   Unicellular algae.
8.   Multicellular and autotrophic.

9. Unicellular and prokaryotic.
10. Multicellular, heterotrophic, mostly ingestive.
11. Multicellular algae.
12. Unicellular and eukaryotic.
13. Multicellular, heterotrophic, mostly absorptive.
14. Sponges.
15. Mosses.
16. Bacteria.

17.-28.   **MATCHING TEST:** types of animals

Choices: A. Sponges   E. Arthropods
           B. Coelenterates F. Echinoderms
           C. Annelid worms G. Chordates
           D. Molluscs

17. Snails and octopuses.
18. Segmented worms.
19. Have a spiny internal skeleton.
20. Pass water through pores in the body wall.
21. Have a bony endoskeleton covered with skin and muscle.
22. Have a common opening serving as both mouth and anus.
23. Have waterproof external skeleton with jointed appendages.
24. Jellyfish.
25. Fish.
26. Spiders.
27. Leeches.
28. Starfish.

----

29. Scientific principles are _____ essential to biology.
    A. proven facts
    B. unproven assumptions

30.-35.   **MATCHING TEST:** Which scientific principles do each of the following violate?

Choices: A. Natural Causality
           B. Uniformity in Time and Space
           C. Common Perception

30. In biblical times, humans lived to be 900 years old.
31. God created all life forms on earth in six days.
32. A 6 foot tall man may look 10 feet tall to you.
33. Gravity did not affect the dinosaurs as much as it does us.
34. Miracles.
35. An anorexic woman sees herself as being fat.

36.	Define the four interrelated operations which make up the "Scientific Method".
37.	State the three natural processes which result in evolution, according to Darwin and Wallace.
38.	Which of the following are <u>not</u> evidence for the occurrence of evolution?
	A.	Similarities between related modern organisms.
	B.	Variations seen among white humans living in the US.
	C.	The fossil record shows gradual change over time in the structure of organisms.
	D.	Mice can be taught to run through mazes without making mistakes.
	E.	Development of pesticide resistance in insects.

****

ANSWERS TO REVIEW QUESTIONS:

1.	False; [see section 1.
	in the outline]

2.	False; [see 2.A.]

3.	False; [see 2.A.1)-2)]

4.	True; [see 2.B.]

5.	False; [see 2.D.]

6.	C and E; [see 3.A.]

7.-16.	[see 3.B.-F.]

7.	B	12.	B
8.	D	13.	C
9.	A	14.	E
10.	E	15.	D
11.	D	16.	A

17.-28.	[see 3.F.3)a.-g.]

17.	D	23.	E
18.	C	24.	B
19.	F	25.	G
20.	A	26.	E
21.	G	27.	C
22.	B	28.	F

29.	B; [see 4.]

30.-35.	[see 4.A.-C.]

30.	B	33.	B
31.	A	34.	A
32.	C	35.	C

36.	see 5.A.-D.

37.	see 7.A.

38.	B and D; [see 7.B.]

****

UNIT I: THE LIFE OF THE CELL

CHAPTER 2:   INORGANIC CHEMISTRY

CHAPTER OUTLINE.

This chapter is concerned with matter and energy. The structure of atoms and molecules is covered as well as the three major types of chemical bonding. Also, important inorganic molecules, especially water, are discussed.

1.   **Matter and energy**.

    A.   <u>Matter</u> is the physical material of the universe.

    B.   <u>Energy</u> is the capacity to do work such as moving pieces of matter around.
        1)   Forms of energy:
            a.   <u>Kinetic</u> energy: energy of movement, including electrical energy (movement of electrons in a wire), and heat energy (movement of atoms).
            b.   <u>Potential</u> energy: "stored" energy which can become kinetic.
        2)   Transfer of energy regulates all interactions among bits of matter.

2.   **Structure of matter**: 92 naturally occurring <u>elements</u> which can be combined with each other to make chemical compounds.

    A.   An <u>element</u> is a substance, with specific properties, that cannot be broken down or converted into another substance under ordinary conditions.

    B.   <u>Common elements</u>.
        1)   In living organisms, the more important elements can be remembered by using the phrase "<u>C</u> HOPKINS <u>CaFe</u>, <u>Mighty Good</u> <u>salt</u>" for the elements Carbon (<u>C</u>), Hydrogen (<u>H</u>), Oxygen (<u>O</u>), Phosphorus (<u>P</u>), Potassium (<u>K</u>), Iodine (<u>I</u>), Nitrogen

(N), Sulfur (S), Calcium (Ca), Iron (Fe), Magnesium (Mg for "Mighty Good"), Sodium and Chlorine (Na and Cl which make up "salt").

2)   The most common elements found in biological molecules can be remembered using the nonsense word CHNOPS (pronounced "cha-nops").

3)   Other definitions involving elements:

    a.   Compound: A substance composed of precise proportions of two or more elements in a specific geometric pattern (i.e., CO_2).

    b.   Mixture: A substance with two or more elements in variable proportions (i.e., soda).

C.   Atomic structure.

1)   An atom is the smallest particle of an element.

2)   Atoms contain:

    a.   Nucleus: central and dense; resists disturbances by outside forces.

       i.   Nuclei contain equally heavy sub-atomic particles of two types:

          a)   Protons (positively charged).

          b)   Neutrons (uncharged).

       ii.   Atomic number: numbers of protons present; constant for any particular element (i.e., every Hydrogen atom has one proton).

       iii.   Atomic weight: total number of protons and neutrons; can vary within atoms of an element due to differing numbers of neutrons, resulting in isotopes (atoms of the same element, differing in the number of neutrons present). Example: some Hydrogen atoms have one neutron while others have 2 neutrons, but all Hydrogen atoms have one proton.

    b.   Orbitals: outer regions containing light, negatively charged electrons which zip around the nucleus at great speed. Isolated atoms are electrically neutral, having equal numbers of protons and electrons. Electron orbitals are more easily modified than the nucleus.

       i.   Each orbital contains two electrons at most.

       ii.   Large atoms have several layers of electron shells at increasing distances from the nucleus, each shell containing one or more orbitals.

       iii.   Electrons in a shell close to the nucleus have less energy than those in a shell further away. Thus, electron shells have different energy levels.

       iv.   Number of orbitals in various electron shells:

          a)   The closest electron shell to the nucleus has one orbital and 2 electrons at most.

          b)   The second closest shell has at most 4 orbitals and 8 electrons.

          c)   The third closest shell has at most 9 orbitals and 18 electrons.

       v.   Electrons usually occupy orbitals in the shells closest to the nucleus before occupying shells further away. Example:

Neutral Atoms (equal numbers of protons & electrons)	Atomic Number	Arrangement of Electrons		
		First Shell	Second Shell	Third Shell
C (carbon)	6	2,	4	
Na (sodium)	11	2,	8,	1
Fe (iron)	26	2,	8,	16

3)   Atoms can use energy absorbed from the sun to move electrons from lower energy levels to higher ones. Later, these electrons can move back to a lower energy level again, releasing the absorbed energy which may be chemically trapped by the organism or given off as heat.

D.   <u>Atomic reactivity</u>: depends on the number of electrons in the outermost shell.

    1)   A <u>stable</u> atom is one with completely full or completely empty outer electron shells.

        a.   It will not react with other atoms.

        b.   Example: "<u>Inert gases</u>" such as Helium (He, completely full first and completely empty second electron shell) and Neon (Ne, completely full first and second and completely empty third electron shell) are very stable.

    2)   An atom with a partially full outer electron shell is reactive and can become more stable by losing, gaining, or sharing electrons.

        a.   Most atoms react in ways which either fill or empty the outer electron shell.

        b.   Example: Carbon has 6 electrons, arranged 2, 4. Carbon is more stable with 8 electrons in the outer shell, so it forms 4 bonds with other atoms to gain the 4 electrons it needs to become more stable (see Table 2-2 in text).

3.   **Chemical Bonds: Joining Atoms to make Molecules**: Atoms become more stable by having completely full or empty outer electron shells. Atoms with partially full outer shells will react with other atoms to become more stable. These reactions create chemical bonds, energy relationships that bind the atoms together into <u>molecules</u>. Three types of chemical bonds are possible.

A.   <u>Ionic bonds</u>: occur when one atom releases an electron and another atom gains the electron.

    1)   This reaction usually occurs between atoms like Sodium, which has only one electron in its outer shell, and Chlorine, which needs only one more electron to fill its outer shell.

    2)   Both become more stable if Sodium donates its outer electron (to empty its outer shell) and Chlorine accepts the electron (to fill its outer shell). This reaction causes neutral atoms to become electrically charged <u>ions</u>:

Atoms before ionic bonding	Atoms after ionic bonding
Sodium (Na):	**Sodium Ion (Na^+):**
11 protons	11 protons
11 electrons (2, 8, 1)	10 electrons (2, 8, 0)
	(positive ion)
Chlorine (Cl):	**Chloride Ion (Cl^-):**
17 protons	17 protons
17 electrons (2, 8, 7)	18 electrons (2, 8, 8)
	(negative ion)

3) Na^+ and Cl^- ions attract each other electrically, forming crystals of "salt." The attraction between oppositely charged ions is called an <u>ionic bond</u>.

4)   Remember, ions are atoms with unequal numbers of protons and electrons while neutral atoms have equal numbers of protons and electrons.

B.   <u>Covalent bonds</u>: occur when two atoms share a pair of electrons with each other.

1)   Such "time sharing" helps both atoms become more stable at least part of the time by temporarily filling up their outer electron shells.

2)   Some atoms may form several covalent bonds with other atoms:

   a.   <u>Single covalent bond</u>: one pair of electrons is shared between 2 atoms. Example: in hydrogen gas, we have H_2 or H-H molecules.

   b.   <u>Double covalent bonds</u>: two pairs of electrons are shared between 2 atoms. Example: in oxygen gas, we have O_2 or O=O molecules.

   c.   <u>Triple covalent bonds</u>: three pairs of electrons are shared between 2 atoms. Example: in nitrogen gas, we have N_2 or N≡N molecules.

3)   Different atoms form covalent bonds with each other. Example: a water molecule is made of 2 Hydrogen atoms and one Oxygen atom, forming 2 covalent bonds:

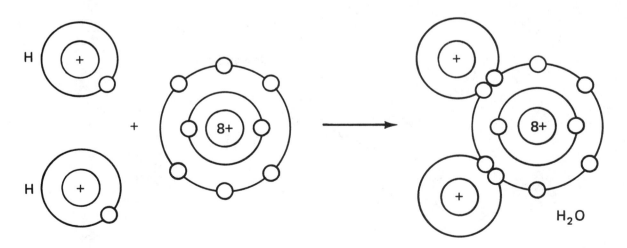

The covalent bonds allow each H and the O to have completely filled outer electron shells part of the time. This increases their stability.

4) Most biological molecules are made of atoms joined by covalent bonds. Types of covalent bonds found in biological molecules:

a. <u>Non-polar</u>: electrically symmetrical bonds formed when two similar atoms share electrons. Examples: H_2 gas or O_2 gas.

b. <u>Polar</u>: covalent bonds between atoms of different sizes, where the larger atom pulls more strongly on the electron pair than the smaller atom does. This produces a molecule with slightly charged ends:

   i. The end with the stronger atom becomes slightly negative in charge since the shared electrons are more often near this atom.

   ii. The end with the weaker atom becomes slightly positive since the shared electrons are less often near this atom.

   iii. So, a polar molecule has both a slightly positive and a slightly negative end even though the molecule as a whole is electrically neutral.

   iv. Example: in water, the stronger O end becomes slightly negative while the weaker H end becomes slightly positive:

slightly positive end      slightly negative end

C. <u>Hydrogen bonds</u>: electrical attraction between the polar parts of molecules, often containing Hydrogen. Example: water molecules form hydrogen bonds between them.

4.   Important **inorganic molecules**, especially water which is essential to life.

A.   Types of molecules:
1)   <u>Organic</u>: any molecule containing both C and H.
2)   <u>Inorganic</u>: CO_2 and all molecules lacking C.

B.   <u>Water</u> (H_2O): an extremely important inorganic molecule. Between 60-90% of a typical cell is water. Water has several important properties:
1)   Water is a good <u>solvent</u> (a substance that can dissolve many other molecules) because it is a polar molecule having both positively and negatively charged ends.
   a.   It can dissolve salts by surrounding the positively and negatively charged salt ions with "water envelopes" formed by hydrogen bonding between the polar ends of water and the salt ions:

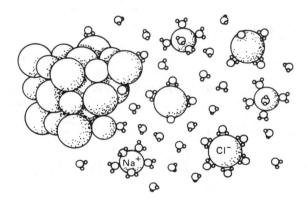

   b.   Water can dissolve other polar molecules like sugars and amino acids by forming hydrogen bonds with their slightly positive and negative regions.
   c.   Water can react with gases like O_2 and CO_2 and with larger molecules like proteins and sugars.
   d.   Charged or polar molecules are <u>hydrophilic</u> ("water loving") since they electrically attract water.
   e.   Nonpolar molecules are <u>hydrophobic</u> ("water hating") and will aggregate in water.
2)   Water has a slight tendency to become ionized, forming H^+ (<u>hydrogen ions</u>) and OH^- (<u>hydroxyl ions</u>).
   a.   If the concentration of H^+ > OH^- , the solution is <u>acidic</u>, but if the concentration of H^+ < OH^- , the solution is <u>basic</u>.
   b.   Chemists measure the degree of acidity by using the <u>pH scale</u>: a pH value below 7 is acidic, one above 7 is basic, while a Ph of 7 is neutral.
   c.   The pH scale is logarithmic: a change of one pH unit represents a 10-fold change in the amount of H^+ ions.
   d.   Ionization of water is crucial, since slight changes in cellular pH may cause various proteins to malfunction.

3) Water is a good <u>temperature</u> <u>moderator</u> in cells (helps cells resist sudden large changes in temperature) due to three properties:

    a. Water has <u>high</u> <u>specific</u> <u>heat</u>: It takes lots of energy to raise the temperature of water. Because of hydrogen bonding among water molecules, much energy is needed to speed up the movement (raise the temperature) of water.

    b. Water has <u>high</u> <u>heat</u> <u>of</u> <u>vaporization</u>: It takes lots of energy to convert liquid water to water vapor due to the hydrogen bonding among water molecules. Lots of heat is needed to evaporate water.

    c. Water has <u>high</u> <u>heat</u> <u>of</u> <u>fusion</u>: Much heat must be lost before water turns to ice.

4) Ice. Water is unusual because its solid form (ice) is <u>less</u> dense than liquid water.

    a. This is why ice floats on water.

    b. In winter, as ponds freeze, ice forms an insulating layer over the water, allowing aquatic plants and animals below to live.

5) Cohesion. Due to hydrogen bonding, liquid water shows <u>cohesion</u>, a solid-like property in which water molecules tend to stick together, especially at the surface.

    a. This produces <u>surface</u> <u>tension</u>, the tendency for heavy objects, like fallen leaves, to float on the thin surface film of water formed by cohesion of water molecules.

    b. Cohesion allows water molecules to rise up narrow cellular tubes in the trunks of trees, reaching up to the leaves.

****

REVIEW QUESTIONS:

1. Define matter.
2. Define energy.

3.-10. **MATCHING TEST**: types of energy

        Choices: A.   Potential energy   B.   Kinetic energy

3. Stored energy.
4. Energy of movement.
5. A rock resting at the top of a cliff.
6. Heat energy.
7. Electrical energy.
8. A rock falling from a cliff.
9. A man at the top of a ski lift.
10. A man skiing down a hill.

----

11. List, from memory, the more important elements in living systems.
12. List, from memory, the most common elements found in biological molecules.

13.   The nucleus of an atom <u>never</u> contains:
       A.   Protons   B.   Neutrons   C.   Electrons

14.-18. **MATCHING TEST**: atomic weight and atomic number

           Choices: A.   Atomic weight   C.   Neither of these
                       B.   Atomic number

14.   Number of neutrons in a nucleus.
15.   Number of protons in a nucleus.
16.   Number of protons and neutrons in a nucleus.
17.   May differ among atoms of the same element.
18.   Never differs among atoms of the same element.

----

19.   Isotopes are atoms of the same element:
       A.   Differing in number of neutrons present.
       B.   Differing in number of protons present.
       C.   Differing in number of electrons present.
       D.   With same atomic weight but different atomic number.
       E.   With same atomic number but different atomic weight.

20.-26. **TRUE-FALSE STATEMENTS** about orbitals and electron energy levels.

20.   Orbitals are regions containing up to 2 electrons moving at high speed.
21.   Orbitals are regions less easily modified than the nucleus of an atom.
22.   Electrons in a shell closer to the nucleus have more energy than those in a shell farther away.
23.   An atom with 16 electrons will have them arranged in shells as follows: nucleus--2--6--8.
24.   Atomic reactivity depends on the number of electrons in the innermost shell.
25.   An atom is most stable when its outermost electron shell is either completely full of electrons or completely empty.
26.   Most atoms react in ways that either fill up or empty out the outer electron shell.

27.-36. **MATCHING TEST**: chemical bonding

           Choices: A.   Ionic bonding   C.   Hydrogen bonding
                       B.   Covalent bonding

27.   Two atoms share a pair of electrons.
28.   One atom donates an electron and another atom accepts it.
29.   Holds a crystal of salt together.
30.   Attraction between the polar regions of molecules.
31.   Results in atoms with unbalanced electrical charges.
32.   Holds a molecule of water together.
33.   Attraction between two water molecules.
34.   Electrical attraction between oppositely charged atoms.
35.   Holds most biological molecules together.
36.   Can result in polar molecules.

----

37.   What is the difference between an organic and an inorganic molecule?

38.-46. **MATCHING TEST**: properties of water

Choices: A.   A polar molecule   D.   High heat of fusion
            B.   High specific heat   E.   Solid less dense than liquid
            C.   High heat of vaporization   F.   Cohesion

38.   Why leaves "float" on water.
39.   Why much energy is needed to convert liquid water to water vapor.
40.   Why water is such a good solvent.
41.   Why much energy is needed to raise the temperature of water.
42.   Why water can break down salt crystals.
43.   Why fish can live in lakes in the wintertime.
44.   Why much heat must be lost before water turns to ice.
45.   Why evaporation cools us in the summertime.
46.   Why belly-flopping dives hurt.

****

ANSWERS TO REVIEW QUESTIONS:

1.   see 1.A.

2.   see 1.B.

3.-10. [see 1.B.1)a.and b.]
        3.   A   7.   B
        4.   B   8.   B
        5.   A   9.   A
        6.   B   10.   B

11.   see 2.B.1)

12.   see 2.B.2)

13.   C   [see 2.C.2)a.i.]

14.-18. [see 2.C.2)a.ii.-iii.]
        14.   C   17.   A
        15.   B   18.   B
        16.   A

19.   A and E; [see 2.C.2)a.iii.]

20.-26. [see 2.C. and D.]
        20.   True   24.   False
        21.   False   25.   True
        22.   False   26.   True
        23.   False

27.-36. [see 3.A.-C.]
        27.   B   32.   B
        28.   A   33.   C
        29.   A   34   A
        30.   C   35   B
        31.   A   36.   B

37.   see 4.A.

38.-46. [see 4.B.1)-5)]
        38.   F   43.   E
        39.   C   44.   D
        40.   A   45.   C
        41.   B   46.   F
        42.   A

****

CHAPTER 3:   BIOLOGICAL MOLECULES

CHAPTER OUTLINE.

This chapter covers the structure, synthesis, and function of the four major types of large organic molecules (carbohydrates, lipids, proteins, and nucleic acids).

1. **General facts**.

   A.   All biological molecules are organic, having C and H and usually other atoms as well.

   B.   Carbon allows organic molecules to become large and diverse since each carbon can form up to 4 covalent bonds with other atoms. This allows molecules with many carbons to assume complex shapes and grow quite large due to their carbon "skeletons" or "backbones".

   C.   Attached to the carbons of most organic molecules are <u>functional</u> <u>groups</u> (see Table 3-1 in the text) that determine the molecules' characteristics.

2. **Making organic molecules**:

   A.   Cells use a "modular approach" to construct large molecules: they hook together smaller pre-assembled subunits just as a large train is assembled by coupling smaller cars together. The table at the top of the next page presents the smaller subunits that go into making up each of the important types of large molecules found in cells.

PRE-ASSEMBLED SUBUNITS (MONOMERS)	LARGE MOLECULE FORMED (POLYMERS)
amino acids	proteins
single sugars (monosaccharides)	carbohydrates
fatty acids } glycerol }	fats
nitrogen bases } phosphoric acid } 5-carbon sugars }	nucleic acids

B.   Types of reactions:
1)   The smaller subunits are hooked together through <u>condensation reactions</u> in which one subunit loses a hydrogen ion (H^+) and the other subunit loses a hydroxyl ion (OH^-). These "condense" together to form H_2O and the subunits join together by forming a covalent bond.
2)   The process is reversed through <u>hydrolysis reactions</u> in which H_2O breaks a larger molecule into 2 smaller subunits, with the water's H^+ joining to one subunit and its OH^- joining to the other subunit.
3)   Example:

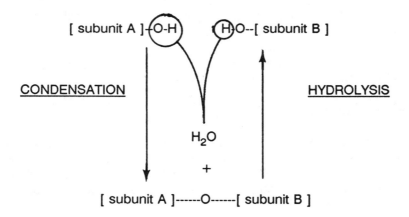

a larger molecule

3. **Carbohydrates**.

    A.   <u>General facts</u> about carbohydrates:
        1)   They are various sized sugars with the general formula $(CH_2O)_n$ where n = the number of carbons in the molecule, usually 3 to 7.
        2)   When dissolved in water, they usually are found in a circled-up ring form, with the carbon "backbone" forming the circle.
        3)   Small carbohydrates are water soluble, since the -OH groups are polar.

    B.   <u>Categories</u> of Carbohydrates:
        1)   <u>Monosaccharides</u> ("single sugars"): one sugar molecule per carbohydrate
            a.   Most common is <u>glucose</u> $(C_6H_{12}O_6)$, made by plants using sunlight energy to join $6CO_2$ and $6H_2O$ together. Glucose molecules are often linked to form larger carbohydrate molecules.
            b.   Other important monosaccharides:
                i.   6-carbon types: <u>fructose</u> (corn sugar) and <u>galactose</u> (found in milk).
                ii.   5-carbon types: <u>ribose</u> and <u>deoxyribose</u> (found in nucleic acids).
        2)   <u>Disaccharides</u> ("double sugars"): two sugar molecules per carbohydrate.
            a.   Used for short-term energy storage and transport in plants.
            b.   Examples:
                i.   <u>Sucrose</u> (table sugar), made of glucose and fructose.
                ii.   <u>Lactose</u> (milk sugar), made of glucose and galactose.
                iii.   <u>Maltose</u> (from starch), made of 2 glucose molecules.
            c.   Made by condensation and split by hydrolysis:

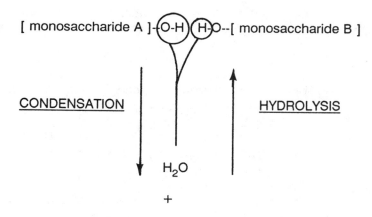

[monosaccharide A]-(O-H)(H-O)--[monosaccharide B]

CONDENSATION   HYDROLYSIS

H_2O

+

[monosaccharide A]--------O-------[monosaccharide B]

<u>a disaccharide</u>

       3)   Polysaccharides ("many sugars"), many sugar (usually glucose) molecules per carbohydrate. Uses:
- a.   For long term storage of energy. Examples:
  - i.   <u>Starch</u> in plants.
  - ii.   <u>Glycogen</u> in animals.
- b.   For structural materials. Examples:
  - i.   <u>Cellulose</u> in plant cell walls.
  - ii.   <u>Chitin</u> in insect skeletons and cell walls of fungi.

4. **Lipids**.

- A.   Lipids are diverse in structure and function, but all:
  - 1)   are insoluble in water (are non-polar) and soluble in ether.
  - 2)   contain large regions composed almost entirely of C and H.

- B.   <u>Types</u> of lipids:
  - 1)   <u>Oils</u>, <u>Fats</u>, and <u>Waxes</u>:
    - a.   Contain mostly C and H, with some O.
    - b.   Exist as long chains, not rings.
    - c.   Fats and oils:
      - i.   Are <u>triglycerides</u>, each molecule containing one <u>glycerol</u> and three <u>fatty</u> <u>acids</u>, joined by condensation reactions.
      - ii.   Have a high concentration of chemical energy and are used for semi-permanent storage of energy in animals (fats) and plants (oils).
      - iii.   Fats are solid at room temperature since their fatty acids are <u>saturated</u> with H atoms, allowing the molecules of fat to stack more densely.
      - iv.   Oils are liquid at room temperature since their fatty acids are <u>unsaturated</u> with H (some of the carbons form bent or "kinky" double bonds with adjacent carbons).
        - a)   The bent double bonds do not allow oils to stack very densely.
        - b)   Oils become fats when Hs are added to the oil's fatty acids, breaking the double bonds between carbons and producing "<u>hydrogenated oils</u>".
    - d.   Waxes have fatty acids joined to alcohol molecules instead of glycerol. Waxes are waterproof molecules made by plants and some animals, like bees and mammals.
  - 2)   <u>Phospholipids</u>: found in cell membranes.
    - a.   Are like oils except one fatty acid is replaced by a phosphate group with a short polar subunit often containing nitrogen.
    - b.   Have "contradictory" ends regarding affinity to water:
      - i.   Heads (phosphate-N end) are <u>hydrophilic</u> ("water loving", mix with water).
      - ii.   Tails (fatty acid end) are <u>hydrophobic</u> ("water fearing", insoluble in water).
  - 3)   <u>Steroids</u>: complex lipids having 4 fused rings of carbons with various functional groups attached. Many steroids are made from <u>cholesterol</u>.

5.   **Proteins**:

A.   <u>Uses</u> of proteins:
1)   Enzymes (control chemical reactions in cells).
2)   Structural components (e.g., elastin, which gives skin its elasticity).
3)   Energy and material storage.
4)   Transport (e.g., hemoglobin, which carries oxygen).
5)   Cell movement.
6)   Hormones.

B.   <u>Structure</u>:
1)   Proteins are chains of amino acids covalently attached by condensation reactions:

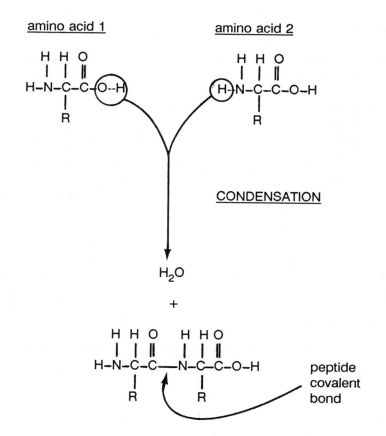

<u>dipeptide subunit with 2 amino acids</u>

2)   Amino acids come in 20 different types, differing in the chemical composition of the R groups.
3)   Levels of protein structure:

        a.   <u>Primary</u> <u>structure</u>: the sequence of amino acids in the peptide chain.

        b.   <u>Secondary</u> <u>structure</u>: either a helical or a pleated sheet shape of the peptide chain due to hydrogen bonding between C=O and N-H regions of different amino acids.

        c.   <u>Tertiary</u> <u>structure</u>: contortion of the helix or pleated sheet into a complex 3-dimensional shape due to interactions between R groups of different amino acids in the peptide chain with each other and with the cellular environment.

        d.   <u>Quarternary</u> <u>structure</u>: the joining of several peptide chains to form "super-protein" complexes. Example: hemoglobin has 4 peptide chains in its structure.

    C.   <u>Function</u>: determined by the sequence of amino acids in the peptide chain or chains, since this sequence determines the exact 3-dimensional shape of the protein.

6.   **Nucleic acids**.

    A.   <u>Genes</u> are nucleic acids with sets of instructions that determine the amino acid sequences of proteins made by cells.

    B.   <u>Subunits</u> of nucleic acids:

        1)   Small molecules of 3 types:

              a.   5-carbon sugars (<u>ribose</u> and <u>deoxyribose</u>).

              b.   <u>Phosphate</u> <u>groups</u>.

              c.   <u>Nitrogen-containing</u> <u>bases</u> of 4 types.

        2)   One small molecule of each category combines to form <u>nucleotides</u>:

              a.   <u>Ribose</u> nucleotide: [phosphate group--ribose sugar--nitrogen base].

              b.   <u>Deoxyribose</u> nucleotide: [phosphate group--deoxyribose sugar--nitrogen base].

        3)   Many nucleotides strung together by condensation reactions form a chain of <u>nucleic</u> <u>acid</u>:

              a.   <u>Deoxyribonucleic</u> <u>acid</u> (<u>DNA</u>): 2 chains of deoxyribose nucleotides arranged in a helix.

              b.   <u>Ribonucleic</u> <u>acid</u> (<u>RNA</u>): a chain of ribose nucleotides.

        4)   <u>Summary</u> of DNA and RNA components:

	DNA	RNA
Pentose sugar	deoxyribose	ribose
Phosphate group	present	present
Nitrogenous bases	4 types	4 types

    C.   Nucleotides not found in genes:

        1)   <u>Cyclic</u> <u>nucleotides</u>: intracellular messengers that signal if hormones come in contact with the cell membrane.

        2)   Nucleotides with extra phosphate groups (Example: <u>ATP</u>: <u>adenosine</u> <u>triphosphate</u>): carry energy from place to place in a cell.

3)   Nucleotides combined with vitamins form "<u>coenzymes</u>" that aid in the function of certain enzymes.

****

REVIEW QUESTIONS:

1.   True or False: Some biological molecules are not organic.

2.   True or False: Organic molecules can become large and complex in structure due to carbon atoms which can form four covalent bonds with other atoms.

3.-9.   **MATCHING TEST:** subunits making up large organic molecules

Choices:	A.	Carbohydrates
	B.	Nucleic acids
	C.	Proteins
	D.	Fats

3.   Nitrogen bases.
4.   Amino acids.
5.   Glycerol.
6.   Phosphoric acid.
7.   5-carbon sugars.
8.   Monosaccharides.
9.   Fatty acids.

10.-13. **MATCHING TEST:** chemical reactions

Choices:	A.	Hydrolysis reactions
	B.	Condensation reactions

10.   Water breaks a large molecule into two smaller ones.
11.   Each subunit loses part of a water molecule and a covalent bond joins the subunits.

[subunit]--O-H   H-O--[subunit]

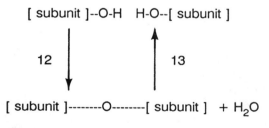

12   13

[subunit]--------O--------[subunit]   + H$_2$O

14.-20. **MATCHING TEST**: carbohydrates

Choices:
A. Disaccharide
B. Monosaccharide
C. Polysaccharide

14. One sugar molecule per carbohydrate.
15. Two sugar molecules per carbohydrate.
16. Many sugar molecules per carbohydrate.
17. Glucose, fructose, and galactose.
18. Starch, glycogen, and cellulose.
19. Sucrose, lactose, and maltose.
20. Ribose and deoxyribose.

21.-28. **MATCHING TEST**: specific carbohydrates

Choices:
A. Glucose   E. Starch
B. Sucrose   F. Glycogen
C. Lactose   G. Cellulose
D. Deoxyribose   H. Chitin

21. Energy storage polysaccharide in plants.
22. Plant cell wall component.
23. Milk sugar.
24. Found in insect skeletons.
25. Most common sugar.
26. Table sugar.
27. Found in some nucleic acids.
28. Energy storage polysaccharide in animals.

29.-35. **MATCHING TEST**: large organic molecules

Choices:
A. Lipids   C. Carbohydrates
B. Proteins   D. Nucleic acids

29. Have the general formula $(CH_2O)_n$.
30. Genes.
31. Enzymes.
32. All are insoluble in water.
33. Usually found in a coiled-up ring form in cells.
34. Determine the amino acid sequences of proteins.
35. Cholesterol is an example.

36.-44. **MATCHING TEST:** lipids

Choices: A. Fats   D. Phospholipids
          B. Oils   E. Steroids
          C. Waxes

36.   Structure contains alcohol.
37.   Cholesterol.
38.   Liquid at room temperature.
39.   Contains all saturated fatty acids.
40.   Contains unsaturated fatty acids.
41.   Energy storage in plants.
42.   Have hydrophilic and hydrophobic ends.
43.   Have four fused rings of carbon atoms.
44.   A major component of cell membranes.

45.-48. **MATCHING TEST:** protein structure

Choices: A. Primary structure
          B. Secondary structure
          C. Tertiary structure
          D. Quarternary structure

45.   Interaction of "R" groups of different amino acids, causing contortions in shape.
46.   Sequence of amino acids in a polypeptide chain.
47.   Joining of several polypeptide chains.
48.   Helical or pleated sheet shape of a peptide chain due to hydrogen bonding.

----

49.   **True or False:** A ribose nucleotide has the structure
       [ribose sugar--nitrogen base--phosphate group].

****

****

ANSWERS TO REVIEW QUESTIONS:

1.	False [see 1.A.]		29.-35.	[see 3. through 6.]	
			29.	C	33. C
2.	True [see 1.B.]		30.	D	34. D
			31.	B	35. A
3.-9.	[see 2.A.1)]		32.	A	
3.	B	7. B			
4.	C	8. A	36.-44.	[see 4.B.1)-3)]	
5.	D	9. D	36.	C	41. B
6.	B		37.	E	42. D
			38.	B	43. E
10.-13.	[see 2.B.]		39.	A	44. D
10.	A	12. B	40.	B	
11.	B	13. A			
			45.-48.	[see 5.B.3)a.-d.]	
14.-20.	[see 3.B.1)-3)]		45.	C	47. D
14.	B	18. C	46.	A	48. B
15.	A	19. A			
16.	C	20. B	49.	F [see 6.B.2)a.]	
17.	B				
21.-28.	[see 3.B.1)-3)]				
21.	E	25 A			
22.	G	26. B			
23.	C	27. D			
24.	H	28. F			

****

CHAPTER 4: ENERGY FLOW IN THE LIFE OF A CELL

CHAPTER OUTLINE.

This chapter focuses on energy flow through the universe and particularly through living systems. The basic laws of thermodynamics are discussed and the basic types of biochemical reactions are outlined. How cells use enzymes to control chemical reactions is emphasized.

1. **Energy flow in the Universe**.

   A. All interactions of material particles are influenced by the flow of energy between them.
      1) Energy is the ability to do work.
      2) Energy flow within "isolated systems" (the concept of closed systems in which energy cannot enter or leave) is governed by the laws of thermodynamics.
      3) In dealing with isolated systems, we need to know:
         a. How much energy is present.
         b. How useful the energy is.

   B. First Law of Thermodynamics (law of conservation of energy): within an isolated system, energy is neither created nor destroyed but can be changed in form.
      1) The total quantity of energy remains constant.
      2) Example: driving a car. The chemical energy of gasoline (100%) is converted to kinetic energy of movement (25%) and heat energy (75%).
      3) Energy always flows "downhill" along gradients from high energy places to low energy places.

   C. Second Law of Thermodynamics: any change in an isolated system causes organized useful energy to decrease and disorganized less useful energy to increase.
      1) Example: driving a car. Some of the concentrated useful energy of organized gasoline molecules is converted into heat (disorganized movement of molecules

27

useless in running the car).

2)   Energy conversion results in <u>entropy</u> (increase in random, disorganized low-level energy movement of molecules). Example: the arrangement of the 8 carbon chain in a gasoline molecule is more orderly than the random movement of 8 CO_2 molecules produced when the car engine burns the gasoline.

3)   Due to entropy, eventually all energy in the universe will be dispersed as heat.

4)   <u>Entropy and life</u>: Living systems are not "isolated systems" because autotrophs (photosynthesizers) absorb sunlight energy and use it to organize CO_2 and water molecules into food sugars and to maintain orderly cell structure. The sun increases its entropy while living organisms maintain order.

2.   Energy Flow in chemical reactions.

A.   In chemical reactions, <u>reactant molecules</u> are converted into <u>product molecules</u>. Types of reactions:

1)   <u>Exergonic</u> ("energy out") <u>reactions</u>: the reactants have more energy than the products.

     a.   Energy is released during the reaction.

     b.   Example: burning coal (pure carbon).

$$\text{C-C-C-C-C} + O_2 \rightarrow CO_2 + \text{energy}$$
$$\text{high energy} \qquad\qquad \text{low energy}$$

2)   <u>Endergonic</u> ("energy in") <u>reactions</u>: the products have more energy than the reactants.

     a.   Energy is required from outside the reaction.

     b.   Example: making sugar by photosynthesis in plants.

$$CO_2 + H_2O + \text{sunlight} \rightarrow \text{sugar} + O_2$$
$$\text{low energy} \qquad \text{energy} \qquad \text{high energy}$$

B.   <u>Exergonic reactions</u>: release energy.

1)   Once started, will continue without adding energy. Example: a coal fire, once started, will continue to burn by itself until the coal is used up.

2)   Require an initial input of energy (<u>activation energy</u>) only to get started, since the molecules must be forced to move closer together for the reaction to begin (the outer electron shells of the atoms must overlap).

3)   Heat (kinetic energy of movement) is the usual source of activating energy (gets the molecules to move faster and bump into each other more violently).

C.   <u>Endergonic reactions</u>: require energy.

1)   Energy is required not only to start ("activate") the reaction but also to continue the reaction.

2)   In photosynthesis, sunlight energy is constantly needed to produce more sugar

molecules. A plant placed in a dark box ceases photosynthesis.

D. <u>Coupled reactions</u>.
1)   In cells, endergonic reactions occur because, in a coupled fashion, they obtain energy released from exergonic reactions.
2)   Example: photosynthesis (endergonic) and solar reactions (exergonic) are coupled:

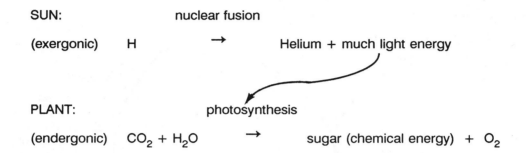

SUN: nuclear fusion

(exergonic)   H   →   Helium + much light energy

PLANT: photosynthesis

(endergonic) $CO_2 + H_2O$ →   sugar (chemical energy) + O_2

The overall process is exergonic since more sunlight energy is produced than plants can use.

3)   In coupled reactions within cells, energy is usually transferred from exergonic reactions to endergonic ones by <u>energy carrier molecules</u>, especially <u>ATP</u> (adenosine triphosphate).

E. <u>Chemical equilibria</u>.
1)   Many chemical reactions are <u>reversible</u>, being capable of occurring in either or both directions depending on the relative concentrations of the molecules involved. Example: the binding of oxygen (O_2) to hemoglobin (Hb) in blood.

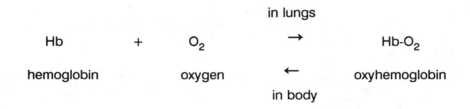

in lungs

Hb   + O_2 →   Hb-O_2

hemoglobin   oxygen   ←   oxyhemoglobin

in body

As blood hemoglobin circulates through the lungs, it picks up oxygen to become oxyhemoglobin. Then, as it passes through other body organs, the Hb-O_2 gives off oxygen.
2)   Reversible reactions reach a <u>chemical equilibrium</u> when the reaction proceeds in both directions at the same rate.
3)   Chemical equilibria are affected by the concentration of molecules involved. In

the hemoglobin example, $Hb-O_2$ molecules tend to give up their O_2 at a low rate:

a.   In the lungs (rich in O_2), most O_2 released by $Hb-O_2$ will be quickly replaced, so that most Hb remains bound to O_2.

b.   In exercising muscles (low in O_2), most O_2 released by the $Hb-O_2$ will not be replaced with another O_2 (since the muscles use up oxygen fast and the oxygen concentration in muscles is low), so most $Hb-O_2$ eventually becomes deoxygenated Hb.

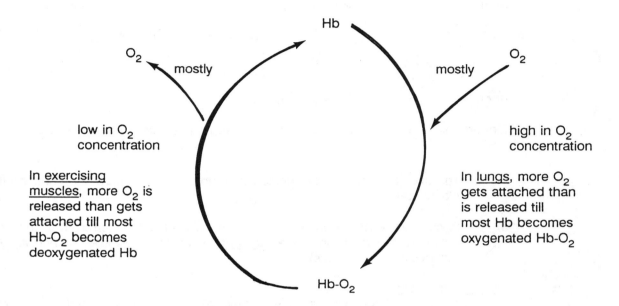

Thus, cells drive reversible reactions back and forth as required by the metabolic demands of the organism.

3.   **Controlling metabolism** (the sum of all chemical reactions) within cells.

A.   <u>General facts</u>.
1)   Cells use <u>enzymes</u> (protein catalysts) to regulate chemical reactions.
2)   Cells allow endergonic reactions to occur by <u>coupling</u> them with exergonic reactions.
3)   Cells use <u>energy-carrier</u> molecules (especially ATP) to store, transport, and regulate energy.
4)   The rate of an exergonic reaction is determined by its <u>activation energy</u> which allows the reactants to collide fast enough to force their outer electron shells together. Heat is one way to provide activation energy. Using enzymes is another.

B.   <u>Catalysts</u>: molecules that reduce activation energy and speed up the rate of reactions

without being permanently changed themselves. Four facts about catalysts:
1) They speed up reactions by reducing activation energy.
2) They cannot cause energetically unfavorable reactions to occur since they can only speed up the rate of slowly occurring spontaneous reactions.
3) They do not change the equilibrium point of reactions.
4) They are not consumed in the reaction they speed up.

C. <u>Enzymes</u>: catalysts, usually proteins, made by living cells.
   1) Features of enzymes:
      a. The four listed above for catalysts.
      b. Unlike catalysts, enzymes are <u>very</u> <u>specific</u>, usually working in only a single type of reaction.
      c. Unlike catalysts, enzyme activity can be <u>regulated</u> (enhanced or suppressed).
   2) <u>Structure and function</u>.
      a. Enzymes are proteins with complex 3-dimensional shapes, including an <u>active</u> <u>site</u> area where reactant molecules ("<u>substrates</u>") fit into based on the substrate's shape and electrical charge. Thus, enzyme function is related directly to enzyme structure (shape).
         i. The substrate must be able to enter the active site.
         ii. Substrate and enzyme must have correct molecular structures to interact with each other, like a lock and its key.
         iii. <u>Poisons</u> can enter the active site but lack the correct chemical composition to react so that they permanently clog up the enzyme.
      b. "<u>Induced</u> <u>Fit</u>" model of enzyme function: once substrate enters the enzyme's active site, both substrate and active site change shape, causing stress on the chemical bonds of the substrate so that a chemical reaction occurs more readily, changing the substrate into one or more products.
      c. When the reaction occurs, the products formed do not fit well in the enzyme's active site and are expelled. The enzyme resumes its normal shape and is ready to accept new substrate molecules.
   3) Various mechanisms of <u>enzyme regulation</u>.
      a. Cells can <u>limit</u> <u>the</u> <u>amount</u> of enzyme present.
      b. Cells can make the enzyme in an <u>inactive</u> <u>form</u> (with a blocked active site) and activate it only when needed. Example: digestive enzymes in the stomach.
      c. Cells can temporarily activate or inactivate some enzymes by <u>end-product</u> <u>inhibition</u>. If too much product is produced, some may temporarily bind to the enzyme, inactivating it. Types of end-product inhibition:
         i. <u>Competitive</u> <u>inhibition</u>. The end-product competes with the substrate for entry into the active site.
         ii. <u>Allosteric</u> ("other shape") <u>inhibition</u>. The end-product binds to some non-active site region (inhibitor site) of the enzyme, changing its shape somewhat. The active site's shape changes due to this, keeping the substrate from entering.
      d. Both competitive and allosteric binding involve reversible reactions with the chemical equilibrium influenced by the concentration of end-product present.

      i.   As end-product becomes scarce, more enzyme molecules become active.

      ii.   Concentration of active enzyme is controlled by concentration of end-product (inhibitor) present.

D.   <u>Coupled reactions</u> and energy-carrier molecules.

    1)   <u>Energy-carrier molecules</u> act like rechargeable batteries, picking up energy from an exergonic reaction, moving it to another cellular location, and releasing the energy there to drive an endergonic reaction.

    2)   <u>ATP</u> (adenosine triphosphate): the most common short-lived energy carrier in living cells.

       a.   **Structure**.

       b.   ATP has many ordinary covalent bonds and two <u>high energy bonds</u> (** above) that are fairly unstable (easily broken to release the energy). Usually, only the outer high energy bond is used by cells to transport energy.

       c.   <u>ATP metabolism</u>: made from ADP (adenosine diphosphate) and inorganic phosphate (P_i) in a reversible reaction. ATP "<u>couples</u>" these reactions: the exergonic reaction releases energy needed by the endergonic one, and ATP transports the energy between them.

energy from an exergonic reaction

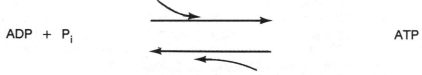

ADP + P_i ATP

energy released to power an endergonic reaction

    3)   <u>Other energy carriers</u>. Some exergonic reactions transfer energy to electrons which are captured by <u>electron-carrier molecules</u> and carried to other cellular regions to supply endergonic reactions. Some electron-carriers:

       a.   <u>NAD</u>: nicotinamide adenine dinucleotide.

       b.   <u>FAD</u>: flavine adenine dinucleotide.

****

REVIEW QUESTIONS:

1.-6.   **True or False** statements on thermodynamics:

1.   With an "isolated (closed) system", the total amount of energy may change as energy is converted from one form to another.
2.   Energy is neither created nor destroyed and energy cannot be converted from one form to another.
3.   Within an isolated system, useful energy tends to change into less useful forms.
4.   Energy conversion results in entropy (increase in the random movement of molecules).
5.   The universe is slowly growing more organized as more complex organisms evolve on earth.
6.   Since living organisms on earth remain highly organized, they violate the second law of thermodynamics.

7.-15. **MATCHING TEST**: chemical reactions

        Choices: A.   Exergonic reactions
                               B.   Endergonic reactions
                               C.   Both of these
                               D.   Coupled reactions

7.   Once started, will continue by themselves.
8.   Need "activation energy" to get started.
9.   Reactants (substrates) have more energy than products.
10.   Photosynthesis and solar reactions.
11.   Products have more energy than reactants (substrates).
12.   Energy is released from the reaction.
13.   Energy is needed to continue the reaction after it begins.
14.   Usually involves "energy-carrier molecules".
15.   Burning wood in a fireplace.

----

16.   Most hemoglobin molecules become saturated with O_2 in the lungs but lose most of the O_2 to exercising muscles. Explain this in terms of chemical equilibrium.

17.-22. **MATCHING TEST**: catalysts and enzymes

        Choices: A.   Catalysts   C.   Both
                               B.   Enzymes   D.   Neither

17.   All are protein molecules.
18.   Can speed up chemical reactions.
19.   Are not changed in the reactions they affect.
20.   Generally are very specific as to the reaction they affect.
21.   Cannot cause energetically unfavorable reactions to occur.
22.   Their activity can be regulated.

----

23.   Explain why the 3-dimensional shape of an enzyme is important to its function. Use the term "active site" in your explanation.

24. How do enzyme poisons work?
25. Explain the "induced fit" model of enzyme function.
26. Name the three ways that enzymes are regulated by cells.
27. What is the difference between competitive inhibition and allosteric inhibition of enzyme action?
28. Explain why ATP (adenosine triphosphate) is considered an "energy-carrier molecule".
29. How can ATP "couple" an exergonic reaction and an endergonic reaction?

****

ANSWERS TO REVIEW QUESTIONS:

1.-4.	[see 1.A.-C.]				23.	see 3.C.2)a.
	1.	False	4.	True		
	2.	False	5.	False	24.	see 3.C.2)a.iii.
	3.	True	6.	False		
					25.	see 3.C.2)b.
7.-15.	[see 2.A.-E.]					
	7.	A	12.	A	26.	see 3.C.3)
	8.	C	13.	B		
	9.	A	14.	D	27.	see 3.C.3)c.
	10.	D	15.	A		
	11.	B			28.	see 3.D.1) and 2)
16.	see 2.E.1)-3)				29.	see 3.D.2)c.
17.-22.	[see 3.A.-C.]					
	17.	B	20.	B		
	18.	C	21.	C		
	19.	C	22.	B		

****

CHAPTER 5: CELLS: UNITS OF LIFE I. CELL STRUCTURE AND FUNCTION

CHAPTER OUTLINE.

Prokaryotic and eukaryotic cells are compared in this chapter. Also, the organization of eukaryotic cells is covered, with emphasis on the various organelles found in these cells.

1. **Development of the Cell Theory.**

    A. Early observations:
        1) Robert Hooke, English, saw microscopic "cells" in thin pieces of cork in 1665.
        2) Antonie van Leeuwenhoek saw microscopic blood cells, sperm, and "animalcules" in the late 1600s.
        3) Theodor Schwann in 1839 stated that cells are the elementary particles of both plants and animals.
        4) Mattias Schleiden in the mid 1800s stated that cells are the absolutely indispensable fundamental basis of life.
        5) Rudolf Virchow in the late 1850s predicted that all cells come from cells.

    B. Modern Cell Theory has three principles:
        1) Every living organism is made up of one or more cells.
        2) Cells are the smallest functional units of life.
        3) All cells arise from pre-existing cells.

2. **Overview of Cell Structure and Function.**

    A. All cells have at least three components:
        1) Plasma (cell) membrane. Double layer of phospholipids with embedded proteins. Three functions:

a. Separate cytoplasm from external environment.
b. Regulate flow of material between cytoplasm and outside environment.
c. allows interactions with other cells.

2) Genetic material (DNA). Hereditary blueprint or instructions for making other cell parts and new daughter cells.
a. Found in nuclear chromosomes of eukaryotic cells.
b. Found in nucleoid cytoplasmic region of prokaryotic cells.

3) Cytoplasm. All material inside the plasma membrane and outside the nuclear membrane; contains molecules and organelles.

B. Cell function limits cell size.
1) All living organisms exchange nutrients and wastes with the external environment through the plasma membrane.
2) If cells grow too large, exchange would be limited by:
a. Distance from the cell's center to the surface, since diffusion would occur too slowly; i.e., oxygen would take 200 days to move 8 inches (=20 cm).
b. Amount of available surface area: Cell volume increases much faster than does cell surface area as cells enlarge.

3. **Types of Cells** (see Table 5-1 in the textbook).

A. Prokaryotic ("before the nucleus") cells: resemble the earliest cells to evolve on earth.
1) Small in size (< 5 micrometers).
2) Surrounded by stiff porous cell wall.
3) Relatively homogeneous cytoplasm and simple internal structure.
4) Simple circular DNA chromosome in the nucleoid region of the cytoplasm.

B. Eukaryotic ("true nucleus") cells: more complex modern cells.
1) Usually > 10 micrometers in diameter.
2) Cytoplasm contains many types of membrane-bound organelles within a fluid matrix (the cytosol).
3) Have a membrane-bound organelle, the nucleus, containing several chromosomes (DNA and proteins).
4) Have cytoplasmic cytoskeleton: network of protein fibers providing shape and organization.

4. **Nucleus**: Control Center of the Cell.

A. Nuclear Envelope:
1) Consists of two porous membranes that isolate the nuclear contents from the rest of the cell.
2) Pores: complex structures of protein and RNA with a small channel through the middle; regulate flow of proteins and RNA to and from the nucleus.

B. Chromatin: the hereditary material.
1) Granular staining strands of DNA and protein called chromosomes ("colored bodies").
2) Cellular processes are governed by information encoded in nuclear DNA.
a. Genetic information is copied from DNA into RNA in the nucleus.

  b.   RNA moves through nuclear envelope pores into the cytoplasm.
  c.   Ribosomes in cytoplasm use RNA information to make proteins.
  d.   Other molecules, especially proteins, move into the nucleus to regulate which DNA regions make RNA.

C.   Nucleolus ("little nucleus"): site of ribosome assembly.
  1)   Ribosomes are organelles of protein and ribosomal-RNA (r-RNA) that serve as "workbenches" for protein synthesis.
  2)   r-RNA genes cluster at nucleolus and make r-RNA; proteins pass into nucleus and also enter nucleolus.
  3)   Large and small ribosomal subunits are assembled in nucleolus and then pass into the cytoplasm through nuclear envelope pores.
  4)   Nucleolus disintegrates during cell division and reforms afterwards in each new daughter cell's nucleus.

5. **Membrane System of Eukaryotic Cells**: an interconnected series of membrane types.

A.   General Facts:
  1)   Includes plasma membrane, endoplasmic reticulum, nuclear envelope, Golgi complex, vesicles, and lysosomes.
  2)   All are composed of lipids and proteins made in the endoplasmic reticulum.

B.   Membrane Structure (more details in Chapter 6).
  1)   All cell membranes have a bilayer of phospholipids with several kinds of embedded proteins and choleserol.
  2)   Structural Framework = phospholipids and cholesterol:
    a.   Phospholipids have polar hydrophilic "heads" and nonpolar hydrophobic "tails."
    b.   Phospholipids line up in a double layer ("bilayer") with heads facing water inside and outside the cell and tails hiding inside the bilayer.
    c.   Cholesterol, a hydrophobic steroid, is found among the phospholipid tails.
  3)   Membrane proteins carry out specialized functions like transport and catalysis; the function of these proteins determines the functions of the membranes.

C.   Principal Components of Membrane Systems.
  1)   Plasma Membrane:
    a.   Outer boundary of the living portion of cells.
    b.   Separates cytoplasm from outside environment.
    c.   Controls transport of selected substances in and out of cells.
  2)   Endoplasmic Reticulum (ER):
    a.   A series of interconnected membrane tubes and channels in the cytoplasm.
    b.   **Types**.
      i.   Smooth ER: lacks ribosomes but has enzymes for lipid synthesis.
      ii.   Rough ER: has ribosomes attached for protein synthesis. Some proteins like hormones, made by secretory cells to be expelled into their surroundings, are made by rough ER and transported into channels where they move through the ER and accumulate

in "pockets" which bud off to form membrane-bound sacs (<u>vesicles</u>) that move to the Golgi complexes.

    3)   <u>Golgi Complexes</u>: processing and packaging.

        a.   Stacks of ER-derived membranes. Vesicles from smooth ER may fuse with Golgi membranes (emptying their contents into the Golgi sac) as other vesicles bud off the Golgi membranes.

        b.   <u>Major functions</u>.

            i.   <u>Sorts out</u> various proteins and lipids received from ER vesicles by type and destination.

            ii.   <u>Modifies</u> some molecules by adding chemical groups to them.

            iii.   <u>Packages</u> these materials into new vesicles that are transported to other cell regions or sent out of the cell.

    4.   <u>Lysosomes</u>: intracellular digestion.

        a.   Membrane-bound vesicles made in the Golgi complex.

        b.   Contain digestive enzymes originally made in ER.

        c.   <u>Functions</u>:

            i.   Digest food particles by fusing with food vacuoles in the cytoplasm.

            ii.   Digest defective organelles.

D.   <u>Flow of Membranes through a cell</u>.

    1)   All membranes are made by the ER and move back and forth among the nuclear and cell membranes, Golgi complex, lysosomes and food vacuoles.

    2)   **Steps:**

        a.   Membrane molecules are made by rough and smooth ER and pinched off as vesicles.

        b.   The vesicles fuse with Golgi complexes which sort out their various components and modify some for special uses.

        c.   Golgi complexes process and repackage the membrane materials into new vesicles which move to the appropriate cellular areas for use.

6.   **Chloroplasts and Mitochondria**: energy capture and extraction.

A.   <u>Basic facts</u>.

    1)   Both are thought to have evolved from bacteria long ago.

    2)   Both are oblong in shape, 1-5 micrometers long, and surrounded by a double membrane.

    3)   Both make ATP for different uses.

    4)   Both have their own DNA.

    5)   Chloroplasts capture light energy and use it to make sugar (<u>photosynthesis</u>).

    6)   Mitochondria convert sugar energy into ATP energy for cell use.

B.   **Chloroplasts**: in plants and unicellular algae.

    1)   <u>Function</u>: photosynthesis.

    2)   <u>Structure</u>: sac bound by a pair of membranes.

        a.   Outer and inner membranes with little space between.

        b.   <u>Stroma</u>: semi-fluid material surrounded by the inner membrane.

        c.   <u>Thylakoids</u>: disc-shaped membrane sacs in the stroma containing green chlorophyll and other pigment molecules that capture sunlight energy.

          d.   <u>Granum</u>: a stack of thylakoid sacs.

    C.   **Mitochondria**: "cellular powerhouses," found in both plant and animal cells.
        1)   <u>Function</u>: cellular respiration (energy extraction from food molecules and storage as ATP).
            a.   In the cytosol, a small amount of food energy is extracted by <u>anaerobic</u> ("without oxygen") metabolism.
            b.   In mitochondria, <u>aerobic</u> ("with oxygen") metabolism completes the food energy extraction, releasing about 18 times as much energy than occurs in the cytosol.
        2)   <u>Structure</u>: sac bound by a pair of membranes.
            a.   Outer membrane is smooth.
            b..   Inner membrane loops to form deep folds called <u>cristae</u>.
            c.   Intermembrane compartments (between the membranes).
            d.   <u>Matrix</u> space surrounded by inner membrane.

7.   **Plastids and Vacuoles**: storage and elimination.

    A.   **Plastids**: double-membrane organelles found in plants. <u>Functions</u>:
        1)   Photosynthesis (chloroplasts).
        2)   Storage of carotenoid pigments (yellow, red, and orange colors in fruits).
        3)   Storage of photosynthetic products (starches made from sugars) during the winter (Example: potatoes).
    B.   **Vacuoles**: single-membrane bound sacs. <u>Functions</u>:
        1)   Storage of food or waste products.
        2)   Elimination of waste products.
        3)   Cell support.

8.   **Cytoskeleton**: shape, support, and movement.

    A.   <u>Components</u> (network of protein fibers in the cytoplasm):
        1)   <u>Microtubules</u>: thick protein fibers (25 nanometers in diameter).
        2)   <u>Intermediate filaments</u> (10 nm in diameter).
        3)   <u>Microfilaments</u>: thin protein fibers (5-7 nm in diameter).

    B.   <u>Functions</u>.
        1)   Determines <u>cell shape</u>, if no cell wall is present.
        2)   Causes <u>cell movement</u> by assembly/disassembly or sliding of fibers past each other.
        3)   <u>Move organelles</u>.
        4)   <u>Move chromosomes</u> into daughter cells during eukaryotic cell division.
        5)   In animal cells, <u>cytoplasmic division</u> occurs by cell pinching caused by a ring of microfilaments that contract.

    C.   **Microfilaments**: contain the proteins <u>actin</u> and <u>myosin</u> which slide past each other to allow for muscle contraction and organelle movement.

    D.   **Intermediate Fibers:**
        1)   At least five types, each made of a different protein.

        2)   Are permanent frameworks providing shape to cells and anchoring cell parts.

E.   **Microtubules:** hollow cylinders of the protein <u>tubulin</u>. <u>Facts:</u>

1)   Some are permanent parts of flagella and cilia.

2)   Some are transitory, appearing and disappearing during different stages of cell division (Example: the spindle apparatus).

3)   Made by <u>microtubular</u> <u>organizing</u> <u>centers</u>, like the <u>centrioles</u> in animal cells (make spindle fibers or act as <u>basal bodies</u> which produce microtubules in cilia).

4)   Cause cellular movement in two ways:

     a.   Different microtubules can move in opposite directions, pulling objects away from each other.

     b.   Microtubules can increase or decrease in length by adding or removing tubulin subunits, or can "move" by adding tubulin to one end while removing tubulin from the other end.

F.   **Cilia** and **Flagella**: whip-like microtubular appendages of the plasma membrane:

1)   They have a "<u>9 + 2 arrangement</u>" of microtubular pairs (as seen in cross-section), produced by basal body centrioles.

2)   <u>Cilia</u> ("eyelash") are short, numerous, and bend parallel to the cell membrane; show a fairly stiff "rowing" motion.

3)   <u>Flagella</u> ("whip") are long, few in number, and bend perpendicular to the cell membrane; show a continuous bending motion.

4)   They are powered by energy from ATP made by mitochondria found near the basal bodies.

5)   Used for food gathering and movement in unicellular organisms and small aquatic invertebrates.

6)   Some animal cells have cilia to move eggs along (in the oviducts of female mammals) or clearing debris and microorganisms (in the respiratory tracts of land vertebrates).

7)   Prokaryotes may have slender projections that undulate or spin, but these are not true flagella since they contain no microtubules.

****

****

REVIEW QUESTIONS:

1.-3. **True or False** statements on the Cell Theory:

1. There are some living organisms that do not have cells.
2. The cell nucleus is the smallest unit of life.
3. Some bacteria arise spontaneously, not from other bacteria.

4.-9. **MATCHING TEST**: cell types

        Choices: A.   Eukaryotic cells
                             B.   Prokaryotic cells
                             C.   Both

4. Lack a membrane-bound nucleus.
5. Have many chromosomes with DNA and protein.
6. Lack most cytoplasmic organelles.
7. Larger and more complex.
8. Have nucleoid regions in the cytoplasm.
9. Have DNA.

10.-20. **MATCHING TEST**: mitochondria and chloroplasts

        Choices: A.   Mitochondria
                             B.   Chloroplasts
                             C.   Both

10. Make ATP using energy.
11. Capture sunlight energy to make sugar.
12. Convert sugar energy into ATP energy.
13. Have DNA.
14. Have thylakoid membranes and semi-fluid stroma.
15. Extract energy from food molecules.
16. Found in plants.
17. Have cristae membranes and semi-fluid matrix.
18. Have chlorophyll.
19. Function in photosynthesis.
20. Function in cellular respiration.

21.-33. **MATCHING TEST**: organelles that manufacture or digest proteins and lipids

        Choices: A.   Ribosomes   D.   Lysosomes
                          B.   Endoplasmic reticulum (ER)   E.   None of these
                          C.   Golgi complexes

21. Digest food particles.
22. Interconnected membrane tubes and channels in the cytoplasm.
23. Stacks of membranes in the cytoplasm.
24. Made of RNA and proteins.
25. Function in the nucleus.
26. Membrane-bound vesicles.
27. "Workbenches" for protein synthesis.
28. Sort out various lipids and proteins.
29. May be rough or smooth in appearance.
30. Sites of lipid synthesis.
31. Large and small subunits are assembled in the nucleolus.
32. Packages proteins and lipids into vesicles for transport out of the cell.
33. Digests defective organelles.

-----

34. Describe membrane flow through cells.

35.-41.   **MATCHING TEST**: vacuoles and plastids

> Choices: A.   Vacuoles   C.   Both
>   B.   Plastids   D.   Neither

35. Have single membranes.
36. Have double membranes.
37. May contain carotenoid pigments.
38. May collect toxic wastes that cells can't excrete.
39. May be sites of protein synthesis.
40. May store starches during the winter.
41. May be sites of photosynthesis.

-----

42. Define the following: nuclear envelope, chromatin, and nucleolus.
43. Name two ways that microtubules can cause movement.

44.-49.   **MATCHING TEST**: cilia and flagella

> Choices: A.   Cilia   B.   Flagella   C.   Both

44. Microtubular extensions through the cell membrane.
45. Shorter and more numerous per cell.
46. Have a "9 + 2" arrangement of microtubular pairs.
47. Bend perpendicular to the cell membrane.
48. Bend parallel to the cell membrane.
49. Used for food gathering and for movement.

-----

50. Name the three essential components of all cells.
51. Why can't cells grow large enough so that a large organism could be made of one cell?

****

****

ANSWERS TO REVIEW QUESTIONS:

1.-3.	[see 1.B.]				34.	see 5.D.2)		
	1.	False						
	2.	False			35.-41.	[see 7.A. and B.]		
	3.	False				35.	A	39. D
						36.	B	40. B
4.-9.	[see 3.A. and B.]					37.	B	41. B
	4.	B	7.	A		38.	A	
	5.	A	8.	B				
	6.	B	9.	C	42.	see 4.A.-C.		

1.-3.   [see 1.B.]
　　1.　　False
　　2.　　False
　　3.　　False

4.-9.   [see 3.A. and B.]
　　4.　　B　　　7.　　A
　　5.　　A　　　8.　　B
　　6.　　B　　　9.　　C

10.-20. [see 6.A.-C.]
　　10.　　C　　　16.　　C
　　11.　　B　　　17.　　A
　　12.　　A　　　18.　　B
　　13.　　C　　　19.　　B
　　14.　　B　　　20.　　A
　　15.　　A

21.-33. [see 4.C. and 5.C.]
　　21.　　D　　　28.　　C
　　22.　　B　　　29.　　B
　　23.　　C　　　30.　　B
　　24.　　A　　　31.　　A
　　25.　　E　　　32.　　C
　　26.　　D　　　33.　　D
　　27.　　A

34.   see 5.D.2)

35.-41. [see 7.A. and B.]
　　35.　　A　　　39.　　D
　　36.　　B　　　40.　　B
　　37.　　B　　　41.　　B
　　38.　　A

42.   see 4.A.-C.

43.   see 8.E.4)

44.-49. [see 8.F.]
　　44.　　C　　　47.　　B
　　45.　　A　　　48.　　A
　　46.　　C　　　49.　　C

50.   see 2.A.1)-3)

51.   see 2.B.1)-2)

****

CHAPTER 6: CELLS: UNITS OF LIFE II. MEMBRANE STRUCTURE AND FUNCTION

CHAPTER OUTLINE.

In this chapter, the characteristics and functions of cell walls and cell membranes are discussed. Transport of molecules across cell membranes is emphasized, especially the processes of diffusion and osmosis. Finally, cell connections and communications are covered.

1. **Cell walls**: outer cell coverings of bacteria, plants, fungi and some protists.

    A. Characteristics.
        1) Stiff and non-living, containing mainly cellulose in plants or chitin in bacteria and fungi.
        2) Provides support and protection for otherwise fragile cells.
        3) Porous, permitting easy passage of small molecules.

    B. Formation in plants: new cells secrete vesicles containing sticky polysaccharides such as pectin that glue adjacent cells together, forming the middle lamella. The cells then secrete cellulose, forming the primary cell wall. Then, cellulose is secreted beneath the primary cell wall to produce the secondary cell wall.

2. **Cell (Plasma) membranes**.

    A. **Major functions**.
        1) Isolates (via phospholipids) the cytoplasm from the external environment.
        2) Regulates (via proteins and glycoproteins) the flow of materials into and out of the cell. Most water-soluble molecules (salts, amino acids, sugars) cannot pass through the lipid bilayer.

4) <u>Identifies</u> the cell as belonging to a particular member of a particular species.

B. **Structure.** "<u>Fluid mosaic model</u>": a sea of phospholipids (mosaic "grout") in which proteins (mosaic "tiles") are embedded. <u>Components</u>:
1) A double layer of phospholipids ("<u>lipid bilayer</u>") which moves around in the membrane, with:
a. <u>Hydrophilic heads</u> facing the watery edges of the bilayer (the cytoplasm and extracellular fluids), and
b. <u>Hydrophobic tails</u> tucked inside the bilayer.
2) <u>Cholesterol</u>: makes the bilayer stronger, more flexible but less fluid, and less permeable to water-soluble molecules.
3) <u>Membrane proteins</u>:
a. Embedded within or attached to the surface of the plasma membrane bilayer.
b. Held in place by interactions between hydrophobic amino acids and hydrophobic tails of the phospholipids.
c. Many have carbohydrates attached, forming <u>glycoproteins</u>.

C. **Functions** of membrane proteins:
1) <u>Transport Proteins</u>: regulate movement of most water-soluble molecules through the plasma membrane.
a. Some have pores ("<u>channel proteins</u>"), allowing small water-soluble molecules to penetrate.
b. Some allow ions to freely pass through.
c. Others bind to certain molecules and help them to pass through.
2) <u>Receptor Proteins</u>: molecules that trigger cellular responses when specific molecules (like hormones or nutrients) in the extracellular fluid bind to them.
a. Dozens of types per cell.
b. May trigger increased metabolic rates, cell division or movement, or secretion of hormones.
c. May trigger passage of ions through pores in transport proteins (as occurs in nerve cells).
3) <u>Recognition Proteins and Glycoproteins</u>: Identification tags and cell surface attachment sites. They allow cells of the immune system to avoid attacking normal body cells.

3. **Transport** across membranes.

A. <u>Definitions</u>.
1) A <u>fluid</u> is a liquid or a gas.
2) The <u>concentration</u> of a fluid is the number of molecules per unit volume.
3) A <u>gradient</u> is a physical difference between two regions (i.e., high vs. low concentration).

B. Movement of particles in fluids.
1) Movement of <u>individual</u> particles in a fluid is random.
2) <u>Collectively</u>, particles in a fluid move directionally in response to gradients from regions of high to regions of low concentration, pressure, or electrical charge

("down the gradient").

C.   Movement across membranes: may be active or passive.
   1)   <u>Passive transport</u>: movement <u>down</u> gradients. No energy is necessary to move molecules through the membrane down the gradient (like coasting downhill on a bike).
   2)   <u>Active</u> (or energy-requiring) <u>transport</u>: movement <u>up</u> or <u>against</u> gradients. Energy is required from cellular metabolism to move molecules through the membrane against a gradient (like pedaling a bike uphill). Proteins in the membrane control the direction molecules move across a membrane.

4.   **Passive transport** down concentration gradients (no energy required).

   A.   **Diffusion**: net movement of <u>all</u> molecules in a fluid from regions of higher concentration to regions of low concentration. Example: a mixture of dye in water becomes uniformly dispersed since the dye <u>and</u> the water molecules both move randomly. Specifically:
      1)   Diffusion requires a difference in concentration.
      2)   Net movement of <u>all</u> molecules is down a gradient from high to low concentration.
      3)   The steeper the gradient, the faster the diffusion.
      4)   Diffusion cannot rapidly move molecules over long distances.
      5)   Diffusion continues until the concentration gradients are eliminated.

   B.   Diffusion across membranes.
      1)   Membranes are <u>differentially</u> <u>permeable</u> since they allow some molecules to more easily permeate or pass through than others.
      2)   <u>Simple</u> <u>diffusion</u>: the passive diffusion of some molecules through the cell membrane's lipid bilayer. Examples: H_2O, O_2, CO_2, lipid soluble molecules. Rates of simple diffusion depend on:
         a.   The concentration gradient.
         b.   The size of the molecule.
         c.   The lipid solubility of the molecule.
      3)   <u>Facilitated</u> <u>diffusion</u>: the movement of water-soluble molecules (like ions, amino acids, monosaccharides) across the membrane, requiring the aid of membrane-transport proteins with pores.
         a.   <u>Channel proteins</u> have pores of specific size and characteristics that allow only particular ions to pass through.
         b.   <u>Carrier proteins</u> have amino acid regions that bind to certain molecules, then change shape and allow the molecule to pass through the membrane.
         c.   Facilitated diffusion occurs more slowly than simple diffusion. Rates depend on the concentration gradient and the density of the transport proteins in the membrane.
      4)   Molecules moving by either simple or facilitated diffusion can go in either direction across the membrane but always <u>down</u> the concentration gradient. Example: O_2 usually diffuses into cells while CO_2 usually diffuses out.

   C.   **Osmosis**: effect of diffusion of water across a membrane that does not allow some other molecules to cross due to the membrane's differential permeability.
      1)   Example: A thin plastic sac with tiny pores in it acts like a differentially permeable

membrane, allowing water but not sugar molecules to pass through.

    a.   Place a closed-off sac filled with 10% sugar-water solution in a glass of pure water and the sac will expand until it bursts. Why?

    b.   Since the concentration of pure water (100%) in the glass is greater than the concentration of water in the sac (90% H_2O and 10% sugar initially), more water enters the bag than leaves it at any time due to simple diffusion down the water gradient.

    c.   Thus, the sac expands until it bursts.

2)   <u>Osmotic</u> <u>pressure</u>: the amount of force or pressure necessary to keep osmosis from causing the solution with the lower water concentration from expanding.

3)   Osmosis is driven by <u>water</u> <u>concentration</u> <u>differences</u> between solutions, not by the type of dissolved molecules.

D.   Effects of osmosis on cells.

1)   Most cell membranes are highly permeable to water but much less permeable to other molecules. Osmosis in cells depends on differences between water concentration in the cytoplasm and in the extracellular fluid.

2)   Terms used in comparing concentrations of water and dissolved molecules in cytoplasm and extracellular fluid:

    a.   <u>Isotonic</u> ("equal strength") <u>solution</u>: cytoplasm and extracellular fluids have equal concentrations of water and dissolved molecules.

    b.   <u>Hypertonic</u> ("greater strength") <u>solution</u>: has the greater dissolved molecule concentration and the lesser water concentration.

    c.   <u>Hypotonic</u> ("lesser strength") <u>solution</u>: has the lesser dissolved molecule concentration and the greater water concentration.

    d.   Example: a cell's cytoplasm has a 1% salt concentration and the extracellular fluid has a 10% salt level. The cytoplasm is hypotonic and the extracellular fluid is hypertonic.

3)   Effect of osmosis on cells.

    a.   A cell placed in an <u>isotonic</u> salt solution will neither expand nor contract since water enters and leaves the cell at equal rates (equal water concentration outside and inside the cell).

    b.   A cell placed in <u>hypertonic</u> solution (like strong salt water) will shrink since more water leaves than enters (lower water concentration outside than inside the cell).

    c.   A cell placed in <u>hypotonic</u> solution (like pure water) will swell up since more water enters than leaves (greater concentration of water outside than inside the cell).

	RED BLOOD CELL	PLANT CELL
cell placed in hypotonic solution	HEMOLYSIS (cell expands and bursts)	TURGIDITY (cell becomes quite rigid)
cell placed in hypertonic solution	CRENATION (cell shrinks & shrivels up)	PLASMOLYSIS (cell gets soft & flaccid)

4) <u>Vacuoles</u> and cell volume.
a. <u>Contractile vacuoles</u>, found in some freshwater protozoans, collect water entering by osmosis, then contract to squirt the water out through exit pores in the plasma membrane.
b. <u>Central vacuoles</u>, found in many plant cells, have several functions and effects:
    i. Store hazardous wastes.
    ii. Store sugars and amino acids.
    iii. Store colored pigments.
    iv. As water enters, swelling of central vacuoles occurs and this creates <u>turgor pressure</u>, making the cell quite rigid.

5. **Energy-requiring transport** across membranes: movement of molecules against diffusion gradients ("uphill", from regions of lower to regions of greater concentration):

A. **Active Transport**: use of energy by specific membrane proteins to move individual molecules across the plasma membrane (allows cells to acquire or expel substances).
    1) These proteins have two active sites: one binds to the molecule being transported, the other binds to an energy-supplying molecule (usually ATP).
    2) The ATP donates the energy that allows the protein to move the molecule through the membrane, much like a pump uses energy to pull water from a well.

B. **Endocytosis** ("into the cell"): plasma membrane engulfs a particle and pinches off a membrane-bound sac (or <u>vesicle</u>), with the particle inside, into the cytoplasm. Types of endocytosis:
    1) <u>Pinocytosis</u> ("cell drinking" or fluid phase endocytosis): taking small molecules

48

into a cell by the dimpling in of a region of the plasma membrane, forming a vesicle in the cytoplasm.

2) <u>Receptor-mediated endocytosis</u>: acquiring particles by binding them to receptor proteins concentrated in areas of the plasma membrane called <u>coated pits</u>. Once bound to molecules, coated pits deepen into a U-shaped pocket that pinches off to form a vesicle in the cytoplasm.

3) <u>Phagocytosis</u> ("cell eating"): taking large particles (like bacteria) into a cell by forming extensions of the cell membrane called <u>pseudopods</u> ("false feet") which trap the particle inside a vesicle brought into the cytoplasm. This is how an <u>Amoeba</u> eats and how white blood cells kill bacteria.

C. **Exocytosis** ("out of the cell"): the reverse of endocytosis; how cells dispose of waste materials or secrete hormones into the extracellular fluid. Steps:
1) A cytoplasmic vesicle containing particles to be excreted moves to the inner cell surface;
2) The vesicle fuses with the plasma membrane and opens towards the extracellular fluid;
3) The vesicle contents diffuse out of the cell.

6. **Cell connections and communication**: plasma membranes also function in holding clusters of cells together, providing means of communication. Four types of connections are known:

A. <u>Desmosomes</u> (cell-to-cell adhesion): attachment between cells in tissues that are stretched, compressed, or bent as the organism moves. In a desmosome, proteins and carbohydrates glue adjacent cells together and intermediate fibers add strength to the connections.

B. <u>Tight Junctions</u> (leak-proofing): near-fusion of adjacent cells along a series of ridges, forming waterproof gaskets between cells in tubes or sacs which must hold their contents without leaking (Example: the urinary bladder).

C. <u>Gap Junctions</u> (cell-to-cell communication) : clusters of protein channels directly connecting the insides of adjacent cells in tissues where cell-to-cell communication is necessary.

D. <u>Plasmodesmata</u> (cell-to-cell communication): thousands of large tubes, lined with plasma membrane, that penetrate through cell walls and connect the cytoplasm of adjacent plant cells. These channels allow water, nutrients, and hormones to pass between cells.

****

REVIEW QUESTIONS.

1. Compare primary and secondary cell walls.
2. What are the four major functions of cell membranes?

3.-10. **MATCHING TEST**: cell walls and cell membranes

        Choices: A.   Cell wall
                               B.   Cell membrane

3. Contains cellulose in plants.
4. Isolates the cytoplasm from the external environment.
5. Regulates flow of materials into and out of cells.
6. Contains chitin in fungi.
7. Communicates with other cells.
8. Stiff, porous, and non-living.
9. "Fluid-mosaic model".
10. Has a lipid bilayer.

---

11. **True or False**: Phospholipids in the cell membrane have hydrophobic tails facing the watery edge of the bilayer.
12. Compare active and passive transport of molecules across cell membranes.
13. **True or False**: Proteins in a membrane control the direction molecules move across a membrane.

14.-17. **True or False** statements on diffusion:

14. Diffusion does not require a difference in concentration of molecules.
15. With diffusion, some molecules move down a concentration gradient and others move up the gradient since all movement is random.
16. The steeper the concentration gradient, the faster the rate of diffusion will be.
17. Diffusion continues even after all concentration gradients are eliminated.

---

18. Compare simple diffusion and facilitated diffusion.
19. **True or False**: Molecules must always diffuse into a cell from the outside.

20.-24. **MATCHING TEST**: diffusion and osmosis

        Choices: A.   Osmosis
                               B.   Diffusion

20. Effect of movement of all molecules down the concentration gradient.
21. Effect of water moving down its concentration gradient across a differentially permeable membrane.
22. Movement of O_2 into a cell and CO_2 out of a cell.
23. A cell expands when placed in pure water.
24. Can cause cells to shrink.

25.-32. **MATCHING TEST**: comparing concentrations

Choices: A.   Isotonic solution
                B.   Hypertonic solution
                C.   Hypotonic solution

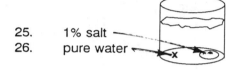

25. 1% salt
26. pure water

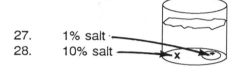

27. 1% salt
28. 10% salt

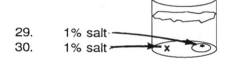

29. 1% salt
30. 1% salt

31. higher H_2O concentration
32. lower H_2O concentration

33.-38. **MATCHING TEST**: effect of osmosis on cells

Choices: A.   Cells placed in hypotonic solution
                B.   Cells placed in hypertonic solution
                C.   Cells placed in isotonic solution

33. Animal cells will expand.
34. Animal cells will shrivel up.
35. Red blood cells will burst.
36. Celery will wilt.
37. Lettuce leaves will become turgid (rigid, crisp) in fluid.
38. Red blood cells will neither shrivel up nor swell up.

-----

39. Explain the differences between active transport and facilitated diffusion.
40. Explain the differences between pinocytosis and phagocytosis.

41.-44. **MATCHING TEST**: cell connections and communications

Choices: A.   Tight junctions   C.   Gap junctions
                B.   Plasmodesmata   D.   Desmosomes

41. Clusters of protein channels for communication between cells.
42. Waterproof gaskets between cells.
43. Attachments between cells that are stretched, compressed, or bent as organisms move.
44. Large, membrane-bound tubes for water passage between cells.

****

ANSWERS TO REVIEW QUESTIONS:

1. see 1.B.

2. see 2.A.1)-4)

3.-10. [see 1. and 2.]

3.	A		7.	B
4.	B		8.	A
5.	B		9.	B
6.	A		10.	B

11. False; [see 2.B.1)a.-b.]

12. see 3.C.1) and 2)

13. True; [see 3.C.3)]

14.-17. [see 4.A.1)-5)]

14.	False	16.	True
15.	False	17.	False

18. see 4.B.2)-4)

19. False; [see 4.B.4)]

20.-24. [see 4.A.-C.]

20.	B	23.	A
21.	A	24.	A
22.	B		

25.-32. [see 4.D.2)a.-d.]

25.	B	29.	A
26.	C	30.	A
27.	C	31.	C
28.	B	32.	B

33.-38. [see 4.D.3)a.-c.]

33.	A	36.	B
34.	B	37.	A
35.	A	38.	C

39. see 5.A. and 4.B.

40. see 5.B.1)-3)

41.-44. [see 6.A.-D.]

41.	C	43.	D
42.	A	44.	B

****

CHAPTER 7: PHOTOSYNTHESIS: TAPPING SOLAR ENERGY

CHAPTER OUTLINE.

The process of photosynthesis is detailed in this chapter. The light dependent and light independent steps of photosynthesis in chloroplasts are outlined, and the two major ways used by plants to trap carbon dioxide are presented.

1. **General facts.**

   A.   The flow of energy through life on today's earth begins with the <u>sun</u>.

   B.   However, when the earth formed, 4.5 billion years ago, matter collided and fused to release heat.
      1)   Storms and volcanos released more energy, but no organisms existed to trap the energy.
      2)   Energy rich molecules formed, using heat and lightning energy, and accumulated.
      3)   As the earth cooled, living cells arose, feeding on the chemical "soup."
      4)   The cells gradually consumed most of the organic molecules in the soup, and sources of organic energy became scarce.
      5)   Through chance, some cells began to use sunlight energy to make organic molecules like glucose (rich in chemical energy) from inorganic molecules of CO_2 and H_2O (poor in chemical energy), a process called <u>photosynthesis</u>.
      6)   The most common type of photosynthesis releases O_2 into the atmosphere, breaking down organic molecules.
      7)   When cells metabolize glucose without oxygen ("glycolysis"), only a small amount of the light energy trapped in glucose by photosynthesis is extracted again to drive cellular reactions.
      8)   When oxygen is used to break down glucose ("cellular respiration"), 18 to 19 as much energy is released as occurs during glycolysis. Cells that respire could

grow and reproduce faster.

C.   Photosynthesis and cellular respiration are complementary processes which evolved about 2 billion years ago, driving the flow of energy and the cycling of carbon through organisms and ecosystems:

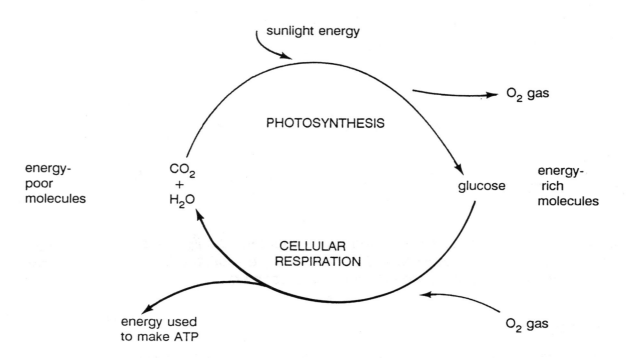

2.   **Photosynthesis: an overview.**

A.   <u>Photosynthesis</u>: uses the energy of sunlight to make energy-rich products (glucose and O_2) from energy poor reactants (CO_2 and H_2O)
       1)   Photosynthesis converts electromagnetic sunlight energy into chemical energy stored in the chemical bonds of glucose and O_2.
       2)   Overall chemical reaction of photosynthesis:

$$6\ CO_2 + 6\ H_2O + energy \rightarrow C_6H_{12}O_6 + 6\ O_2$$

       3)   In plants, photosynthesis occurs within chloroplasts, most of which are in leaf cells.

B.   <u>Leaves, chloroplasts, and photosynthesis.</u>
       1)   Leaves generally are only a few cell layers thick.
            a.   Upper and lower leaf surfaces consist of a layer of transparent cells (<u>epidermis cells</u>).
            b.   The outer epidermal cell layers are covered by a waxy <u>cuticle</u> layer to reduce water evaporation from the leaf.

      i.   Leaves obtain CO_2 for photosynthesis from the air.

      ii.   Adjustable epidermal pores, called <u>stomata</u>, open and close and this affects CO_2 movement into leaves.

    c.   Inside of leaves are cells called <u>mesophyll</u> ("middle of leaf").

      i.   These contain most of a leaf's chloroplasts.

      ii.   Vascular bundles (veins) supply water and minerals to mesophyll cells and carry away sugars made there.

2)   <u>Chloroplasts</u>: organelles with a double outer membrane enclosing a semi-fluid medium (<u>stroma</u>).

    a.   Stroma contains <u>thylakoids</u> (disc-shaped, interconnected membranous sacs).

    b.   The thylakoids are piled atop each other in stacks called <u>grana</u>.

3)   Photosynthesis involves many enzymes catalyzing many reactions.

    a.   Conceptually, it can be seen as two reactions coupled together by energy carrier molecules (ATP and NADPH):

      i.   <u>Light dependent reactions</u>: chlorophyll and other molecules in thylakoid membranes capture sunlight energy, converting some into chemical energy of ATP and NADPH.

      ii.   <u>Light independent reactions</u>: stroma enzymes use chemical energy from ATP and NADPH to make glucose and other organic molecules.

3.   **The light dependent reactions**: Converting light to chemical energy.

    A.   <u>Light, chloroplast pigments, and photosynthesis.</u>

      1)   Solar radiation comes in many wavelengths, from short (high energy) gamma rays through ultraviolet, visible and infrared light, to long (low energy) radio waves.

      2)   Solar radiation is composed of energy packets (or "<u>photons</u>") ranging from high energy photons (short wavelength radiation) to low energy photons (long wavelength radiation).

      3)   <u>Visible light</u> contains the wavelengths with energies most useful to living cells.

      4)   One of 3 processes may occur when light strikes an object like a leaf; the light may be:

        a.   <u>Absorbed</u> (its energy drives biological processes).

        b.   <u>Reflected</u> or bounced back (gives color to an object).

        c.   <u>Transmitted</u> or passed through (gives color).

      5)   Several types of molecules absorb different wavelengths of light in chloroplasts:

        a.   <u>Chlorophyll</u>: the key light-capturing molecule in thylakoid membranes; absorbs blue and red light but reflects green and thus, appears green to us.

        b.   <u>Accessory pigments</u>: capture light energy and transfer it to chlorophyll:

          i.   <u>Carotenoids</u>: absorb blue and green light, and appear yellow or red.

          ii.   <u>Phycocyanins</u>: absorb green and yellow, and appear blue or purple.

        c.   Because of these three types of pigments, all wavelengths can drive photosynthesis to some extent.

B.   <u>The light dependent reactions</u>.

    1)   Thylakoid membranes (in stacks called "grana") contain chlorophyll, accessory pigments, and electron-carrier molecules assembled into thousands of <u>photosystems</u> of 2 types (I and II).

    2)   Each photosystem has 2 major parts:
       a.   A <u>light harvesting complex</u>, and
       b.   An <u>electron transport system</u> (ETS).

    3)   Steps in energy capture:
       a.   The light harvesting complex has about 300 chlorophyll and accessory pigment molecules (called "<u>antenna molecules</u>") which absorb light energy and pass it to
       b.   The "<u>reaction center</u>" (a specific chlorophyll molecule). An electron in this molecule absorbs the energy, leaves the molecule, and jumps over to the ETS.
       c.   The electron is then passed through a series of ETS carrier molecules (embedded in the thylakoid membrane) which slowly extract its energy and use it to make ATP and NADPH, a process called <u>photophosphorylation</u>.

    4)   <u>Photosystem II</u>: generates ATP.
       a.   STEP 1: Photons of light are absorbed by antenna molecules of <u>photosystem II</u>:
       b.   STEP 2: The energy is passed to the reaction center where it boosts an electron out of the chlorophyll molecule.
       c.   STEP 3: The reaction center's energized electron is accepted by the first electron carrier of the ETS.
       d.   STEP 4: The electron is passed from carrier to carrier in the ETS, releasing energy used to make ATP by a process called <u>chemiosmosis</u>. Meanwhile,

    5)   <u>Photosystem I</u>: generates NADPH.
       a.   STEP 5: Photons of light are absorbed or harvested by antenna molecules of <u>photosystem I</u>:
       b.   STEP 6: The energy is passed to the reaction center where it boosts an electron out of the chlorophyll molecule.
       c.   STEP 7: This energized electron is accepted by the ETS.
       d.   STEP 8: To replace it's lost electron, photosystem I's reaction center receives the electron leaving photosystem II:
          i.   Photosystem I's high energy electron moves through its ETS to an $NADP^+$ molecule.
          ii.   Eventually, each $NADP^+$ molecule picks up 2 electrons from photosystem I's ETS and one H^+ ion to form NADPH.
       e. $NADP^+$ and NADPH both are water soluble molecules dissolved in the stroma.

    6)   <u>Splitting water</u> maintains the flow of electrons through the photosystems.
       a.   Overall electron flow: Reaction center of photosystem II → Photosystem II's ETS → Reaction center of photosystem I → photosystem I's ETS → NADPH.
       b.   To maintain this one-way flow of electrons, photosystem II's reaction center must get new electrons to replace the ones it gives up.
       c.   STEP 9: Photosystem II's reaction center chlorophyll attracts electrons

from water molecules within the thylakoid compartment, causing the water molecules to split apart:

$$2 H_2O \rightarrow O_2 \text{ gas} + 4 H^+ + 4 \text{ electrons}$$

d.   The fate of water:

$$2 H_2O \rightarrow \begin{cases} O_2 \text{ gas (given off to the atmosphere or used in the cell)} \\ 4 \text{ electrons (taken up by photosystem II's reaction center)} \\ 4 H^+ \text{ (attached to 4 NADP}^+\text{)} \end{cases}$$

7)   Diagram of the light dependent reactions (photosystems I and II):

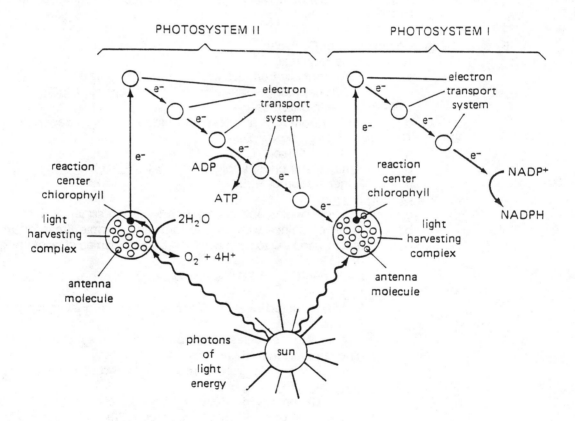

8)   <u>Simplified summary</u> of light dependent reactions: using sunlight energy, chlorophyll splits water, releasing oxygen gas, and hydrogen ions and energy used to make ATP and NADPH.

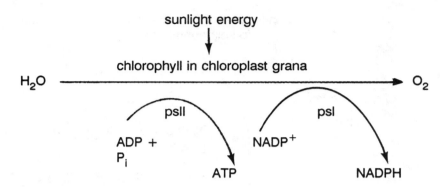

4.   **The light independent reactions.** Securing chemical energy in glucose molecules.

A.   The ATP and NADPH made in the light-dependent reactions are dissolved in the stroma.
1)   They will provide energy for making glucose from CO_2 and H_2O.
2)   The light-independent reactions can occur independently of light if ATP and NADP are available.

B.   Calvin-Benson cycle or C_3 cycle:
1)   The C_3 cycle requires:
a. CO_2, normally from the air.
b.   A CO_2-capturing sugar, ribulose bisphosphate (RuBP).
c.   Catalytic enzymes.
d.   Energy from ATP and NADPH made in the light-dependent reactions.
2)   The three parts of the C_3 cycle:
a.   Carbon fixation.
b.   Synthesis of phosphoglyceraldehyde (PGAL).
c.   Regeneration of RuBP.
3)   Carbon fixation.
a. C_3 cycle begins and ends with a 5-carbon sugar, RuBP.
b.   RuBP combines with CO_2 to form an unstable 6-carbon molecule that reacts with H_2O to form two 3-carbon molecules of phosphoglyceric acid (PGA).
c.   Capturing CO_2 in PGA is called carbon fixation.
4)   Synthesis of PGAL.
a.   Energy donated by ATP and NADPH is used to enzymatically convert PGA to PGAL.
b.   Two PGAL molecules (3 carbons each) may combine to become a glucose molecule (6-carbons).
c.   PGAL may also be used to make lipids and amino acids.
5)   Regeneration of RuBP.
a.   Using ATP energy, 10 PGAL molecules (10 x 3-carbons = 30) can make 6 molecules of RuBP (6 x 5-carbons = 30).
b.   Overall, to capture 6 CO_2 molecules as one glucose molecule, 6 molecules of RuBP enter and are regenerated in each "turn" of the C_3 cycle.

C.   <u>Simplified summary</u> of light independent reactions: using ATP and NADPH generated in the light dependent reactions, chloroplast stroma enzymes use CO_2 molecules to make glucose.

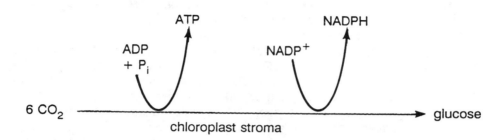

5.   **Relation between light-dependent and light-independent reactions.**

A.   <u>Relationship between light dependent and light independent reactions</u>: they are closely coordinated and coupled together by the energy-carriers ATP and NADPH: both reaction series are necessary for photosynthesis.

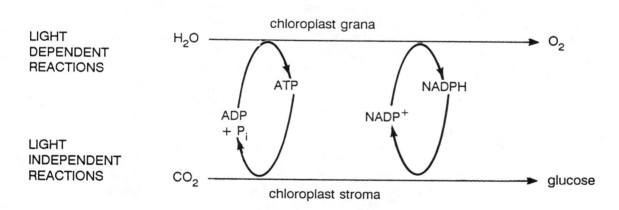

B.   Water, carbon dioxide, and the <u>C_4 pathway</u>:
1)   Factors limiting the rate of photosynthesis are the amounts of light, water, and carbon dioxide.
2)   To allow sufficient CO_2 to enter a leaf, plants must keep <u>stomata</u> (adjustable pores in the waterproof leaves of broadleaf plants) open.
   a.   But in dry weather this allows too much water to escape.
   b.   When plants close their stomata in very hot, dry weather, the C_3 cycle may shut down due to lack of enough CO_2 to combine with RuBP.
3)   <u>Photorespiration</u>: what happens to carbon fixation when the stomata close, allowing O_2 to rise and CO_2 levels to fall within leaves.

        a.   Depending on the relative concentrations of gases, RuBP can combine with either CO_2 or O_2.

           i.   The O_2 + RuBP reaction is the first step in <u>photorespiration</u>, ultimately generating CO_2 from RuBP.

           ii.   Photorespiration not only prevents CO_2 fixation but causes carbon loss from RuBP.

        b.   During hot, dry weather, stomata seldom open.

           i. CO_2 in leaves declines as photosynthesis occurs.

           ii.   New CO_2 can't get in and O_2 from photosynthesis can't get out.

           iii.   Result: Photorespiration predominates, and plants consume all their carbohydrate supplies and die.

4)   <u>Two-stage carbon fixation in C_4 plants</u>: a way to reduce photorespiration and increase photosynthesis in dry, hot weather.

        a.   In C_3 plant leaves, almost all chloroplasts reside in mesophyll cells, while bundle sheath cells around veins lack chloroplasts.

        b. C_4 plants (e.g., corn and crabgrass) are adapted to hot, dry weather since both mesophyll and bundle sheath cells have chloroplasts, allowing 2-step carbon fixation.

        c.   <u>First stage</u>: initial carbon fixation.

           i.   Mesophyll cells of C_4 plants contain 3-carbon phosphoenolpyruvate (<u>PEP</u>) molecules instead of RuBP.

           ii. CO_2 reacts with PEP to make 4-carbon oxaloacetic acid molecules (hence, C_4 plants).

           iii.   This reaction is not hindered by high O_2 levels.

        d.   <u>Second stage</u>: shuttling carbon into the C_3 cycle.

           i.   Oxaloacetic acid moves into bundle sheath cells and breaks down, releasing CO_2 there.

           ii.   This high CO_2 concentration in bundle sheath cells allows the RuBP + CO_2 reaction to fix carbon in the regular C_3 cycle.

           iii.   The remnants of the shuttle molecules return to mesophyll cells where ATP energy is used to regenerate PEP.

5)   <u>Comparison</u> of the C_3 and C_4 pathways:

        a. C_4 uses more energy (PEP must be regenerated) than C_3.

        b. C_4 is better when lots of light but little water is present since "expensive" photosynthesis is better than none at all.

        c. C_3 is better when lots of water (stomata stay open, lots of CO_2 enters) but dim light (little ATP is made) is present.

        d.   Thus, C_3 plants such as Kentucky bluegrass have an advantage in cool, wet, cloudy climates (spring and fall) while C_4 plants such as crabgrass have the advantage in hot, dry, sunny climates (like mid-summer in the temperate zone).

****

REVIEW QUESTIONS:

1.   Review the steps in energy capture and energy transfer that developed during the early history of the earth.

2.   Diagram how photosynthesis and cellular respiration are complementary processes.

3.   Name and describe the functions of the three major types of pigments found in chloroplasts.

4.-17. **MATCHING TEST:** photosynthesis

        Choices: A. Light dependent reactions   B. Light independent reactions

4. CO_2 is captured and converted into sugars.
5.   Light energy is converted into chemical energy of ATP and NADPH.
6.   Occurs in chloroplast grana.
7.   Uses chemical energy to make glucose.
8.   Uses chlorophyll, carotenoids, and phycocyanins to trap light energy.
9.   Calvin-Benson, or C_3 cycle.
10.   Energy obtained from NADPH and ATP.
11.   Produces oxygen gas.
12.   Thylakoid membranes.
13.   Photosystems I and II.
14.   Carbon fixation occurs.
15.   Involves electron transport.
16.   Occurs in chloroplast stroma.
17.   Water is split into oxygen and hydrogen.

----

18.   Draw a diagram to summarize the workings of photosystems I and II.
19.   Explain what occurs during the light independent reactions of photosynthesis.
20.   Diagram a simplified summary of the light dependent and light independent reactions of photosynthesis, showing how they are interrelated by energy-carrier molecules.
21.   Compare the C_4 and C_3 pathways for carbon fixation during photosynthesis.

****

ANSWERS TO REVIEW QUESTIONS:

1.   see 1.B.1)-8)

2.   see 1.C.

3.   see 3.A.5)

4.-17. [see 3.B.-4.B.]

4.	B	11.	A
5.	A	12.	A
6.	A	13.	A
7.	B	14.	B
8.	A	15.	A
9.	B	16.	B
10.	B	17.	A

18.   see 3.B.7)

19.   see 4.A.-C.

20.   see 5.A.

21.   see 5.B.1)-5)

****

CHAPTER 8: GLYCOLYSIS AND CELLULAR RESPIRATION: HARVESTING ENERGY FROM FOOD

CHAPTER OUTLINE.

The processes of glycolysis and cellular respiration are detailed in this chapter. Glycolysis and fermentation are explained, and the role of mitochondria in converting the chemical energy of sugars into ATP energy during aerobic respiration is presented.

1. **General facts.**

   A. Photosynthesis converts sunlight energy into chemical energy stored in molecules like sugars and fats. However, plants and animals use ATP, not sugar or fat, as their immediate energy source.

   B. Most cells produce ATP using energy from other molecules.
       1) This chapter focuses on the metabolism of glucose since every cell metabolizes glucose for energy at least part of the time.
       2) When other molecules are used as energy sources, they are converted either into glucose or into molecules that enter into the process of glucose metabolism.

2. **Glucose metabolism**: An overview.

   A. The chemical equations for glucose synthesis by photosynthesis and complete metabolism of glucose are symmetrical:

PHOTOSYNTHESIS

$$6\ CO_2\ +\ 6\ H_2O\ +\ \text{sunlight energy} \rightarrow C_6H_{12}O_6\ \text{(glucose)}\ +\ 6\ O_2$$

COMPLETE GLUCOSE METABOLISM

$$C_6H_{12}O_6\ \text{(glucose)}\ +\ 6\ O_2 \rightarrow 6\ CO_2\ +\ 6\ H_2O\ +\ \text{heat and chemical energy}$$

---

    1)   Most of the energy released from complete glucose metabolism is heat.

    2)   Cells, however, can extract enough chemical energy from complete glucose metabolism to make a great deal of ATP.

  B.   <u>Steps in glucose metabolism</u> in eukaryotic cells:

    1)   <u>Glycolysis:</u>

        a.   Is <u>anaerobic</u> (no O_2 required) but also occurs in aerobic conditions (with oxygen present).

        b.   Occurs in the cytosol.

        c.   <u>Glucose</u> is split in half, forming 2 molecules of <u>pyruvic acid</u> and releasing a little energy to make 2 ATP molecules.

        d.   If no oxygen is present, <u>fermentation</u> occurs in the cytosol, converting the pyruvic acid into either ethanol or lactic acid; no additional ATP is made.

        e.   If oxygen is present, <u>cellular respiration</u> occurs.

    2)   <u>Cellular respiration:</u>

        a.   Is <u>aerobic</u> (O_2 required).

        b.   Occurs in the mitochondria, where the pyruvic acids are completely broken down to CO_2 and water, releasing enough useable energy to make 34 to 36 ATP molecules.

3.   **Glycolysis** ("to break apart a sweet"):

  A.   <u>General steps.</u>

    1)   <u>Glucose activation:</u> use of energy from 2 ATPs to change stable glucose into a highly reactive molecule of fructose diphosphate.

    2)   <u>Energy harvest:</u>

        a.   Fructose diphosphate splits into 2 molecules of PGAL (<u>p</u>hospho<u>g</u>lycer<u>a</u>ldehyde) which eventually become 2 pyruvic acids, generating 4 ATPs in the process.

        b.   Also, electrons and H atoms given off in these reactions are picked up by the electron carrier NAD^+ to make 2 energized NADH molecules.

        c.   Thus, there is a net gain of 2 ATP molecules for each glucose molecule broken apart during glycolysis.

    3)   The electron energy carried in NADH can only be used to make ATP when oxygen is available.

        a.   Under anaerobic conditions, NADH production is not a method of energy capture. Instead, it's a way of getting rid of hydrogens and electrons during glycolysis.

b. NAD^+ is used up as it accepts electrons and hydrogens to become NADH.

c.   Cells must dispose of the hydrogens and electrons, and regenerate NAD^+, or else glycolysis will stop.

4.   **Fermentation.**

A.   Under anaerobic conditions, pyruvic acid can build up to toxic levels and NAD^+ molecules can be depleted. Cells convert pyruvic acid and regenerate NAD^+ (for further glycolysis) by one of two fermentation processes.

B.   Lactic acid fermentation (in some bacteria and animal cells, especially muscles):

1)   During vigorous exercise, muscle cells use up O_2 in cellular respiration faster than it can be supplied by breathing.

2)   Under these conditions, pyruvic acid is converted to lactic acid, regenerating NAD^+:

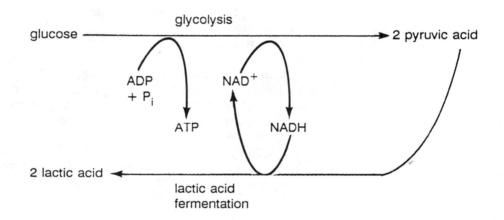

3)   The regenerated NAD^+ can again accept electrons during glycolysis, and energy production can continue.

4)   Lactic acid, however, is toxic in high concentrations, leading to fatigue.

5)   After exercise ceases, the lactic acid is transported to the liver, where it is converted back to pyruvic acid.

6)   Many microorganisms use lactic acid fermentation, including the bacteria that produce yogurt, sour cream, and certain cheeses.

C.   Alcoholic fermentation (in many microorganisms, including baker's and brewer's yeast):

1)   Under anaerobic conditions, pyruvic acid is converted to ethyl alcohol (ethanol) and CO_2, regenerating NAD^+:

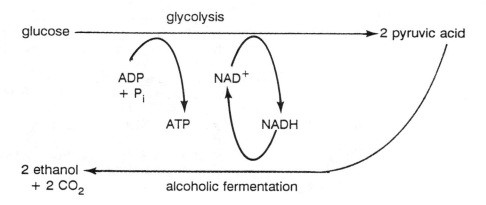

2)   This is useful in both the baking industry (CO_2 is used to make bread "rise") and the alcohol industry (stills explode if the CO_2 is not allowed to escape).

5.   **Cellular respiration**: occurs in eukaryotic cells under aerobic conditions in mitochondria.

A.   Relation to mitochondrial structure:
   1)   In the <u>matrix</u>, pyruvic acid is converted into CO_2 with much energy captured in a few ATP and many electron carriers.
   2)   In the <u>inner</u> <u>membrane</u>, the energized electron carriers are used to make most of the ATP produced by glucose metabolism.

B.   The main events of cellular respiration (see Text Figure 8-4).

   1)   <u>Step 1</u>. Glycolysis in the cell cytosol: glucose → 2 pyruvic acids
   2)   <u>Step 2</u>. The pyruvic acids are transported into the mitochondrial matrix.
   3)   <u>Step 3</u>. Each pyruvic acid → CO_2 + a 2-carbon acetyl group → citric acid cycle → 2 CO_2 + 1 ATP + energetic electrons for several electron carrier molecules.
   4)   <u>Step 4</u>.
      a.   Electron carriers donate their energetic electrons to the electron transport system of the inner membrane.
      b.   The energy of the electrons transports H^+ ions from the matrix to the intermembrane compartment, creating a concentration gradient.
      c.   Electrons combine with O_2 and H^+ to make H_2O.
   5)   <u>Step 5</u>.   Chemiosmosis: the H^+ ion gradient discharges through ATP-synthesizing enzymes in inner membrane, and the energy is used to make much ATP.
   6)   <u>Step 6</u>. The ATP is transported into the cytosol.

C.   <u>Pyruvic acid transport</u>.
   1)   The outer membrane of mitochondria contains many large pores, and is highly permeable to pyruvic acid.
   2)   Facilitated diffusion through proteins in inner membrane allows pyruvic acid to enter mitochondrial matrix.

D.   Matrix reactions:
1)   Each pyruvic acid reacts with a coenzyme A (CoA) molecule, producing a CO_2, an acetyl-CoA, and an NADH molecule containing an energetic electron and an H^+ from the pyruvic acid.
2)   The 2 acetyl-CoAs enter the citric acid or Krebs cycle:
   a.   Each acetyl-CoA combines with oxaloacetic acid to form citric acid and coenzyme A.
   b.   In a series of reactions, each citric acid is eventually converted to oxaloacetic acid and 2 molecules of CO_2, with most of the released energy captured as one ATP and 4 electron carriers (one $FADH_2$ [flavin adenine dinucleotide] and 3 NADH).
   c.   Simplified summary of the matrix reactions:

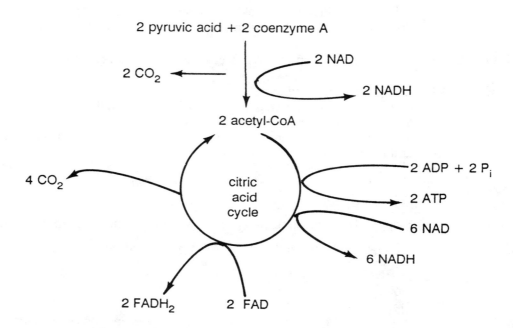

E.   Electron Transport.
1)   During glycolysis and the matrix reactions, one glucose molecule has produced:
   a.   2 NADH and 2 ATP during glycolysis.
   b.   8 NADH, 2 $FADH_2$, and 2 ATP during the matrix reactions.
2)   The 10 NADH and 2 $FADH_2$ electron carriers move to the inner mitochondrial membrane and deposit their electrons into an electron transport system (ETS) there.
   a.   As the electrons move from molecule to molecule along the ETS, the energy they release is used to make ATP by chemiosmosis.
   b.   Each NADH makes 3 ATP molecules.
   c.   Each $FADH_2$ makes 2 ATP molecules.
   d.   Thus, 34 ATPs are made during the inner membrane reactions:

10 NADH	x	3 ATP each	= 30 ATP
2 FADH$_2$	x	2 ATP each	= 4 ATP
			TOTAL 34 ATP

3) <u>Oxygen</u> accepts electrons and H$^+$ ions at the end of the ETS chain, forming water molecules. This clears out electrons from the ETS chain, allowing more to enter and release their energy to make more ATP. If no oxygen is present, the ETS chain gets clogged with electrons and ceases its function, stopping cellular respiration.

F. <u>Chemiosmosis</u>.
1) H$^+$ ion pumping across the inner membrane generates a concentration gradient (concentration of H$^+$ high in intermembrane compartment and low in the matrix).
2) During chemiosmosis, H$^+$ ions move down their gradient from intermembrane compartment to matrix, through ATP-synthesizing enzymes.
3) Flow of H$^+$ ions provides energy to make 34 ATP molecules from ADP and P$_i$.

G. <u>ATP transport</u>.
1) The outer mitochondrial membrane is very permeable to ATP, ADP and P$_i$.
2) Proteins in the inner mitochondrial membrane exchange matrix ATP and cytosol ADP and P$_i$.

H. <u>Summary</u> of glucose metabolism under aerobic conditions:

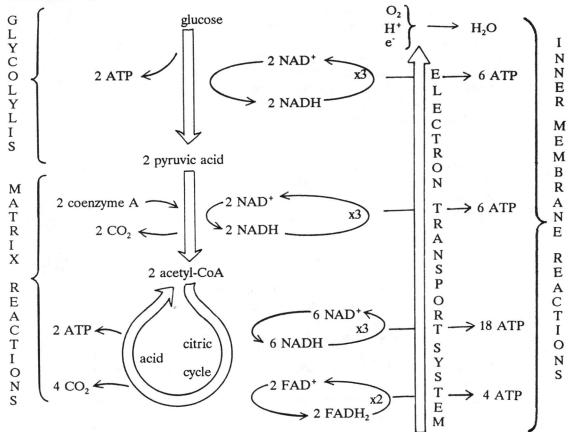

So, under aerobic conditions, for each glucose molecule broken down into CO_2 and H_2O:

Glycolysis generates	2 ATPs
Citric acid cycle generates	2 ATPs
Electron transport system generates	34 ATPs
TOTAL:	38 ATPs

****

REVIEW QUESTIONS:

1.-17. **MATCHING TEST**: glucose metabolism

           Choices:

- A. Glycolysis
- B. Fermentation
- C. Both A. and B.
- D. Cellular respiration
- E. None of these

1.   Most of the ATP is made.
2.   Occurs only under anaerobic conditions.
3.   Occurs only under aerobic conditions.
4.   Photophosphorylation.
5.   Occurs under either anaerobic or aerobic conditions.
6.   Can occur under anaerobic conditions.
7.   Glucose is split into 2 pyruvic acid molecules.
8.   Occurs in mitochondria.
9.   PGAL (phosphoglyceraldehyde).
10. Lactic acid.
11. Occurs in the cytoplasm.
12. Produces CO_2 and ATP.
13. Requires some ATP energy to get started.
14. Ethanol.
15. Acetyl-CoA.
16. Citric acid cycle.
17. Fructose diphosphate.

----

18. Sequence the following to illustrate the flow of energy:
   - A. ATP energy from photophosphorylation.
   - B. Sunlight energy.
   - C. ATP energy from metabolism of glucose.
   - D. glucose energy.
19. Explain, using chemical equations, how photosynthesis and aerobic cellular respiration are "symmetrical."
20. How is NAD^+ useful and necessary in glucose metabolism?

21. How do lactic acid fermentation and alcoholic fermentation differ? How are they similar?
22. What are the main events of cellular respiration?
23. Diagram a simplified summary of lactic acid fermentation.
24. Diagram a simplified summary of alcoholic fermentation.
25. Diagram a simplified summary of the matrix reactions of cellular respiration.
26. Explain what is meant by "chemiosmosis."
27. Diagram a simplified summary of glucose metabolism under aerobic conditions (glycolysis and cellular respiration).

****

ANSWERS TO REVIEW QUESTIONS:

1.-17.

1.	D	9.	A		20.	see 3.A.2)-3)
2.	B	10.	B			
3.	D	11.	C		21.	see 4.B.-C.
4.	E	12.	D			
5.	A	13.	A		22.	see 5.B.1)-6)
6.	C	14.	B			
7.	A	15.	D		23.	see 4.B.2)
8.	D	16.	D			
		17.	A		24.	see 4.C.1)
					25	see 5.D.2)c.

18. B → A → D → C

19. see 2.A.

20. see 3.A.2)-3)

21. see 4.B.-C.

22. see 5.B.1)-6)

23. see 4.B.2)

24. see 4.C.1)

25 see 5.D.2)c.

26. see 5.F.1)-3)

27. see 5.H.

****

UNIT I EXAM:

<u>Chapters</u> <u>1</u> *(Introduction)* - <u>8</u>. *All questions are dichotomous. Circle the correct choice in each case.*

Chapter <u>1</u>.

1.   Biology is (basically different from / essentially similar to) other sciences.
2.   The basic assumptions of science (can / cannot) be proven.
3.   The conclusions of science are (permanent / temporary).
4.   Science accepts only (natural / supernatural) explanations for natural processes.
5.   Creationism (is / is not) a science.
6.   Organisms that produce their own food are (autotrophic / heterotrophic).
7.   Monera are (eukaryotic / prokaryotic) organisms.
8.   Prokaryotic forms (do / do not) possess distinct nuclei.
9.   Fungi are (autotrophic / heterotrophic).
10. Redi's experiments (disproved / supported) the theory that life can occur by spontaneous generation.

Chapter <u>2</u>.

1.   The smallest unit of matter in an element is the (atom / molecule).
2.   A positive unit in an atom is the (neutron / proton).
3.   An electron is (heavier / lighter) than a proton.
4.   The number of protons in an atom determines its atomic (number / weight).
5.   (Helium / Hydrogen) is more likely to explode.
6.   Electrons close to a nucleus have (less / more) energy than those farther away from the nucleus.
7.   In salt, sodium and chlorine atoms attract each other by forming (covalent / ionic) bonds.
8.   The sodium in salt tends to (give up / take on) an electron.
9.   In a polar water molecule, the hydrogen region has a (negative / positive) electrical charge.
10. When atoms share electrons, they form (covalent / ionic) bonds.

Chapter 3.

1. Sucrose is a (disaccharide / monosaccharide).
2. In carbohydrates, the amount of carbon is equal to the amount of (hydrogen / oxygen).
3. When two monosaccharides become a disaccharide, a water molecule is (added / taken away).
4. Two glucose molecules join together to form (maltose / sucrose).
5. Animals store their food in the form of (glycogen / starch).
6. (Carbohydrates / Fats) have more energy per gram.
7. Fats are made of fatty acids and (cholesterol / glycerol).
8. The (hydrophilic / hydrophobic) portion of a phospholipid is located in the middle of the cell membrane.
9. The acid portion of an amino acid is the ($COOH$ / NH_2) group.
10. The sequence of amino acids is the (primary / secondary) structure of a protein.

Chapter 4.

1. Within a closed system, the amount of energy is (constant / variable) through time.
2. A fire (changes / creates) energy.
3. The (first / second) law of thermodynamics is concerned with entropy.
4. Photosynthesis and similar reactions (decrease / increase) entropy.
5. Eventually, all molecules in the universe will become (randomly dispersed / totally concentrated).
6. Reactions that release energy are (endothermic / exothermic).
7. Enzymes (decrease / increase) the activation energy needed for chemical reactions to occur.
8. ATP contains (deoxyribose / ribose) sugar.
9. Conversion of ATP to ADP (releases / stores) energy.
10. In coupled reactions, the "downhill" reaction liberates (less / more) energy than the "uphill" reaction.

Chapter 5.

1. More primitive types of cells are called (eukaryotic / prokaryotic) cells.
2. The presence of large numbers of ribosomes is characteristic of (rough / smooth) endoplasmic reticulum.
3. The majority of the hereditary material is found in the (cytoplasm / nucleus).
4. (Chloroplasts / Mitochondria) are associated with release of energy from sugar.
5. (Chloroplasts / Mitochondria) are associated with the storage of energy in sugar.
6. Lipid-producing enzymes are more common in (rough / smooth) endoplasmic reticulum.
7. (Animals / Plants) store food in the form of starch.
8. (Animals / Plants) are more likely to have vacuoles.
9. Higher (animals / plants) lack ciliated or flagellated cells.
10. Cilia are (longer / shorter) than flagella.

Chapter 6.

1. As a cell increases in size, its (surface area / internal volume) increases more rapidly.
2. Red blood cells will (burst / shrink) when placed in fresh water.
3. The water-loving portion of a compound is (hydrophilic / hydrophobic).
4. The rate of diffusion can be increased by (decreasing / increasing) the temperature.
5. In diffusion, molecules move toward regions of (higher / lower) concentration.
6. More water will enter a cell if it is placed in a (hypertonic / hypotonic) solution.
7. Solutions with higher salt concentrations than a cell are (hypertonic / hypotonic) when compared to the cell.
8. Fresh-water organisms deal constantly with the tendency of their cells to (gain / lose) water.
9. Endocytosis is the movement of substances (into / out of) cells.
10. The movement of a solid substance into a cell is called (phagocytosis / pinocytosis).

Chapter 7.

1. (Aerobic respiration / Photosynthesis) probably evolved first.
2. Chloroplasts are associated with (grana / stroma).
3. Glucose is synthesized in the (grana / stroma).
4. The most energetic wavelengths of light are (blue / red).
5. Ultraviolet light is visible to (humans / insects).
6. Light energy is first captured in (photosystem I / photosystem II).
7. As electrons are transferred from one carrier to another, the electrons (gain / lose) energy.
8. Photosynthesis (produces / uses) O_2 and (produces / uses) CO_2.
9. In the light dependent reactions, chlorophyll in the (thylakoid membranes / stroma) captures sunlight energy and uses it to make (glucose / ATP).
10. The electron transport system is part of the light (dependent / independent) process of photosynthesis.

Chapter 8.

1. (Aerobic / Anaerobic) forms of life evolved first.
2. Aerobic respiration (produces / uses) O_2 and (produces / uses) CO_2.
3. Glycolysis (requires / does not require) oxygen in order to function.
4. Glycolysis occurs in the (mitochondria / cytoplasm) of a cell.
5. Pyruvic acid is produced by (glycolysis / the citric acid cycle).
6. The chemical energy in sugar is used to make (O_2 / ATP).
7. When NADPH becomes $NADP^+$ the hydrogens are used to make (water / sugar).
8. Lactic acid fermentation occurs when oxygen is (present / absent) from muscle cells.
9. When each pyruvic acid is completely broken down, (3 / 6) CO_2 molecules are released.
10. Each $FADH_2$ molecule releases enough energy to make (2 / 3) molecules of ATP.

****

ANSWER KEY:

Chapter 1:

1.   essentially similar to
2.   cannot
3.   temporary
4.   natural
5.   is not
6.   autotrophic
7.   prokaryotic
8.   do not
9.   heterotrophic
10. disproved

Chapter 2:

1.   atom
2.   proton
3.   lighter
4.   number
5.   hydrogen
6.   less
7.   ionic
8.   give up
9.   positive
10. covalent

Chapter 3:

1.   disaccharide
2.   oxygen
3.   taken away
4.   maltose
5.   glycogen
6.   fats
7.   glycerol
8.   hydrophobic
9.   COOH
10. primary

Chapter 4:

1.   constant
2.   changes
3.   second
4.   decrease
5.   randomly dispersed
6.   exothermic
7.   decrease
8.   ribose
9.   releases
10. more

Chapter 5:

1.   prokaryotic
2.   rough
3.   nucleus
4.   mitochondria
5.   chloroplasts
6.   smooth
7.   plants
8.   plants
9.   plants
10. shorter

Chapter 6:

1.   volume
2.   burst
3.   hydrophilic
4.   increasing
5.   lower
6.   hypotonic
7.   hypertonic
8.   gain
9.   into
10. phagocytosis

Chapter 7:

1.   photosynthesis
2.   grana
3.   stroma
4.   blue
5.   insects
6.   photosystem II
7.   lose
8.   produces, uses
9.   thylakoid
     membranes, ATP
10. dependent

Chapter 8.

1.   anaerobic
2.   uses, produces
3.   does not require
4.   cytoplasm
5.   glycolysis
6.   ATP
7.   water
8.   absent
9.   3
10. 2

****

UNIT II. INHERITANCE

CHAPTER 9:   CELLULAR REPRODUCTION

CHAPTER OUTLINE.

After briefly discussing the cell cycle in prokaryotes, this chapter covers mitosis, a basic type of eukaryotic cell division. The structure of eukaryotic chromosomes is presented and the typical chromosome numbers in body cells and sex cells is reviewed. The eukaryotic cell cycle is described, as well as interphase, mitosis, and cytokinesis.

1.   **General Facts**.

    A.   All cells come from other living cells.

    B.   Since all organisms consist of one or more cells, cellular reproduction is absolutely essential for life to continue on earth.

    C.   Most cells reproduce according to the simple repeating phrases: <u>enlarge, divide in two, enlarge, divide in two</u>, etc.
        1)   This process, called cell division, produces two daughter cells that are roughly identical copies of the original cell.
        2)   Each round of cell growth and division is called a <u>cell cycle</u>.
        3)   For unicellular organisms, cell cycle = life cycle.
        4)   Multicellular organisms begin life as a fertilized egg.
            a.   Repeated cell divisions, called <u>mitosis</u>, produce the many cells of the adult.
            b.   Specialized cells in reproductive organs undergo <u>meiosis</u>, producing sex cells (sperm and eggs).

2.   **Essentials of cellular reproduction**.

    A.   Every cell must pass on to its offspring two essential requirements for life:
        1)   Hereditary information (DNA) to direct life processes.
        2)   Other materials needed by the offspring to survive and use the hereditary information.

    B.   DNA (Deoxyribonucleic Acid): the heredity information of all living cells.
        1)   DNA has four different types of subunits, called nucleotides.
        2)   The sequence of nucleotides encodes a message.
        3)   Genes are segments of DNA from 100s to 1000s of nucleotides long that act as units of heredity. They specify the proteins made by the cell.
        4)   A cell needs a complete set of genes to survive. Before dividing, each cell must first duplicate its DNA so that each daughter cell receives a complete set.

    C.   Other necessary materials that must be passed on:
        1)   Molecules needed to read the genetic instructions and to stay alive until new materials are acquired from the environment.
        2)   Mitochondria and chloroplasts in the cytoplasm.
        3)   Other organelles, nutrients, and enzymes.

    D.   Cell cycle: The complete sequence of activities of a cell from one cell division to the next. Steps:
        1)   Cell growth and metabolism.
        2)   Cell division.

3.   **Prokaryotic cell cycle**.

    A.   For most of the cell cycle:
        1)   Growth occurs by absorbing nutrients and making more of its own molecules.
        2)   Then, its DNA in the form of a single circular chromosome replicates.
        3)   Then, the cell divides by "fission".

    B.   Cell division (fission).
        1)   One point on the chromosome is attached to the plasma membrane.
        2)   The chromosome replicates, and each piece attaches to nearby distinct points on the plasma membrane.
        3)   The plasma membrane grows between the chromosomes, pushing them apart.
        4)   The plasma membrane around the middle of the cell grows inward.
        5)   Two new daughter cells are formed, each with one chromosome and about half of the cytoplasm with necessary materials for growth.

4.   **Eukaryotic cell cycle**.

    A.   Stages.
        1)   Interphase with three subphases.
        2)   Cell division with four phases, and division of the cytoplasm.

    B.   Interphase (periods between nuclear divisions).

    1)   The longest period of cell division.

    2)   The subphases of interphase:

        a. G_1 phase (first time gap): the period before DNA synthesis occurs.

            i.   Cells acquire nutrients from the environment.

            ii.   Cells carry out their specialized functions.

            iii.   Cells grow.

            iv.   Late in G_1, at the "restriction point," each cell evaluates its ability to complete the cell cycle. Only if the evaluation is positive will the cell go on and divide.

        b. $\underline{S}$ phase (synthesis of DNA phase): the DNA replicates, as do the centrioles of animal cells.

        c. G_2 phase (second time gap): the period after DNA synthesis and before the next cell division. Materials needed for cell division are made.

    3)   Differences often occur relative to G_1:

        a.   Embryonic cells divide rapidly with virtually no G_1 so that almost no growth occurs between divisions.

        b.   Nerve cells in the adult brain no longer divide and remain in the G_1 phase for life.

  C.   <u>Cell division</u>.

    1)   Consists of two events:

        a.   Nuclear division (<u>mitosis</u>): chromosome copies migrate so that identical complete sets are packaged in two new nuclei.

        b.   Cytoplasmic division (<u>cytokinesis</u>): the cytoplasm is split into two daughter cells, each receiving one nucleus and approximately half the cytoplasm.

        c.   So, the parent cell and the daughter cells are essentially identical to each other.

    2)   Exceptions:

        a.   Some cells (vertebrate skeletal muscle and some fungi) undergo mitosis without cytokinesis, producing single cells with many nuclei.

        b.   Sex cells receive only half the chromosomes of the body cells.

    3)   Since a multicellular organism grows from a single fertilized egg by repeated cel! divisions, every cell of a multicellular organism is genetically identical (with the exceptions noted above).

    4)   Much of biology is the study of cellular activities during interphase, such as photosynthesis, muscle movement, and thought.

5.   **The eukaryotic chromosome**.

  A.   The complicated events of eukaryotic cell division are largely an evolutionary solution for sorting out large numbers of long chromosomes.

    1)   Prokaryotic cells have one circular DNA chromosome about 1-2 mm in circumference.

    2)   Each human (eukaryotic) cell has 46 chromosomes with total length of about 2000 mm, made of equal amounts of DNA and proteins.

  B.   <u>Chromatin and chromosomes</u>.

    1)   In each nucleus during interphase, chromosomes are highly uncoiled and they

are referred to collectively as <u>chromatin</u> ("color").

2) During cell division, the chromatin becomes highly condensed or coiled <u>chromosomes</u> ("colored bodies"), being easily distinguishable from each other.

3) Each eukaryotic chromosome is a single long molecule of DNA complexed with special <u>histone</u> proteins.

   a. During interphase, the cell can "read" the genes of the DNA since the chromosomes are uncoiled and extended.

   b. During cell division, the chromosomes must be sorted out and moved equally into two daughter nuclei, and condensed chromosomes are easier to organize and move around.

4) <u>Chromosome terminology.</u>

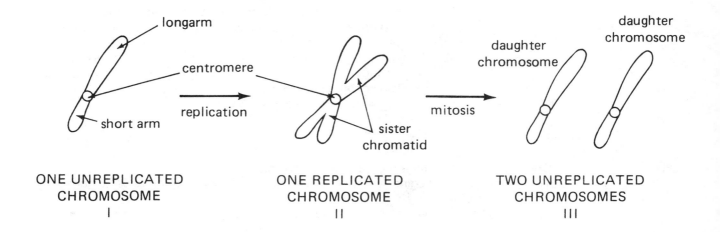

ONE UNREPLICATED CHROMOSOME
I

ONE REPLICATED CHROMOSOME
II

TWO UNREPLICATED CHROMOSOMES
III

   a. The number of chromosomes is equal to the number of functional <u>centromeres</u> present. Hence, I and II above show one chromosome each while III shows 2 chromosomes.

   b. After the centromeres split during mitosis, each "sister chromatid" becomes a "daughter chromosome".

5) The number of chromosomes in body cells:

   a. In many species, non-reproductive body cells have pairs of chromosomes that are the same length and stain in the same patterns.

      i. The chromosomes of each pair have similar genetic content.

      ii. The members of each pair are called <u>homologues</u> ("to say the same thing").

   b. Cells with pairs of homologous chromosomes are called <u>diploid</u> <u>cells</u>.

   c. Human body cells have 23 pairs of homologous chromosomes (a total of 46 chromosomes).

   d. Not all cells have pairs of homologous chromosomes:

      i. During sexual reproduction, cells with only one copy of each type of chromosome (instead of pairs of these) are produced by meiosis. These are <u>haploid</u> <u>cells</u>.

ii. In animals, these haploid cells are the <u>sex</u> <u>cells</u> or <u>gametes</u> (sperm and eggs).

iii. Human gametes have 23 chromosomes, one from <u>each</u> homologous pair.

iv. Sperm and egg fuse to form a diploid cell, the fertilized egg or <u>zygote</u>. So, 23 (sperm) + 23 (egg) → 46 zygote in humans.

v. Thus, one chromosome of each pair in a diploid body cell is inherited from each parent.

e. <u>Polyploid</u> organisms have more than 2 homologues of each type of chromosome.

f. <u>Terminology</u> for various chromosome numbers:

i. <u>Haploid</u> <u>number</u> (n): The number of different kinds of chromosomes; the number of chromosomes in the sex cells. In human sex cells, n = 23 chromosomes.

ii. <u>Diploid</u> <u>number</u> (2n): 2 homologues of each type per cell; the chromosome condition in animal body cells. In human body cells, 2n = 46 chromosomes.

iii. <u>Triploid</u> <u>number</u> (3n): 3 homologues of each type per cell; <u>tetraploid</u> <u>number</u> (4n): 4 homologues of each type of chromosome per cell.

iv. <u>Polyploid</u> cells are 3n, 4n, 5n, etc.

6. **Mitosis**.

A. Chromosome contents are duplicated during interphase.

1) As mitosis begins, each chromosome has 2 sister chromatids attached by a centromere.

2) We divide mitosis into four phases, but it really is a continuous process.

B. <u>Prophase</u>.

1) Replicated chromosomes coil up and thicken (condense), becoming distinctly visible under the microscope.

2) The nuclear envelope disintegrates.

3) The nucleolus breaks up.

4) The <u>spindle</u> <u>apparatus</u> is assembled:

a. During interphase, microtubules radiate out from a <u>microtubule organizing center</u> (MOC) near the nucleus.

b. During prophase, the MOC splits in two, and each daughter center forms its own array of microtubules.

i. The two MOCs move to opposite sides of the nucleus becoming the <u>poles</u>.

ii. Microtubules extending from each pole nearly surround the nucleus and are called the <u>spindle apparatus</u>.

5) The spindles of plant and animal cells differ.

a. In animal cells, each MOC contains a pair of <u>centrioles</u>.

i. During interphase, the two centrioles separate and a new centriole develops near the base of each "parent" centriole (the MOC now has 2 pairs of centrioles).

        ii.   When the MOC splits, each daughter center contains one pair of centrioles.

    b.   In animal cells, spindle microtubules extend
        i.   Across the nucleus towards the other pole.
        ii.   Outward in all directions from the MOC towards the plasma membrane, forming a star-shaped <u>aster</u>.

    c.   Plant cells lack centrioles and asters.

6)   Capture of the chromosomes.

    a.   After the spindle forms, the nuclear envelope disintegrates: the double membrane breaks up into vesicles that resemble pieces of endoplasmic reticulum.

    b.   Some of the spindle microtubules can now attach to the chromosomes at sites near the centromeres called <u>kinetochores</u>.
        i.   Each sister chromatid has its own kinetochore.
        ii.   The kinetochores of one sister chromatid of each chromosome has microtubules leading to one pole, and the other sister chromatid's microtubules lead to the other pole.

    c.   Other microtubules retain free ends that overlap along the cell's equator. These <u>polar microtubules</u> will push the two spindle poles away from each other during anaphase.

C.   <u>Metaphase</u>.
1)   Each kinetochore microtubules pulls towards its pole.
2)   This pulling aligns each chromosome along the equator of the cell.
3)   Metaphase ends when all the chromosomes are lined up at the equator.

D.   <u>Anaphase</u>.
1)   The centromere of each chromosome divides and the sister chromatids separate to become independent daughter chromosomes.
2)   Daughter chromosomes move to opposite poles as kinetochore microtubules shorten.
3)   Since daughter chromosomes are identical copies of the original chromosomes, the two clusters of chromosomes that form at the poles contain one copy of every chromosome.
4)   Simultaneously, the polar microtubules grow longer and push the spindle fibers away from each other.

E.   <u>Telophase</u>.
1)   Begins when the chromosomes reach the poles of the spindle.
2)   The spindle disintegrates.
3)   Vesicles that formed when the old nuclear envelope broke up during prophase coalesce around each group of chromosomes, forming two new nuclear envelopes.
4)   The chromosomes uncoil, becoming long and thin again.
5)   Usually, cytokinesis occurs.

7.   **Cytokinesis**: Division of the cytoplasm during telophase.

  A.   In <u>animal</u> cells:
    1)   Microfilaments of actin and myosin form rings around the cell's equator.
    2)   The rings contract to pinch in the equator.
    3)   The equator constricts down to nothing, dividing the cytoplasm into 2 new daughter cells, each containing one nucleus.

  B.   In <u>plant</u> cells, the stiff cell wall precludes the cell changing shape. Instead:
    1)   The Golgi complex buds off carbohydrate-filled vesicles that gather at the equator.
    2)   The vesicles fuse, producing a hollow, pancake-shaped structure called the <u>cell plate</u>.
    3)   The edges of the cell plate merge with the original plasma membrane around the circumference of the cell.
    4)   The carbohydrate becomes the beginning of the primary cell wall.

8.   Cell division and **asexual reproduction**.

  A.   In asexual reproduction, offspring are formed by cell division from a single parent without the fusion of male and female gametes.
    1)   Many unicellular organisms reproduce asexually (i.e., <u>Paramecium</u>, <u>Euglena</u>, yeast).
    2)   Some multicellular organisms can also reproduce asexually by "budding" (i.e., <u>Hydra</u>).

  B.   Due to mitosis, asexually reproduced offspring are <u>genetically</u> <u>identical</u> to their parents. A group of such identical offspring derived from one parent or each other is a <u>clone</u>.

****

REVIEW QUESTIONS:

1.   List the steps involved in prokaryotic cell division.

2.-8.   **MATCHING TEST**: interphase of cell division.

  Choices: A.   S phase   B. G_1 phase   C. G_2 phase

2.   Phase in cells that will no longer divide.
3.   Period after DNA synthesis occurs.
4.   Period before DNA synthesis occurs.
5.   Period of DNA replication or synthesis.
6.   Spindle fiber proteins are made.
7.   Period of most cell growth and metabolic activity.
8.   Nearly eliminated in embryonic cells that divide rapidly.
----

9.   Compare the size, number, and composition of prokaryotic and eukaryotic chromosomes.
10.   Why is it advantageous for eukaryotic cells to have long, uncoiled chromosomes during interphase but to have coiled up condensed chromosomes during mitosis?

11.-13. **MATCHING TEST**: chromosome morphology.

Choices:

A.   Chromosome
B.   Chromatid
C.   Centromere

14.-22. **MATCHING TEST**:   chromosome numbers in cells.

        Choices: A.   Haploid cells
                        B.   Diploid cells
                        C.   Polyploid cells

14.   Abnormal human cells with 92 chromosomes.
15.   Have homologous pairs of chromosomes.
16.   Sex cells.
17.   Produced by meiosis.
18.   Have single chromosomes of each homologous type.
19.   Human cells with 46 chromosomes.
20.   Cells with 3 or more sets of homologous chromosomes.
21.   Cells of a zygote.
22.   Human cells with 23 chromosomes.

23.-32. **MATCHING TEST**: stages of mitosis.

        Choices: A. Telophase   C. Metaphase
                        B. Anaphase   D. Prophase

23.   Chromosomes have reached the poles of the spindle.
24.   Chromosomes migrate to the cell's equator.
25.   Replicated chromosomes coil up and condense.
26.   The centromeres divide.
27.   Daughter chromosomes move towards the poles.
28.   The spindle breaks down.
29.   The nuclear envelope disintegrates.
30.   Sister chromatids become daughter chromosomes.
31.   The spindle forms.
32.   Cytokinesis occurs.

----

33.   Compare cytokinesis in plant cells and animal cells.
34.   What happens during asexual reproduction?

****

ANSWERS TO REVIEW QUESTIONS:

1.   see 3.A.-B.

2.-8.   [see 4.B.1)-3)]
        2.   B   6. C
        3.   C   7. B
        4.   B   8. B
        5.   A

9.   see 5.A.-B.

10.   see 5.B.3)a.-b.

11.-13. [see 5.B.4)]
        11.   A
        12.   C
        13.   B

14.-22. [see 5.B.5)a.-f.]
        14.   C   19.   B
        15.   B   20.   C
        16.   A   21.   B
        17.   A   22.   A
        18.   A

23.-32. [see 6.B.-E.]
        23.   A   28.   A
        24.   C   29.   D
        25.   D   30.   B
        26.   B   31.   D
        27.   B   32.   A

33.   see 7.A.-B.

34.   see 8.A.-B.

****

CHAPTER 10: MEIOSIS AND THE LIFE CYCLES OF EUKARYOTIC ORGANISMS

CHAPTER OUTLINE.

After briefly contrasting asexual and sexual reproduction, this chapter covers meiotic cell division. The stages of meiosis I and meiosis II are described, especially as they relate to the production of sex cells, and meiosis I and II are compared. Mitosis and meiosis are then compared. Finally, three basic types of life cycles among eukaryotic organisms are outlined and contrasted.

1. **General facts**.

    A.   Offspring produced through <u>asexual reproduction</u> arise from mitotic cell division, meaning that the offspring are genetically identical to the parents.
       1)   Asexual reproduction is advantageous only if:
          a.   The parental organism is well-adapted to its environment.
          b.   The environment never changes.
          c.   The organisms never move to a new location.
       2)   However, in changing environments or when offspring migrate to new locations, genetic variability in the offspring may allow some to be better adapted to the new or changing situation.

    B.   <u>Sexual reproduction</u> allows offspring to be similar but not identical to their parents.
       1)   Most multicellular organisms are diploid, with pairs of homologous chromosomes in their body cells.
          a.   Segments of DNA ("genes") on the homologous chromosomes influence the same traits.
          b.   However, genes of homologous chromosomes often are not identical and

may influence the same trait in different ways.
2)   In sexual reproduction, 2 haploid sex cells (usually from 2 different parents) fuse to form a diploid cell (zygote) that develops by mitosis into the offspring.
3)   Each gamete contains genetic information from one parent.
4)   Thus, if the parents have dissimilar genes, their offspring will receive a <u>unique</u> combination of genes from both parents.

C.   The 3 features typical of sexual reproduction in eukaryotic organisms:
    1)   The parents have diploid cells at some stage during their life cycles.
    2)   The homologous chromosomes of the parents separate from one another through a special cell division called <u>meiosis</u>, producing haploid cells.
        a.   In animals, haploid cells are gametes (sperm and eggs).
        b.   In plants and fungi, the haploid cells are <u>spores</u> that grow by mitosis into a multicellular haploid body which then produces gametes by mitosis.
    3)   During fertilization, 2 haploid gametes fuse to form a diploid fertilized egg (or zygote), with one copy of each homologous chromosome donated by <u>each</u> parent.

D.   The key to sexual reproduction is <u>meiosis</u>, the production of haploid cells with unpaired chromosomes from diploid cells with paired homologues. Only by separating homologous chromosomes in meiosis and fusing the resulting haploid cells (<u>fertilization</u>) can each new generation remain diploid.

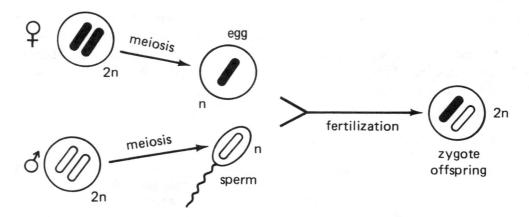

2.   **Overview of meiosis** ("to diminish"): reduces the number of chromosomes by half.

A..   In meiosis, each daughter cell receives one member of each homologous chromosome pair.

B.   Meiosis involves 2 nuclear divisions (<u>meiosis I</u> and <u>II</u>):
    1)   Chromosomes are duplicated during the interphase just prior to meiosis I and do not duplicate again during meiosis.
    2)   During meiosis I, homologous chromosomes pair up, then move to opposite poles with each daughter nucleus receiving one replicated homologue of each pair of chromosomes (sister chromatids remain together since centromeres do

not divide). These daughter nuclei are haploid.

3)   In meiosis II, the centromeres divide and sister chromatids become daughter chromosomes with one of each type going into each daughter nucleus. Meiosis II is similar to mitosis occurring in a haploid cell, with no reduction in number of chromosomes.

4)   Each nuclear division is usually accompanied by cytokinesis.

5)   So, one diploid cell --(meiosis I)--> 2 haploid cells --(meiosis II)--> 4 haploid cells.

3.   Mechanisms of meiosis.

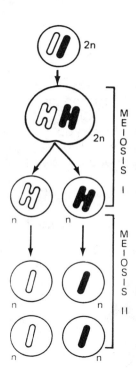

A.   Meiosis I details.

     1)   Prophase I.

         a.   Begins after interphase, when chromosome replication has occurred.

         b.   The ends of homologous maternal and paternal chromosomes attach to the nuclear envelope next to each other:

             i.   Then, the two chromosomes pair up much like zipping the zipper of a jacket.

             ii.   Proteins join the homologous chromosomes.

         c.   Large enzyme complexes assemble at random places along the paired chromosomes:

             i.   They snip apart the chromosomes, and graft the broken ends together so that pieces of maternal and paternal chromosomes are joined.

             ii.   Thus, the maternal and paternal chromosomes now intertwine, forming <u>chiasmata</u> ("crosses").

         d.   Chiasmata (singular = chiasma) serve two functions:

             i.   <u>Crossing over</u>: Maternal and paternal homologous chromosomes swap DNA segments so that neither chromosome is quite the same as before.

             ii.   Chiasmata keep the homologous chromosomes paired during their attachment to the spindle microtubules.

         e.   Eventually, the enzyme complexes leave the chromosomes and the protein zippers disappear.

         f.   The chromosomes coil and condense, becoming visible with a light microscope, with chiasmata connecting homologues.

         g.   The spindle assembles outside the nucleus.

         h.   The nuclear membrane breaks up and the spindle microtubules capture the chromosomes by attaching to their kinetochores.

     2)   Metaphase I.

         a.   The spindle microtubules move the chromosomes to the equator of the cell.

         b.   The chromosomes are aligned in homologous pairs along the equator.

         c.   The two homologues of each pair are attached to kinetochore microtubules leading to opposite poles of the spindle.

         d.   Movement at metaphase of mitosis (unpaired homologues) compared to meiosis I metaphase (paired homologues):

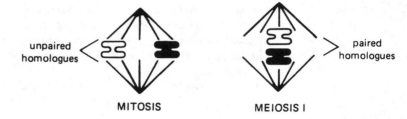

unpaired homologues < | > paired homologues

MITOSIS   MEIOSIS I

e.   <u>Different</u> paired homologues will align at the equator randomly relative to each other:

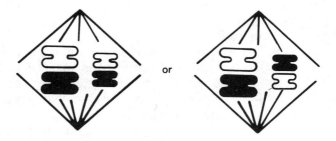

or

3)   <u>Anaphase I</u>: Homologous chromosomes separate from one another and move to opposite poles of the cell.
    a.   Chiasmata holding homologous chromosomes together loosen up.
    b.   Kinetochore microtubules pull the homologues away from each other.
    c.   The centromeres holding sister chromatids together do not split: sister chromatids remain attached to one another.
    d.   So, one chromosome of each homologous pair moves to one pole while its homologue moves to the other pole.
    e.   Comparison of mitosis anaphase (centromeres split) and meiosis I anaphase (centromeres do not split):

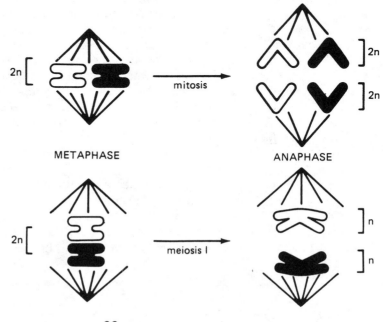

      f.   At the end of anaphase I, there are two groups of chromosomes: each is haploid, containing one member of each pair of homologous chromosomes.

    4)   Telophase I.

        a.   The spindle disappears as the chromosomes reach the poles.

        b.   In plants, cytokinesis usually does not occur, while in animals it usually does.

        c.   No further chromosome replication occurs (chromosomes already have sister chromatids) prior to meiosis II.

B.   Meiosis II details: very much like mitosis in a haploid cell.

    1)   Prophase II: spindle reforms and chromosomes attach as in mitosis.

    2)   Metaphase II: chromosomes migrate to the equator as in mitosis.

    3)   Anaphase II: sister chromatids separate as the centromeres split and daughter chromosomes move to opposite poles as in mitosis.

    4)   Telophase II and cytokinesis: occur as in mitosis. Nuclear membranes reform, chromosomes relax, and cytoplasm divides.

C.   Comparison of mitosis and meiosis.

MITOSIS	MEIOSIS
Chromosomes replicate once.	Chromosomes replicate once.
Cells divide once.	Cells divide twice.
One diploid cell → 2 diploid cells	One diploid cell → 4 haploid cells
Is "nonreductional" division (no reduction in chromosome number: diploid → diploid number of chromosomes)	Meiosis I is "reductional" (diploid → haploid number of chromosomes)
	Meiosis II is "nonreductional" (haploid → haploid number of chromosomes)

3.   **Eukaryotic life cycles.**

A.   Both mitosis and meiosis play crucial roles in the life cycles of most eukaryotic organisms.

    1)   Common patterns to all life cycles:

        a.   Diploid cells → haploid cells by meiosis.

        b.   Fusion of 2 haploid cells (fertilization) → diploid cell with new gene combination.

        c.   Depending on the organism, either haploid cells (from a.) or diploid cells

(from b.) can produce multicellular bodies by <u>mitosis</u>.
2) Organisms with seemingly quite different life cycles may vary in 3 aspects:
    a. The interval between meiosis and fertilization.
    b. When meiosis and mitosis occur.
    c. The relative proportion of the life cycle spent in the diploid and haploid stages.

B. **Haploid life cycles.**
1) Some eukaryotes (like the alga <u>Chlamydomonas</u>) have mainly haploid life cycles.
2) When sexual reproduction occurs, fertilization produces a diploid cell which immediately undergoes meiosis to produce haploid cells.
3) The haploid cells divide by mitosis, but mitosis never occurs in a diploid cell, so:

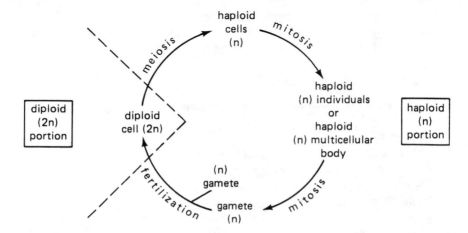

C. **Diploid life cycles:** animals have life cycles just the reverse of the above.
1) Virtually the entire cycle consists of diploid cells.
2) Haploid sperm and eggs are formed by meiosis.
3) Gametes fuse to form diploid zygotes.
4) The diploid zygote divides by mitosis to form the multicellular body but mitosis never occurs in haploid cells. So:

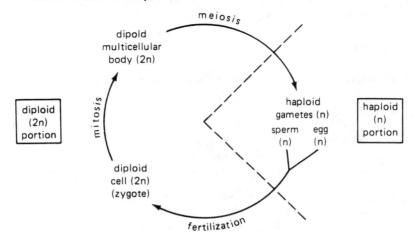

D. **Alternation of generation life cycles**: Most plants have both multicellular diploid and multicellular haploid stages in their life cycles.
1) Diploid cells produce haploid <u>spores</u> by meiosis.
2) The spores divide by mitosis to produce <u>haploid</u> multicellular bodies (the haploid generation).
3) Some cells in the haploid body differentiate into haploid gametes.
4) Two gametes fuse to form a <u>diploid</u> <u>zygote</u>.
5) The zygote divides by mitosis to produce the diploid multicellular body (the diploid generation).
6) In primitive plants (like ferns), both generations are free living and independent plants, but in the flowering plants the haploid generation is reduced to the pollen grains and the embryo sacs in the flowers' ovaries. So,

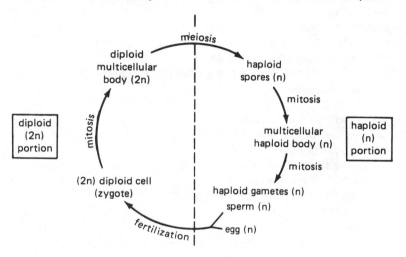

4. **Chromosomes, meiosis, and inheritance**. Major facts to remember:

A. Diploid organisms have pairs of homologous chromosomes.
B. One member of each homologous pair comes from the organism's mother, the other from the father.
C. Homologues contain the same type of genetic information (i.e., hair color).
D. If the parents differ genetically, the homologues will not have the same exact types of genes (i.e., blond hair in one, brown hair in the other).
E. Meiosis separates the paired homologues of diploid cells, producing haploid cells with one chromosome from each pair.
F. The exact chromosome from each pair (from mother or father) included by meiosis in any haploid cell is random.
G. Fusion of gametes produces a diploid zygote with pairs of homologous chromosomes.

****

REVIEW QUESTIONS:

1.-17. **MATCHING TEST:** mitosis and meiosis

Choices:
A. Mitosis   C. Both of these
B. Meiosis   D. Neither of these

1. Production of haploid cells from haploid cells.
2. The mechanism by which unicellular organisms reproduce.
3. Produces genetically variable cells.
4. Produces sperm and egg cells in animals.
5. Allows multicellular organisms to grow.
6. Produces sperm and egg cells in plants.
7. Ensures that each body cell gets a complete set of genes.
8. Can produce haploid cells.
9. Produces spores in plants and fungi.
10. Produces genetically identical cells.
11. Production of haploid cells from diploid ones.
12. Occurs in the human body.
13. Maintains the same number of chromosomes.
14. Doubles the number of chromosomes.
15. Reduces the number of chromosomes by half.
16. Chromosomes replicate once.
17. Cells divide twice.

----

18. When is asexual reproduction advantageous and when is sexual reproduction advantageous?
19. Name the three typical features of organisms reproducing sexually.

20.-28. **MATCHING TEST:** meiosis.

Choices:
A. Prophase I   E. Prophase II
B. Metaphase I   F. Metaphase II
C. Anaphase I   G. Anaphase II
D. Telophase I   H. Telophase II

20. Individual chromosomes migrate to the equator.
21. Chiasmata form.
22. Chromosomes with 2 chromatids each move towards the poles.
23. Centromeres divide.
24. Homologous chromosomes pair up.
25. Daughter chromosomes migrate towards the poles.
26. Homologous pairs of chromosomes move together to the equator.
27. Crossing over occurs.
28. Homologous chromosomes move towards opposite poles.

29. Diagram a typical haploid life cycle.
30. Diagram a typical animal diploid life cycle.
31. Diagram a typical plant alternation of generation life cycle.

****

ANSWERS TO REVIEW QUESTIONS:

1.-17. [see entire chapter]

1.	A	10.	A	
2.	A	11.	B	
3.	B	12.	C	
4.	B	13.	A	
5.	A	14.	D	
6.	A	15.	B	
7.	A	16.	C	
8.	C	17.	B	
9.	B			

18. see 1.A.-B.

19. see 1.C.1)-3)

20.-28. [see 3.A.-B.]

20.	F	25.	G
21.	A	26.	B
22.	C	27.	A
23.	G	28.	C
24.	A		

29. see 3.B.

30   see 3.C.

31. see 3.D.

****

CHAPTER 11: PRINCIPLES OF INHERITANCE

CHAPTER OUTLINE.

This chapter presents Mendel's concepts of inheritance: the segregation and independent assortment of genes, the dominance and recessiveness of different alleles, and the randomness of fertilization. Also, the relationship of genes to chromosomes, especially sex-linkage, sex determination, linkage, and crossing over, are covered. Finally, variations on the Mendelian theme (mutation, incomplete dominance, multiple alleles, codominance, polygenic inheritance, gene interactions, and environmental influences) are enunciated.

1. **General facts**.

    A. Any <u>theory of inheritance</u> must explain the following:
        1) All living things arise from pre-existing living things.
        2) The offspring of 2 parents usually resemble the parents and each other.
        3) Many traits in the offspring are not exactly like those of either parent.

    B. In the 19th century, most scientists explained the above by <u>blending inheritance</u>: traits from the parents blend in the offspring to produce intermediate traits (i.e., tall dad x short mom → intermediate son or daughter).
        1) According to this however, over time all variation would be "blended out" (everyone would be of intermediate height).
        2) Also, evolution could not occur by natural selection if the variations selected for would be reduced each generation by blending.
        3) Gregor Mendel solved these problems by discovering the correct ideas about inheritance.

2.   **Gregor Mendel** and the <u>origin</u> <u>of</u> <u>genetics</u>.

A.   Why did Mendel succeed while others failed?
   1)   He chose the edible pea as his experimental subject.
      a.   Due to flower shape, peas normally <u>self-fertilize</u>, resulting in many <u>true-breeding</u> varieties (all offspring are identical to the parent) available from seed dealers.
      b.   Different pea plants could be forced to <u>cross-fertilize</u> by manually manipulating the flowers (carrying pollen from one plant to another). Mendel was able to mate different true-breeding varieties and examine their offspring.
   2)   At first, he studied one trait at a time and used traits with easily recognizable differences.
   3)   Mendel used a quantitative approach: he <u>counted</u> the numbers of offspring with various differences and converted the numbers to <u>ratios</u>.

B.   Single trait experiments: the <u>Law</u> <u>of</u> <u>Segregation</u>.
   1)   Mendel raised varieties of peas that were true-breeding for different forms of a <u>single</u> trait and cross-fertilized them (the <u>P</u> or <u>parental</u> <u>generation</u>).
      a.   Such single trait crosses are called <u>monohybrid</u> <u>crosses</u>.
      b.   Mendel saved the offspring seeds (the F_1 <u>generation</u>) from these crosses and planted them the following year, this time allowing the plants to produce offspring (the F_2 <u>generation</u>) by self-fertilization.
      c.   A typical set of results:

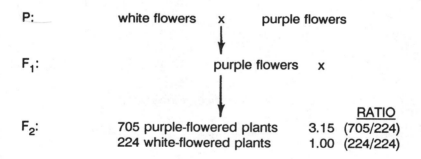

		RATIO
F_2:	705 purple-flowered plants	3.15 (705/224)
	224 white-flowered plants	1.00 (224/224)

      d.   When he planted some F_2 seeds and allowed the plants to self-fertilize, he found:
         i.   All the white flowered F_2 plants bred true, producing only white-flowered offspring in the F_3 generation.
         ii.   About 2/3 of the purple-flowered F_2 plants were hybrids, producing purple and white offspring in a 3:1 ratio.
         iii.   About 1/3 of the purple-flowered F_2 plants bred true, producing only purple offspring.
   2)   <u>Mendel's explanation</u> consisted of a five part hypothesis:
      a.   <u>Each</u> <u>trait</u> <u>is</u> <u>determined</u> <u>by</u> <u>a</u> <u>pair</u> <u>of</u> <u>discrete</u> <u>physical</u> <u>units</u> (now called <u>genes</u>). So, in each pea plant cell, a pair of genes controlling flower color exists.

   b.   <u>Pairs</u> <u>of</u> <u>genes</u> <u>separate</u> (<u>segregate</u>) <u>from each</u> <u>other</u> <u>during</u> <u>gamete</u> <u>formation</u>. This is the <u>Law</u> <u>of</u> <u>Segregation</u>:
      i.   Each gamete gets <u>one</u> gene of each pair.
      ii.   When sperm and egg fuse, the zygote gets one gene of the pair from each parent.
   c.   <u>Which</u> <u>gene</u> <u>of</u> <u>a</u> <u>pair</u> <u>becomes</u> <u>included</u> <u>in</u> <u>a</u> <u>particular</u> <u>gamete</u> <u>is</u> <u>determined</u> <u>by</u> <u>chance</u>.
      i.   If the pea cells have an $\underline{A_1}\underline{A_2}$ gene pair, a particular gamete is equally likely to get gene $\underline{A_1}$ or gene $\underline{A_2}$.
      ii.   However, a gamete will <u>not</u> get both $\underline{A_1}$ and $\underline{A_2}$.
   d.   <u>There</u> <u>may</u> <u>be</u> <u>two</u> <u>or</u> <u>more</u> <u>alternative</u> <u>forms</u> <u>of</u> <u>a</u> <u>gene</u>.
      i.   White flowers and purple flowers are determined by 2 different forms of the flower color gene.
      ii.   These different forms of the same gene are called <u>alleles</u>.
      iii.   A <u>dominant allele</u> masks or hides the expression of a <u>recessive allele</u>. For instance, if <u>P</u> is the allele for purple flower color and <u>p</u> is the allele for white, and a <u>Pp</u> plant has purple flowers, then purple (<u>P</u>) is dominant and white (<u>p</u>) is recessive.
   e.   <u>True-breeding</u> <u>organisms</u> <u>have</u> <u>similar</u> <u>alleles</u> <u>for</u> <u>the</u> <u>trait</u> (<u>PP</u> or <u>pp</u>), <u>and</u> <u>hybrids</u> <u>have</u> <u>different</u> <u>alleles</u> (<u>Pp</u>).
      i.   <u>PP</u> and <u>pp</u> are <u>homozygous</u> individuals and can produce one genetic type of gamete each:

   <u>PP</u> → <u>P</u> gametes only

   <u>pp</u> → <u>p</u> gametes only

      ii.   <u>Pp</u> is <u>heterozygous</u> and can produce 2 genetic types of gametes:

   <u>Pp</u> →   <u>P</u> and <u>p</u> gametes in equal numbers

3)   Mendel's explanation for flower color inheritance:

P: purple (<u>PP</u>)   x   white (<u>pp</u>)   diploid pea cells

    ↓   ↓   <u>Law of Segregation</u>

    (<u>P</u>)   (<u>p</u>)   haploid gametes

random fertilization

F_1:   purple (<u>Pp</u>)   <u>P</u> = dominant purple
              <u>p</u> = recessive white

                      <u>Law of Segregation</u>

1/2 (<u>P</u>)   1/2 (<u>p</u>)   haploid gametes

	1/2 (P)	1/2 (p)
1/2 (P)	1/4 PP purple	1/4 Pp purple
1/2 (p)	1/4 Pp purple	1/4 pp white

Use a <u>Punnett</u> <u>square</u> to determine the probable results of all fertilizations

F_2:   1/4 PP, 2/4 Pp, 1/4 pp   1:2:1 expected ratio of gene types

        3/4 purple,   1/4 white   3:1 expected ratio of color types

        a.   The 1:2:1 ratio refers to gene combinations or <u>genotypes</u>.

        b.   The 3:1 ratio refers to morphological appearances or <u>phenotypes</u>.

        c.   The F_2 generation consists of 3 genotypes:
        1/4 <u>PP</u> and 1/4 <u>pp</u> which are both pure-breeding, and 2/4 <u>Pp</u> which are hybrid.

SUMMARY OF BASIC GENETICS TERMS:

Genotype   genetic composition.

Phenotype   physical appearance.

Homozygous   a pair of identical genes capable of producing only one type of gamete.

Heterozygous   a pair of different genes capable of producing two types of gametes.

Alleles   alternate forms of the same gene.

Locus   site on a chromosome where alleles are located.

Hybrid F_1 offspring of a cross between two pure breeding forms; a heterozygote.

Monohybrid   a cross involving one pair of gene differences.

Dihybrid   a cross involving two pairs of gene differences.

    4)   The <u>test</u> <u>cross</u>.

        a.   Used to determine whether an individual with a dominant phenotype is homozygous or heterozygous in genotype.

        b.   <u>Method</u>: mate the dominant (with the unknown genotype) to a recessive individual (who must be homozygous) and examine the offspring:

            i.   If <u>all</u> the offspring are dominant, the dominant parent must be homozygous. For example:

purple   x   white   →   500 purple offspring

<u>PP</u>   <u>pp</u>   all <u>Pp</u>

ii.   If about <u>half</u> the offspring are dominant and about <u>half</u> are recessive, the dominant parent must be heterozygous. For example:

purple   x   white   →   250 purple + 250 white

<u>Pp</u>   <u>pp</u>   50% <u>Pp</u>   50% <u>pp</u>

C.   Multiple trait experiments: <u>Independent</u> <u>assortment</u>.
       1)   Mendel did <u>dihybrid</u> <u>crosses</u>: he looked at 2 variable traits simultaneously by cross-breeding plants that differed for the 2 traits.
       2)   A typical dihybrid experiment:
            a.   Traits studied and their genetic control:
                   i.   <u>Trait 1</u>: seed shape, with smooth (allele <u>S</u>) and wrinkled (allele <u>s</u>) forms.
                   ii.   <u>Trait 2</u>: seed color, with yellow (allele <u>Y</u>) and green (allele <u>y</u>) forms.
            b.   Results.

P: smooth yellow   x   wrinkled green   phenotypes

<u>SSYY</u>   <u>ssyy</u>   genotypes

↓   ↓

(SY)   (sy)   gametes

                      random fertilization

F₁: smooth yellow <u>SsYy</u>   smooth (<u>S</u>) and yellow (<u>Y</u>) are dominant traits

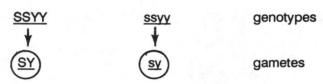

1/4 (SY,) 1/4 (Sy,) 1/4 (sY,) 1/4 (sy)   gametes

Use a <u>Punnett</u> <u>square</u> to determine the outcome of all possible random fertilizations with 4 types of eggs and 4 types of sperm:

	1/4 SY	1/4 Sy	1/4 sY	1/4 sy
1/4 SY	1/16 SSYY smooth yellow ♥	1/16 SSYy smooth yellow ♥	1/16 SsYY smooth yellow ♥	1/16 SsYy smooth yellow ♥
1/4 Sy	1/16 SSYy smooth yellow ♥	1/16 SSyy smooth green ♦	1/16 SsYy smooth yellow ♥	1/16 Ssyy smooth green ♦
1/4 sY	1/16 SsYY smooth yellow ♥	1/16 SsYy smooth yellow ♥	1/16 ssYY wrinkled yellow ♣	1/16 ssYy wrinkled yellow ♣
1/4 sy	1/16 SsYy smooth yellow ♥	1/16 Ssyy smooth green ♦	1/16 ssYy wrinkled yellow ♣	1/16 ssyy wrinkled green ♠

Summary of the F_2 generation taken from the Punnett square:

F_2	EXPECTED RATIOS	"GENERALIZED GENOTYPES"		PHENOTYPES
	9/16	♥	S-Y-	smooth yellow
	3/16	♦	S-yy	smooth green
	3/16	♣	ssY-	wrinkled yellow
	1/16	♠	ssyy	wrinkled green

c.   Independent Assortment: The distribution of alleles for one trait into the gametes does not affect the distribution of alleles for other traits (i.e., whether a gamete gets S or s does not affect whether it gets Y or y).

d.   Aids in determining gamete types.

    i.   Use the formula 2^n (n = the number of heterozygous gene pairs) to determine the number of different types of gametes an organism can produce. Examples:

for AAbb, $2^n = 2^0 = 1$ type   (Ab)

for AaBB, $2^n = 2^1 = 2$ types   (AB, aB)

for AaBb, $2^n = 2^2 = 4$ types   (AB, Ab, aB, ab)

for AaBbDd, $2^n = 2^3 = 8$ types   (can you name them? (ABD, ABd, etc.)

97

ii.   To determine the gene combinations in gametes, use a "tree diagram". Examples:

with <u>AaBb</u>,
      <u>A</u> segregates from <u>a</u>
      <u>B</u> segregates from <u>b</u>
      The <u>Aa</u>s and <u>Bb</u>s show independent assortment

This can be diagrammed as follows:

with <u>AaBb</u>,

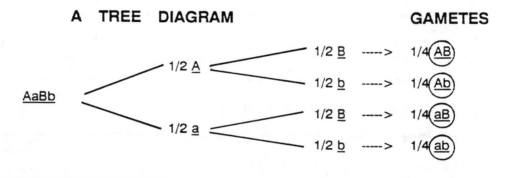

with <u>AaBbDd</u>,

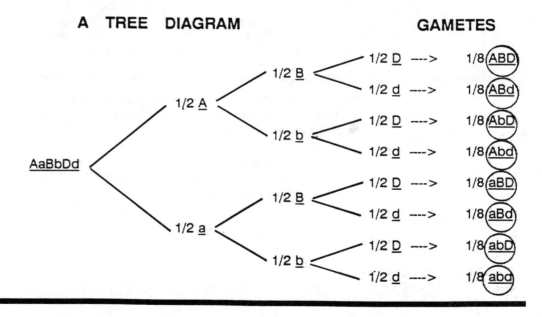

D.   Mendel published his results in an 1866 research paper.
1)   Few read it and apparently no one understood the explanation or appreciated its significance.
2)   It wasn't until 1900 that Mendel's principles and his paper were "rediscovered" and understood by three botanists (Correns, de Vries, and von Tschermak).

3.   **Genes and chromosomes**.

A.   In 1902, Sutton wrote a paper on the correlations between chromosome behavior during meiosis and Mendel's ideas about gene movement:
1)   Diploid cells have pairs of genes and pairs of homologous chromosomes.
2)   Only one gene of each pair is passed to the offspring in each sperm or egg cell; during meiosis, only one of each pair of homologous chromosomes ends up in each haploid daughter cell.
3)   Different genes show independent assortment; different types of chromosomes also assort independently.
4)   Sutton's hypothesis: Genes are part of chromosomes.

B.   Morgan proved that genes are part of chromosomes through his studies with common fruit flies <u>Drosophila</u> <u>melanogaster</u> which grow rapidly, produce many offspring, have 4 pairs of chromosomes, and are small and cheap to raise.

C.   Chromosomal basis of <u>sex determination</u>: the first direct evidence supporting Sutton's hypothesis.
1)   In mammals and many insect species:
a.   Males have a pair of chromosomes that differ in size and shape, called the <u>XY pair</u>. In females, this pair is <u>XX</u>. These are the <u>sex chromosomes</u>.
b.   The other chromosome pairs are homologous and identical looking in males and females. These are the <u>autosomes</u>. Fruit flies have 3 pairs of autosomes while humans have 22 pairs of autosomes.
2)   In organisms with XY males and XX females, the type of sperm that fertilizes the egg determines the sex of an offspring:

XY male $\rightarrow$ 1:1 ratio of X and Y sperm cells

XX female $\rightarrow$ all egg cells have one X

So,

X egg x X sperm $\rightarrow$ XX female offspring

X egg x Y sperm $\rightarrow$ XY male offspring

D.   <u>Sex-linkage</u>.
1)   Morgan discovered that the gene for white eye color in fruit flies is located in the X chromosome alone ("<u>sex-linkage</u>").
2)   Matings showed the following:

P: red eyed females x white eyed males

                                  ↓

F$_1$:   red eyed females and males

                                  x

                                  ↓

F$_2$:   2/4 red eyed females   ⎫
                                        ⎬   3/4 red eyed
                 1/4 red eyed males   ⎭

                 1/4 white eyed males   1/4 white eyed

Although a 3 red:1 white ratio occurred in the F$_2$, only males had white eyes.

3)   Morgan's explanation.

    a.   The allele for white eyes is recessive and the red eye allele is dominant.

    b.   These genes are located in the X chromosome only, not in the Y or the autosomes.

    c.   Using the symbols W for an X chromosome with the red allele and w for an X with the white allele, and Y for the Y chromosome without eye color alleles, we can diagram the matings as follows:

P: female: WW x male: wY

sperm: eggs:	1/2 (w)	1/2 (Y)
1 (W)	1/2 Ww female red eyes	1/2 WY male red eyes

F$_1$:   female: Ww x male: WY

sperm: eggs:	1/2 (W)	1/2 (Y)
1/2 (W)	1/4 WW female red eyes	1/4 WY male red eyes
1/2 (w)	1/4 Ww female red eyes	1/4 wY male white eyes

F$_2$:   2/4   red eyed females
                1/4   red eyed males
                1/4   white eyed males

d. Sex-linked genes in the X chromosome that are not found in the Y are called X-linked genes.

    i. Females can be homozygous for X-linked genes.

    ii. Females can be heterozygous for X-linked genes, masking the effect of recessive alleles.

    iii. Males have only one dose of their X-linked genes, so all of them must be expressed (whether they are dominant or recessive in females).

E. Gene linkage on the autosomes.

    1) Linkage:

        a. Like the X chromosome, the autosomes also contain many genes.

        b. Sex-linkage is merely a special example of the general phenomenon of linkage: 2 or more genes tend to be inherited together because they lie in the same chromosome.

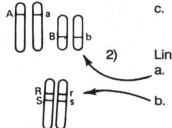

        c. When linkage is complete, instead of obtaining a 9:3:3:1 ratio in the F_2 generation of a dihybrid cross, the ratio is 3:1 since the 2 pairs of genes are inherited together due to their linkage (Figure 11-12 in the text).

    2) Linkage requires that the Law of Independent Assortment be modified as follows:

        a. Genes on different chromosomes assort independently during meiosis (as Mendel demonstrated).

        b. Genes on the same chromosome usually will not assort independently, but instead tend to be inherited together.

F. Linkage, crossing-over, and genetic mapping.

    1) Linked genes do not always stay together, since:

        a. During meiosis, homologous chromosomes often exchange pieces, a process called crossing-over.

        b. At the chiasmata (intertwined chromatid "ropes" of homologous chromosomes during prophase I of meiosis), the chromatids often exchange corresponding pieces with each other, forming new allele combinations in the chromatids.

        c. When crossing-over occurs, the 2 chromatid "ropes" swap corresponding segments so that series of genes are switched over from one chromatid to the other.

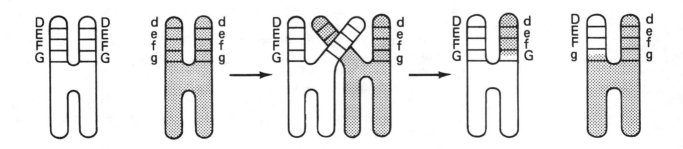

        d.   Crossing-over during meiosis explains why new combinations of traits occur for linked genes. This is an example of <u>genetic recombination</u> (generating new combinations of linked genes due to exchanges of DNA between homologous chromosomes).

    2)   <u>Chromosome mapping</u>.
        a.   Crossing-over occurs more often between linked genes that are farther apart in a chromosome than between genes closer together because, generally, crossing over occurs at random along a chromosome.
            i.   This is similar to saying that the farther apart 2 row houses are along a street, the more likely that a random lightning bolt will hit the ground between them, while the closer together they are, the less likely it is for the lightning to strike between them.
            ii.   Therefore, the number of offspring with recombinant traits is an indication of the spacing of genes along a chromosome: the greater the number of recombinant offspring, the farther apart are the linked genes.
        b.   This idea has been used to "map" genes in chromosomes:
            i.   The "distance" between 2 genes is the percent of recombinant offspring observed during a dihybrid testcross (<u>RrSs</u> x <u>rrss</u>).
            ii.   So, if 10% of the offspring are recombinant, the genes are 10 "map units" apart in the chromosome.

G.   Relationships among genes alleles, and chromosomes.
    1)   <u>Chromosomes</u> are long continuous strands of DNA.
    2)   A <u>gene</u> is a segment of DNA located at a specific spot (<u>locus</u>).
    3)   The different types of chromosomes in the diploid body cells of most multicellular organisms occur in pairs called <u>homologues</u>. Most such cells have:
        a.   One to many pairs of <u>autosomes</u>.
        b.   One pair of <u>sex chromosomes</u>.
    4)   Homologous chromosomes carry the same genes at similar loci.
        a.   The exception is the pair of sex chromosomes.
        b.   <u>X chromosome</u> has many genes, including those for traits unrelated to sex.
        c.   <u>Y chromosome</u> has few genes other than those that determine maleness.
        d.   Males = XY; females = XX.
    5)   Differences in DNA composition at the same gene locus on two homologous chromosomes form different <u>alleles</u> of the gene.
    6)   Homozygous and heterozygous:
        a.   <u>Homozygous genes</u>: both homologous chromosomes have identical alleles for a given gene pair.
        b.   <u>Heterozygous genes</u>: Homologous chromosomes have non-identical alleles for a given gene pair.
    7)   <u>Linked genes</u>: genes located on a single chromosome.
        a.   They tend to be inherited together.
        b.   <u>Crossing over</u> exchanges DNA segments of homologous chromosomes. If the segments have different alleles, crossing over may create new combinations of linked genes.

4.   **Variations** on the Mendelian theme.

A.   Extensions of the gene concept:
1)   Genes mutate at low rates.
a.   A <u>mutation</u> is a change in the molecular composition of a gene.
b.   Different alleles originate as mutations.
2)   Many traits are <u>not</u> inherited in a simple, single gene, dominant vs. recessive manner.

B.   <u>Incomplete dominance</u>: the heterozygote phenotype is intermediate between the 2 homozygous phenotypes.
1)   Example: snapdragon flower color:

P: red (<u>RR</u>)   x   white (<u>R'R'</u>)

F$_1$:   pink (<u>RR'</u>)

2)   This is not blending inheritance since in the F$_2$ generation, red, pink, and white flowered plants are produced in a 1:2:1 ratio, indicating that the <u>R</u> and <u>R'</u> alleles have not changed:

F$_1$:   pink (<u>RR'</u>) x pink (<u>RR'</u>)

	1/2 Ⓡ	1/2 Ⓡ'
1/2 Ⓡ	1/4 <u>RR</u> red	1/4 <u>RR'</u> pink
1/2 Ⓡ'	1/4 <u>RR'</u> pink	1/4 <u>R'R'</u> white

F$_2$:   1/4 <u>RR</u>, 2/4 <u>RR'</u>, 1/4 <u>R'R'</u> : 1:2:1 genotype ratio

1/4 red, 2/4 pink, 1/4 white: 1:2:1 phenotype ratio

3)   Notice that with incomplete dominance, the genotypic and phenotypic ratios are <u>identical</u>.

C.   <u>Multiple alleles</u> and <u>codominance</u>.
1)   <u>One</u> diploid individual can have only 2 alleles for a particular gene.
2)   But if we sample <u>all</u> the individuals of a species, we often find more than 2 allelic forms for a particular gene ("<u>multiple alleles</u>"), for example the ABO blood typing alleles in humans.
3)   The ABO system of multiple alleles:
a.   3 alleles of a given gene (I^A, I^B, i) produce 4 different blood type phenotypes: A, B, AB, and O.

b. The gene involved normally makes a glycoprotein antigen found in the red blood cell surface:
- i. Allele I^A makes antigen A.
- ii. Allele I^B makes antigen B.
- iii. Allele i is defective; makes no antigen.

c. Dominance relationships:
- i. I^A and I^B are both dominant to i which is recessive.
- ii. I^A and I^B are equally dominant to each other ("codominance") so that an $I^A I^B$ person makes both A and B antigens. Both are phenotypically detectable in the $I^A I^B$ heterozygote.
- iii. Relationship between genotypes and phenotypes:

GENOTYPES	PHENOTYPES
$I^A I^A$ and $I^A i$	type A
$I^B I^B$ and $I^B i$	type B
$I^A I^B$	type AB
ii	type O

d. See Figure 11-17 for the relationships between blood types of donors and recipients involving blood transfusions.

D. **Polygenic inheritance:** when a trait is equally influenced by many genes.
1) Example: kernel coloration in wheat. Two pairs of genes (R1 R'1 and R2 R'2) are involved:
- a. Each R1 and R2 actively makes enzymes responsible for one unit of red kernel pigment.
- b. R'1 and R'2 are inactive, making no enzyme for red pigment.
- c. Each gene pair shows incomplete dominance:

R1 R1 = 2 red units,   R1 R'1 = 1 red unit,   R'1 R'1 = 0 red units

R2 R2 = 2 red units,   R2 R'2 = 1 red unit,   R'2 R'2 = 0 red units

- d. Each kernel cell has a pair of each type of gene (types 1 and 2) whose total effects add up to give many gradations of red kernel color intensity:

GENOTYPES	RED UNITS	PHENOTYPES
R1 R1 R2 R2	4	dark red
R1 R'1 R2 R'2	2	medium red
R'1 R'1 R'2 R'2	0	white

Other intensities between 0 and 4 are possible as well (i.e., R1 R'1 R'2 R'2 = 1 unit = pale red)

2) The more genes that contribute to a single trait, the greater the number of categories with increasingly fine gradation between categories. A group of such individuals would show continuous variation for the trait.

3) Many common human characteristics show continuous variation (i.e., height, body build, IQ, and the color of hair, skin, and eyes).

   i. Some of the variation is due to gene differences among people ("nature").

   ii. Some of the variation is due to environmental influences ('nurture").

E. Gene interactions.

1) For some traits, there are genes whose actions are required for other genes to be expressed.

2) Example:

   a. In mice, genes exist for the synthesis of melanin pigment and other genes for distribution of the pigment.

   b. The distribution genes cannot function if no melanin is synthesized:

      If MM and Mm = melanin synthesis, and mm = albino, and AA and Aa = agouti distribution, and aa = black fur, then

      MMAA = agouti mouse   The A and a alleles
      MMaa = black mouse   cannot function if
      mmAA = albino mouse   no M genes are
      mmaa = albino mouse   present.

F. Multiple effects of single genes.

1) Pleiotropy: single genes having multiple phenotypic effects.

2) Example: The genes for albinism in mice cause white fur, pink eyes, and blindness. Such mice:

   a. Lack pigment in the fur.

   b. Lack eye pigmentation (eyes are pink because blood color shows through).

   c. Without pigment, eyes are quite sensitive to normal daylight, which destroys receptors in the eyes, causing blindness.

G. Environmental influences on gene action.

1) Both the genotype of an organism and the environment in which an organism lives profoundly affect its phenotype.

2) Example: Himalayan rabbits have white fur except for black ears, nose, tail, and feet.

   i. Actually, each cell has genes for black fur but the enzyme to make the black pigment is "temperature sensitive", being able to act only at temperatures below 34° C.

   ii. A rabbit's main body temperature is above 34° C (inactive pigment enzyme) but the extremities are less than 34° C (active enzyme in ears, nose, tail, feet).

3) Most environmental influences are more subtle and complicated than this example, especially human characteristics like intelligence and musical ability.

****

REVIEW QUESTIONS AND ANSWERS:

1.   Why was Mendel successful in explaining inheritance while others failed? [see 2.A.1)-3)]

2.-17. **Define** the following terms:

2.   Blending inheritance [see 1.B.]
3.   Segregation [see 2.B.2)b.]
4.   Alleles [see 2.B.2)d.]
5.   Homozygous and heterozygous [see 2.B.2)e.]
6.   Genotypes and phenotypes [see 2.B.3)a.-c.]
7.   Test cross [see 2.B.4)]
8.   Independent assortment [see 2.C.2)c.]
9.   "Tree diagram" [see 2.C.2)d.ii.]
10.   Sutton's hypothesis [see 3.A.1)-4)]
11.   Sex chromosomes and autosomes [see 3.C.1)-2)]
12.   Sex-linkage [see 3.D.1)-3)]
13.   Linkage [see 3.E.1)-2)]
14.   Mutation [see 4.A.]
15.   Incomplete dominance [see 4.B.1)-3)]
16.   Multiple alleles [see 4.C.1)-3)]
17.   Polygenic inheritance [see 4.D.1)-3)]

GENETICS PROBLEMS:

Monohybrid crosses.

1.   When two plants with red flowers are mated together, the offspring always are red, but if two purple-flowered plants are mated together, sometimes some of the offspring have red flowers. Which flower color is dominant?

2.   In sheep, white ($\underline{B}$) is dominant to black ($\underline{b}$). Give the F_2 phenotypic and genotypic ratios resulting from the cross of a pure-breeding white ram with a pure-breeding black ewe.

3.   If you found a white sheep and wanted to determine its genotype, what color animal would you cross it to and why?

4.   Squash may be either white or yellow. However, for a squash to be white, at least one of its parents must also be white. Which color is dominant?

5.   In peas, yellow seed color is dominant to green. What will be the expected proportion of each color in the offspring of the following crosses?

        A.   A heterozygous yellow with a heterozygous yellow.
        B.   A heterozygous yellow with a green.
        C.   A green with a green.

6.   If tall (<u>D</u>) is dominant to dwarf (<u>d</u>), give the genotypes of the parents that produce 3/4 tall plants and 1/4 dwarf plants among their progeny.

Dihybrid crosses.

7.   In pigs, mule hoof (fused hoof) is dominant (<u>C</u>) while cloven foot is recessive (<u>c</u>). Belted coat pattern (<u>S</u>) is dominant to solid color (<u>s</u>). Give the F_2 genotype and phenotype ratios expected from the cross <u>CCSS</u> x <u>ccss</u>.

8.   In the F_2 generation of the previous question, what proportion of the cloven-hoofed, belted pigs would be homozygous

9.   Flat tail (<u>F</u>) is dominant to fuzzy tail (<u>f</u>), and toothed (<u>T</u>) is dominant to toothless (<u>t</u>). Give the results of a cross between two completely heterozygous parents.

10.   In rabbits, black (<u>B</u>) is dominant to brown (<u>b</u>), and spotted coat (<u>S</u>) is dominant to solid coat (<u>s</u>). Give the genotypes of the parents if a black, spotted male is crossed with a brown, solid female and all the offspring are black and spotted.

11.   In the preceding problem, give the genotypes of the parents if some of the offspring were brown and spotted.

12.   In cattle, having horns (<u>p</u>) is recessive to hornless or polled (<u>P</u>). Coat color is controlled by incompletely dominant genes <u>RR</u> for red, <u>rr</u> for white, and <u>Rr</u> for roan. If two heterozygous roan-polled cattle are mated, what kinds of offspring are expected?

13.   If a yellow guinea pig is crossed with a white one, the offspring are cream-colored. What is the simplest explanation for this result? What kinds of offspring are expected if two cream-colored guinea pigs mate?

14.   In carnations, red or white phenotypes are dependent on homozygous genotypes, while the heterozygotes are pink. Give the F_1 and F_2 genotypic and genotypic ratios expected from a cross: red x white.

Sex-linked.

15.   A normal woman whose father was a hemophiliac marries a normal man. What are the chances of hemophilia occurring in their children?

16.   Another woman with no history of hemophilia in her family marries a normal man whose father was a hemophiliac. What are the chances of hemophilia occurring in their children?

17.   Color-blindness (<u>c</u>) is a sex-linked recessive trait, while normal color vision (<u>C</u>) is dominant.

        A.   If two normal-visioned parents have a color-blind son, what are the parents' genotypes?
        B.   What are the chances that their daughter will be color-blind?

18.   In cats, yellow is due to gene <u>B</u>, and black to its allele <u>b</u>. These genes are sex-linked. The heterozygous condition results in tortoiseshell. What kinds of offspring (sex and color) are expected from a cross: black male x tortoise-shell female?

19. In fruit flies, normal long wings is dominant ($\underline{V}$) and vestigial (shortened) wings is recessive ($\underline{v}$). These genes are autosomal. The sex-linked gene controlling red eye color ($\underline{W}$) is dominant to white eyes ($\underline{w}$). A male with red eyes and normal wings mates a white-eyed vestigial-winged female. Give the expected ratio of phenotypes in the F_2 generation.

Gene interactions.

20. In poultry, there are two independently assorting gene loci, each with two alleles that affect the shape of a chicken's comb. One locus has a dominant allele ($\underline{R}$) for rose comb while its recessive allele ($\underline{r}$) produces single comb. The other locus has a dominant gene ($\underline{P}$) for pea comb while its recessive allele ($\underline{p}$) also produces single combs. When the two dominant genes occur together ($\underline{R-P-}$), a walnut comb is produced. So, $\underline{R-P-}$ = walnut, $\underline{R-pp}$ = rose, $\underline{rrP-}$ = pea, and $\underline{rrpp}$ = single. Give the expected phenotypic ratios of offspring from the following matings:

A. $\underline{RRPP}$ x $\underline{rrpp}$

B. $\underline{RrPp}$ x $\underline{rrpp}$

C. $\underline{Rrpp}$ x $\underline{rrPp}$

D. $\underline{RrPP}$ x $\underline{RrPp}$

E. $\underline{rrPp}$ x $\underline{RrPP}$

21. In humans, deafness can be the result of a recessive allele affecting the middle ear ($\underline{dd}$ = deaf), or another recessive allele ($\underline{ee}$ = deaf) that affects the inner ear. Suppose two deaf parents have a child that can hear. Give the genotypes of all three individuals.

22. If two normal people, heterozygous at both loci ($\underline{DdEe}$) for deafness marry, what are the chances that their first child would be normal hearing? What is the chance of deafness in this child?

Multiple alleles.

23. Mallard ducks show a multiple allele pattern of inheritance in which $\underline{M^R}$ produces "restricted mallard" coloring and is dominant over $\underline{M}$ for mallard coloring, and both of these alleles are dominant over $\underline{m}$ for "dusky mallard" coloring. Give the phenotypic ratios expected among offspring from the following crosses:

A. $\underline{M^R M}$ x $\underline{M^R m}$

B. $\underline{M^R M}$ x $\underline{Mm}$

C. $\underline{Mm}$ x $\underline{mm}$

Multiple genes.

24. If there are two pairs of genes involved in producing skin color in black x white crosses, and if the five phenotypic classes are black, dark, medium (mulatto), light, and white, give the expected F_2 results of a white x black mating.

25. Give the darkest phenotype possible among the offspring of the following matings:

A. black x dark

B. black x medium

C. black x white

D. dark x medium

E. dark x white

F. medium x light

G. light x light

H. light x white

****

ANSWERS TO GENETICS PROBLEMS:

1. Purple is dominant.

2. Genotypic ratio: 1 <u>BB</u>: 2 <u>Bb</u>: 1 <u>bb</u>
    Phenotypic ratio: 3 white: 1 black

3. A testcross with a black (<u>bb</u>) sheep.

4. White is dominant.

5. A.   3/4 yellow: 1/4 green   C.   all green
    B.   1/2 yellow: 1/2 green

6. <u>Dd</u> x <u>Dd</u>

7. 9   <u>C-S-</u>   mule-foot, belted pigs
    3   <u>C-ss</u>   mule-foot, solid pigs
    3   <u>ccS-</u>   cloven-foot, belted pigs
    1   <u>ccss</u>   cloven-foot, solid pig

8. There are 3 cloven-foot, belted pigs, but only one is homozygous (<u>ccSS</u>); the other two are heterozygous (<u>ccSs</u>).

9. 9   <u>F-T-</u>   flat and toothed
    3   <u>F-tt</u>   flat and toothless
    3   <u>ffT-</u>   fuzzy and toothed
    1   <u>fftt</u>   fuzzy and toothless

10. black spotted male (<u>BBSS</u>) x brown solid female (<u>bbss</u>).

11. <u>BbSS</u> male x <u>bbss</u> female.

12.

RATIO	GENOTYPES	PHENOTYPES
3	<u>P-RR</u>	polled red
6	<u>P-Rr</u>	polled roan
3	<u>P-rr</u>	polled white
1	<u>ppRR</u>	horned red
2	<u>ppRr</u>	horned roan
1	<u>pprr</u>	horned white

13. A.   Incomplete dominance with cream being heterozygous.
     B.   1 yellow: 2 cream: 1 white.

14. F_1: Rr (pink)
      F_2: 1/4 rr (red): 2/4 Rr (pink): 1/4 rr (white).

15.   1/2 of the sons will inherit the defective gene from mom and get Y from dad. 1/2 of the daughters will also inherit the defective gene from mom but they will be heterozygous since they also inherit a normal gene from dad.

16.   None is expected to have hemophilia or inherit the gene.

17.   A.   Cc (normal carrier woman) x CY (normal man).
      B.   1/2.
      C.   No chance: she inherits a normal C gene from dad as well as either C or c a gene from mom.

18.   Black female, tortoiseshell female, black male, yellow male.

19. F_2 for both sexes: 3/8   red long   3/8   white long
                             1/8   red vestigial   1/8   white vestigial

20.   A.   all walnut
      B.   1/4 walnut: 1/4 rose: 1/4 pea: 1/4 single
      C.   1/4 walnut: 1/4 rose: 1/4 pea: 1/4 single
      D.   3/4 walnut: 1/4 pea
      E.   1/2 walnut: 1/2 pea

21.   DDee x ddEE $\rightarrow$ DdEe.

22.   9/16 D-E- = normal
      3/16 D-ee = deaf
      3/16 ddE- = deaf   so, 9/16 normal; 7/16 deaf.
      1/16 ddee = deaf

23.   A.   3/4 restricted mallard: 1/4 mallard
      B.   1/2 restricted mallard: 1/2 mallard
      C.   1/2 mallard: 1/2 dusky mallard.

24. F_2:   1/16   white
                     4/16   light
                     6/16   medium
                     4/16   dark
                     1/16   black

25.   A.   black   E.   medium
      B.   black   F.   dark
      C.   medium   G.   medium
      D.   black   H.   light

****

110

CHAPTER 12: DNA: THE MOLECULE OF HEREDITY

CHAPTER OUTLINE.

The molecular basis of inheritance, DNA, is covered in this chapter. Proofs that DNA is the hereditary material, the structure of DNA, and how DNA replicates and mutates are emphasized.

1. **The composition of chromosomes.**

   A. In the early 1870s, Miescher analyzed the nucleus biochemically and found a previously unknown acidic material with high phosphorus content which he called <u>nucleic acid</u>.

   B. Eukaryotic chromosomes are made of protein and a type of nucleic acid called <u>DNA</u> (<u>deoxyribonucleic acid</u>). Scientists speculated that the genetic material in chromosomes must be one of these materials.

   C. DNA seemed to be a simple molecule with only 4 kinds of subunits (called <u>nucleotides</u>) strung together in long chains.
      1) Each nucleotide has 3 parts:
         a. A phosphate group ("P").
         b. A 5-carbon deoxyribose sugar ("S").
         c. A nitrogen-containing base ("B").

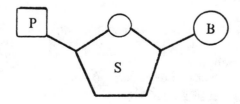

2) All DNA nucleotides have identical P and S subunits, but may have any one of 4 different bases which come in 2 sizes:
   a. Single-ringed <u>pyrimidine</u> <u>bases</u> of 2 types:
      i. <u>Thymine</u> ("<u>T</u>")
      ii. <u>Cytosine</u> ("<u>C</u>")
   b. Double-ringed <u>purine</u> <u>bases</u> of 2 types:
      i. <u>Adenine</u> ("<u>A</u>")
      ii. <u>Guanine</u> ("<u>G</u>")
   c. Aids in remembering these base names:
      i. The smaller bases have the bigger name (single-ring bases are pyrimidines) and the larger bases have the smaller name (double-ring bases are purines)
      ii. The word "purine" begins with the letters P-U. When something smells bad, we say "P-U" and the smell may cause us to gag (GAG are the letters designating the purine bases adenine and guanine). So, purines → P-U → GAG. The other bases (T and C) are pyrimidines.

3) In a single strand of DNA, the P of one nucleotide bonds covalently to the S of the next, forming a long strand of nucleotides with an alternating P-S-P-S-P-S- "backbone". The bases stick out from the backbone:

```
—P-S—P-S—P-S—P-S—P-S—P-S—
   |    |    |    |    |    |
   B    B    B    B    B    B
```

D. Are genes <u>DNA or protein</u>?
   1) Before 1940, scientists erroneously thought that protein was the genetic material. Their flawed reasoning included the following:
      a. DNA is a simple molecule with only 4 kinds of bases. It could not contain the complex information necessary to construct and control an organism.
      b. Protein has 20 different amino acids in its structure and could, therefore, carry large amounts of genetic information.
      c. Since enzymes are proteins, chromosomal proteins act as molds from which copies of enzymes could be made.
   2) Gradually, biologists realized that proteins may not be the genetic material. In some fish, for example, sperm contain DNA plus one protein (protamine) containing one type of amino acid. How could this simple protein contain information for offspring development?
   3) Finally, 2 experiments with microorganisms showed conclusively that DNA is the genetic material.

2. **DNA is the hereditary material**: 2 crucial experiments:

A. **Bacterial transformation.**
   1) In 1928, Griffith demonstrated <u>transformation</u> in pneumonia bacteria (<u>Streptococcus</u> <u>pneumonia</u>).

a. These bacteria occur in 2 genetically determined forms:

    i. Those <u>encapsulated</u> in polysaccharide. These cause pneumonia in mice since the capsule protects the microbes in the blood of mice.

    ii. Those <u>non-encapsulated</u> ("<u>naked</u>"). These are harmless since lack of the capsule allows them to be destroyed in the bloodstream of mice.

b. Griffith showed:

    i. Intense heating kills encapsulated bacteria, rendering them harmless to mice.

    ii. However, injecting a mixture of heat-killed encapsulated bacteria and live naked bacteria causes pneumonia, even though neither type alone could do so.

    iii. The whole process could occur without mice at all. In 1933, Alloway mixed dead encapsulated and live naked bacteria in a test tube and produced live, true-breeding encapsulated bacteria in the tube.

2) In 1944, Avery and co-workers showed that the transforming chemical was DNA:

a. They isolated DNA from the heat-killed encapsulated bacteria and mixed it with live naked bacteria, some of which began to make capsule polysaccharide.

b. Adding protein-destroying enzymes to the extract did not prevent transformation, but adding DNA-destroying enzymes did.

c. So, DNA is the genetic material in bacteria:

    i. A live naked bacterium can take up DNA from its environment and integrate the DNA into its chromosome.

    ii. This incorporated foreign DNA allows the recipient to make capsule molecules.

3) A few skeptics erroneously thought that DNA caused mutations to bacterial "protein genes", allowing them to make the capsule.

B. **Bacteriophage (virus) experiments.**

1) <u>Bacteriophage</u> ("bacteria eaters") or <u>phage</u> are viruses that infect only bacteria.

2) Phage are made of only DNA and protein:

a. When phage encounter bacteria, they attach to the bacterial cell walls and inject their genetic material into the bacteria.

b. The rest of the phage (head, tail, tail fibers) remains outside the bacterium.

c. The phage genetic material uses the bacterial cytoplasm to make many more phage which eventually burst out of the host cell.

3) In 1952, Hershey and Chase showed that, in phage, DNA is the genetic material:

a. DNA has phosphorus but not sulfur while protein has sulfur-containing amino acids but no phosphorus.

b. Hershey and Chase raised one population of phage on radioactive phosphorus ("<u>hot DNA</u>") and another one on radioactive sulfur ("<u>hot protein</u>").

c. When bacteria were infected with phage containing "hot protein":

    i. No "hot protein" entered the bacterial cytoplasm.

    ii. No offspring phage had "hot protein".

d. When bacteria were infected with phage containing "hot DNA":

i.   The "hot DNA" entered the bacterial cytoplasm.

ii.   Some offspring phage had "hot DNA".

e.   Conclusion: The genetic material in phage is DNA, not protein.

3.   **Structure of DNA**: By understanding its structure, we can understand how it functions.

A.   <u>Before 1952</u>, Mirsky had shown that:

      1)   The quantity of DNA is constant in each body cell of a given species, no matter what tissue the cells came from.

      2)   Gametes, however, contain only half as much DNA as do body cells. This is expected if DNA is the genetic material.

B.   <u>In 1950</u>, Chargaff analyzed the amounts of the 4 types of nucleotides in DNA from a variety of species and found curious results:

      1)   The amounts of the 4 bases vary considerably when different species are compared.

      2)   However, in any particular species, the amounts of A and T are equal, and the amounts of G and C are equal.

      3)   The amounts of the purine bases (A + G) = 50% and the amounts of pyrimidines (T + C) = 50%.

C.   In the <u>early 1950s</u>, Wilkins and Franklin used <u>X-ray</u> diffraction to study DNA structure:

      1)   They bombarded crystals of DNA with X-rays and photographed the resulting diffraction patterns.

      2)   The diffraction patterns suggested:

            a.   A DNA molecule is helical (like a corkscrew).

            b.   A DNA molecule has a uniform diameter of 2 nanometers (nm).

            c.   Each DNA subunit is 0.34 nm from its nearest neighbor.

            d.   One full turn of the DNA helix occurs every 3.4 nm.

            e.   The P-S-P-S-P-S-P- backbone is on the outside of the helix while the bases are on the inside.

D.   <u>The double helix</u>. Based on the above information, Watson and Crick made some good guesses about DNA structure:

      1)   Knowing that important biological objects come in pairs, they proposed that each DNA molecule consists of 2 strands, twisted about each other into a "<u>double helix</u>" (like a twisted ladder).

      2)   The S-P-S-P- backbones of the 2 strands (the uprights of the "ladder") are on the outside of the double helix, being parallel to each other but running in opposite directions (like an interstate highway where north and south bound lanes are parallel to each other but oriented so that traffic flows in opposite directions).

      3)   The bases are paired up in the middle (like the rungs of the "ladder").

            a.   Each pairing of bases is composed of a purine and a pyrimidine joined by hydrogen bonds:

                 i.   A pairs with T by 2 hydrogen bonds.

                 ii.   G pairs with C by 3 hydrogen bonds.

            b.   This makes sense of Chargaff's data.

            c.   The <u>base pairing rule</u>: only <u>complementary bases</u> can pair up due to their 3-dimensional structures. A and T are complementary bases, as are

G and C.
4)   The Watson-Crick DNA model, published in 1953, explained several features of inheritance. For instance, each <u>sequence</u> of bases represents a unique set of genetic instructions, coding for a unique protein molecule.

4.   **Semi-conservative DNA replication**: The key to constancy.

A.   According to the base pairing rule (A-T and G-C), the sequence of bases in one strand of a double helix accurately predicts the sequence of bases in the complementary strand.

B.   <u>Steps in DNA replication</u>:
1)   New nucleotides are made in the cytoplasm and enter the nucleus.
2)   The 2 DNA strands of a double helix unwind and separate from each other.
3)   Each parental strand is a <u>template</u>, attracting nucleotides to form a molecule that is complementary to the parental molecule.
4)   So, each new double helix has one parental strand and one newly synthesized daughter strand. Since half of the original double helix is conserved in each new double helix, the process is called <u>semi-conservative</u> <u>replication</u>.
5)   Regarding chromosome replication: Each sister chromatid gets one of the newly synthesized double helix molecules.

C.   <u>Molecular events</u> in DNA replication:
1)   The hydrogen bonds holding base pairs together are broken by the enzyme "DNA helicase", allowing the two DNA strands in a molecule to unwind.
2)   "DNA polymerase" (DNA pol) enzymes bind to each unwound strand at the "replication fork" (the junction between the separated strands and the remaining double helix). DNA pol:
        a.   Causes only free nucleotides (with complementary bases) to base pair in appropriate positions with the original strands.
        b.   Binds adjacent nucleotides together to form the newly synthesized DNA strands.
3)   DNA pol can only travel in one direction (3' → 5') on a DNA strand.
        a.   So, DNA pol appears to move in opposite directions on each strand, since the strands are anti-parallel in orientation.
        b.   After 2 DNA pols enter the replication fork, they will each travel towards the 5' ends of the 2 strands (they appear to move in opposite directions).
4)   Assume the 2 DNA pols settle down on opposite ends of the separated DNA, one at the end of the fork (strand A) and the other at the branch point (strand B):
        a.   The strand A DNA pol builds a new molecule by moving along its parental strand towards the branch point. As the helicase continues to separate the parental DNA strands, this DNA pol follows along on strand A, making one long continuous complementary strand.
        b.   The strand B DNA pol has to bind right behind the helicase and <u>move away</u> from the helicase.
                i.   Thus, as the helicase continues to separate the parental strands, it exposes regions of strand B that the DNA pol can't get to since it's moving away.
                ii.   However, a new DNA pol will soon attach right behind the helicase on strand B and, as it moves away, it will make a

complementary copy of the DNA region it moves along.

iii. When the second strand B DNA pol reaches the place where the first one started, the two short DNA pieces made by the strand B DNA pols are joined up to make a continuous DNA strand by an enzyme "DNA ligase."

iv. Steps i.-iii. above are repeated many times until the entire strand B is replicated.

D. Proofreading.
1) DNA polymerase makes mistakes, due to:
a. The high speed of replication (50-500 nucleotides per second).
b. Spontaneous chemical "flip-flops" in the base structures.
2) DNA polymerase makes about 1 mistake per 10,000 base pairs.
a. But the completely replicated DNA contains about 1 mistake per billion base pairs.
b. This enhanced accuracy is caused by several enzymes that proofread each daughter strand as it is synthesized.

E. Relating DNA replication to chromosome replication and mitosis:
1) During the S phase of interphase:
a. Each chromosome forms 2 sister chromatids.
b. The long DNA molecule of each chromosome replicates semi-conservatively so that each sister chromatid gets one of the newly synthesized double helixes which are exact duplicates of each other.
2) During the following mitosis:
a. At anaphase, sister chromatids become daughter chromosomes and move to opposite ends of the spindle.
b. In this way, each daughter cell formed at telophase receives an exact duplicate of each DNA molecule from the original cell.

5. **Mutation**: The key to variability.

A. Complete constancy is not the best evolutionary strategy:
1) When environments change, species also must change or face extinction.
2) During sexual reproduction, new combinations of existing alleles arise by:
a. Independent assortment of different chromosomes ("sexual recombination").
b. Crossing over of different linked genes ("genetic recombination"), under the precise control of enzymes.
3) These processes can produce new allele combinations if there are a number of different alleles in existence. These alleles arise by mutation.

B. Mutation.
1) A gene mutation is a change in the sequence of bases.
2) Mechanisms of mutation:
a. A mistake in base pairing during replication, usually occurring "spontaneously" (without known cause).
b. Certain chemicals and radiation increase replication errors and can also alter the DNA between replications.

3)   Random changes in DNA composition are unlikely to code for improvements in gene function, and cells have evolved ways to repair most damaged DNA.

C.   <u>Types of gene mutation</u>:
      1)   <u>Point</u> <u>mutations</u>: a pair of bases becomes incorrectly matched.
          a.   Repair enzymes will replace one of the nucleotides with a complementary one.
          b.   But sometimes, the wrong nucleotide is replaced and the mutant sequence replaces the normal one.
      2)   <u>Insertions</u>: new nucleotide pairs are inserted in the midst of a gene.
      3)   <u>Deletions</u>: nucleotide pairs are removed from a gene.

D.   <u>Effects of mutations</u>:
      1)   Most are harmful or neutral and are removed from the population by natural selection or chance.
      2)   Occasionally, a mutation is beneficial to the organism and may become common in the population as its possessors out-reproduce rivals bearing the old allele.

****

REVIEW QUESTIONS:

1.-5.   **MATCHING TEST**: purine and pyrimidine bases

         Choices: A.   Adenine (A)   D.   Thymine (T)
                         B.   Cytosine (C)   E.   A-T and G-C
                         C.   Guanine (G)   F.   A-G and T-C

1.   Purine bases.
2.   Pyrimidine bases.
3.   Base pairs in DNA.
4.   Single ring bases.
5.   Double ring bases.

----

6.   Describe and explain the bacterial transformation experiments of Griffith and Avery.
7.   Describe and explain the bacteriophage experiments of Hershey and Chase.
8.   Describe and explain Chargaff's data in relation to the double helix structure of DNA.
9.   If one strand of a DNA molecule has the base sequence
          A G G T C A A T G C G C,
    what is the base sequence of the other strand?
10.   Describe how DNA replicates in a "semi-conservative" manner.
11.   Define the following types of gene mutations:
          a.   Point mutations.
          b.   Insertion mutations.
          c.   Deletion mutations.

12.-16. MATCHING TEST: DNA structure

Choices:

A.   Nucleotide
B.   Deoxyribose sugar
C.   Phosphate group
D.   Purine base
E.   Pyrimidine base

****

ANSWERS TO REVIEW QUESTIONS:

1.-5.   [see 1.C.2) and 3.D.3)]
    1.   A and C
    2.   B and D
    3.   E
    4.   B and D
    5.   A and C

6.   see 2.A.1)-2)

7.   see 2.B.1)-3)

8.   see 3.B.1)-3); 3.D.1)-4)

9.   [see 3.D.3)a.-c.

    T C C A G T T A C G C G

10.   see 4.B.-C.

11.   see 5.C.

12.-16. [see entire chapter]
    12.   D   15.   A
    13.   B   16.   E
    14.   C

****

CHAPTER 13: FROM GENOTYPE TO PHENOTYPE: GENE EXPRESSION AND REGULATION

CHAPTER OUTLINE.

This chapter covers how genes are expressed and regulated. The concepts of inborn errors of metabolism and one-gene one-enzyme are introduced. The processes of transcription (DNA → RNA) and translation (RNA → protein) are presented along with the three types of RNA and their functions. Transcriptional regulation of genes in prokaryotic and eukaryotic cells is discussed, and the relation of molecular genetics to Mendelian principles is given.

1. **Phenotype** (structure, function, and behavior of an organism).

    A. The phenotype is generated by an organism's genotype as it interacts with the environment.

    B. <u>Questions</u>.
        1) How does a cell use its genotype to create a phenotype?
            a. To go from genotype to phenotype means making the appropriate proteins, both structural and enzymatic.
            b. The sequence of nucleotides in DNA is a code that cells translate into the sequence of amino acids in protein.
        2) How do the genotype and environment interact to produce the many different cell types in an adult organism?
            a. Since all cells in the body contain identical chromosomes and genes,

how can they have such different forms and functions?

b. Cells regulate which genes are used and which are not, depending on the function of the cell and the environmental situation.

2. **Relationship between genes and proteins.**

    A. <u>Early thoughts</u> about how genes act.

        1) Before 1910, Garrod studied the inheritance of human metabolic disorders:

            a. He reasoned that chemical reactions in human cells proceeded in a series of steps, each controlled by a different enzyme.

            b. A defective or absent enzyme would lead to both:

                i. Lack of the end product of the series of reactions.

                ii. Accumulation of large amounts of precursor molecules normally converted to end products.

            c. These situations could lead to disease:

                i. Albinism is the lack of the pigment end product of a series of reactions.

                ii. PKU (phenylketonuria) is the abnormally high accumulation of an amino acid in the blood, causing mental retardation in youngsters.

        2) Garrod referred to these chemical disorders as "<u>inborn</u> <u>errors</u> of <u>metabolism</u>" and proposed that each is due to lack of a different normal enzyme made by its specific gene. Although correct, he had no real proof for his hypothesis.

    B. The <u>one-gene, one-protein (enzyme) hypothesis</u>.

        1) Beadle and Tatum used the common red bread mold <u>Neurospora</u> <u>crassa</u> to study the relationship between genes and enzymes, because this mold:

            a. Can make nearly all the organic compounds it needs.

            b. Can grow on medium containing only sucrose for energy, a few minerals, and the vitamin biotin.

            c. Is haploid for most of its life cycle, so that any mutation must be expressed in the phenotype.

        2) Beadle and Tatum X-rayed <u>Neurospora</u>, producing mutants that lost the ability to make some nutrients needed for growth.

            a. By careful analysis, they could determine exactly which nutrient a particular mutant could no longer make.

            b. They discovered that each specific mutant lost the ability to make an enzyme controlled by that gene.

            c. Their conclusion: Each specific gene somehow encodes the information to make a specific enzyme (the <u>one-gene</u>, <u>one-enzyme</u> hypothesis). Most genes function in precisely this way.

3. From **DNA to protein.**

    A. <u>General facts</u>.

        1) A <u>gene</u> is a segment of DNA with the information to make a protein.

        2) The sequence of nucleotides in DNA encodes the sequence of amino acids in a protein. Different types of proteins have different amino acid sequences.

        3) In eukaryotic cells, the DNA is in the nucleus and proteins are made in the

cytoplasm.

B. An intermediary molecule (RNA) carries the information from nuclear DNA to the ribosomes in the cytoplasm where proteins are made.
    1) RNA differs from DNA in 3 respects:
        a. RNA is single-stranded, not double-stranded.
        b. RNA has ribose sugar, not deoxyribose.
        c. RNA has the pyrimidine base uracil (U), not thymine (T).
    2) Information flow from DNA to protein involves 2 steps:
        a. Transcription: In the nucleus, the DNA code is copied into RNA molecules.
        b. Translation: In the cytoplasm, the RNA is used to make protein. Three types of RNA are involved in translation:
            i. Messenger RNA (mRNA): carries the genetic code to make the protein.
            ii. Transfer RNA (tRNA): picks up and transports amino acids to the mRNA.
            iii. Ribosomal RNA (rRNA): part of the ribosomes that allows the mRNAs and tRNAs to interact, constructing the protein.

4. The **genetic code**: the sequence of bases in the DNA is used to determine the sequence of amino acids in a protein.

A. The genetic code is similar in concept to the Morse code.
    1) Morse code: short sequences of dots and dashes → sequence of letters in a word.
    2) Genetic code: short sequences of bases in DNA → sequence of amino acids in proteins.

B. In 1961, Crick showed that the genetic code is "triplet".
    1) Each sequence of 3 consecutive bases codes for one amino acid.
    2) Thus, since DNA has 4 types of bases, there are $(4)^3 = 4 \times 4 \times 4 = 64$ different code words for the 20 types of amino acids in proteins.

C. Nature of the code.
    1) In order for any language to be understood, the user must know:
        a. What the words mean.
        b. Where words start and stop.
        c. Where sentences start and stop.
    2) Researchers discovered that an RNA with only U bases ("poly U": UUUUUU...) produced a protein composed solely of phenylalanine amino acids (phe-phe-phe...). Thus, the RNA triplet UUU codes for the amino acid phenylalanine.
    3) Eventually, the entire genetic code was deciphered (Table 13-1, not to be memorized!).
    4) The 64 possible mRNA triplets are called codons.
    5) "Punctuations" in the code:
        a. AUG is the first codon in the message, signaling the beginning of the coded information in mRNA.
        b. One of the codons UAG, UAA, or UGA is at the end of each message to

signal that the message is complete (just as a "." or "?" or "!" signals us that a sentence is finished).

    c.   There is no punctuation between codon "words" since each has 3 bases and cells know each message begins with AUG.

    d.   The genetic code is quite <u>redundant</u> or <u>degenerate</u> since about 60 codons code for only 20 amino acids (several codons may specify the same amino acid).

    e.   The code is not ambiguous, since each codon specifies only one amino acid.

5.   **RNA**: intermediary molecules in protein synthesis.

    A.   <u>RNA synthesis</u> (<u>transcription</u>).

        1)   Transcription (DNA → RNA) is restricted in 2 ways:

            a.   Transcription normally copies only the DNA of selected genes into RNA.

            b.   Transcription copies only one strand (the "sense strand" or strand with useful information) into RNA.

        2)   Transcription is a 3-step process:

            a.   Initiation.

            b.   Elongation of the RNA chain.

            c.   Termination.

        3)   <u>Initiation</u>: RNA polymerase locates the beginning of the gene.

            a.   The <u>promotor</u> region of a gene is a short DNA nucleotide sequence located just "upstream" (in the 3' direction) of the body of the gene.

            b.   RNA polymerase recognizes the promotor region and binds to the DNA at that site.

            c.   RNA polymerase covers 50 nucleotides, including the promotor and the first 12 or so in the body of the gene.

        4)   <u>Elongation</u>.

            a.   Once bound to the promotor site, RNA polymerase changes shape, forcing open the DNA double helix.

            b.   RNA polymerase then moves along the DNA sense strand, traveling in the 3' → 5' direction just as DNA polymerase does.

            c.   Using free RNA nucleotides in the nucleus, RNA polymerase makes a single strand of RNA that is complementary to the DNA sense strand.

            d.   DNA-RNA base-pairing rules: A(DNA)-U(RNA), C-G, G-C, T-A.

            e.   After about 10 nucleotides have been added to the growing RNA chain, the beginning of the RNA molecules pulls away from the DNA.

        5)   <u>Termination</u>.

            a.   Not well understood in eukaryotes.

            b.   RNA polymerase may continue moving along the DNA sense strand thousands of bases beyond the body of the gene. Eventually,

                i.   The RNA chain completely detaches from the DNA and from the RNA polymerase.

                ii.   The RNA polymerase detaches from the DNA.

    B.   <u>Types of RNA</u>.

        1)   <u>mRNA</u> (messenger RNA):

            a.   A long, single-stranded molecule including the codons that will be translated into the protein.

        b.   Prokaryotic mRNA is directly transcribed from the gene's DNA, and translation begins even before transcription is completed.

        c.   In eukaryotes:

            i.   The RNA transcribed from DNA contains more nucleotides than will ultimately be translated into protein.

            ii.   In the cytoplasm, mRNA binds to ribosomes where the mRNA codons are translated into the amino acid sequences of proteins.

    2)   rRNA (ribosomal RNA):

        a.   Ribosomes are composites of rRNA and a variety of proteins.

        b.   Each eukaryotic ribosome has 2 subunits:

            i.   A small subunit of one rRNA and about 30 protein molecules. It recognizes and binds mRNA and tRNA.

            ii.   A large subunit of 3 rRNA and 45-50 protein molecules. It contains:

                a)   An enzymatic region that catalyzes the addition of amino acids to the protein being made.

                b)   Two sites (P and A) that bind tRNA molecules.

    3)   tRNA (transfer RNA):

        a.   Binds to amino acids and delivers them to the ribosomes.

        b.   There is at least one type of tRNA for each type of amino acid.

        c.   Acts as "code books" that decipher the mRNA codons and translate them into the amino acids of proteins.

        d.   Are twisted into a "cloverleaf" shape with a "stem":

            i.   Enzymes in the cytoplasm recognize each type of tRNA and attach the correct amino acid to its stem, using ATP as an energy source.

            ii.   The outside bend of the central "leaf" of tRNA has 3 exposed bases (the "anticodon") that deciphers the mRNA code. The anticodon of each tRNA is complementary to the mRNA codon that specifies the amino acid attached to that tRNA.

6.   **Protein synthesis**.

    A.   Protein synthesis occurs in 2 steps:

        1)   mRNA is transcribed from DNA genes in the nucleus, and travels to ribosomes in the cytoplasm.

        2)   A ribosome binds mRNA and appropriate tRNAs, translating the mRNA codons into the amino acid sequence of a protein.

    B.   Translation has 3 steps:

        1)   Initiation of protein synthesis.

        2)   Elongation of the protein chain.

        3)   Termination.

    C.   Initiation of translation.

        1)   Initially, the small ribosomal subunits bind to 2 mRNA codons, the first one always being the "start codon" AUG.

        2)   The tRNA with the "start anticodon" UAC hydrogen bonds to the mRNA "start codon".

3) The large ribosomal subunit then attaches to the small subunit and the "start" tRNA binds to the P site in the large subunit.

D. Elongation of the protein chain occurs one amino acid at a time:
1) The anticodon of a [tRNA-amino acid] complex recognizes the mRNA's second codon and moves into the A site in the large ribosomal subunit. Both the P and A sites are now filled.
2) A ribosome enzyme on the large subunit (the "catalytic site") detaches the "start" amino acid from its tRNA and attaches it to the amino acid attached to the second tRNA by forming a peptide covalent bond. So now,
   a. The "start" tRNA in the P site has no amino acid (is "empty").
   b. The second tRNA in the A site has a short 2 amino acid "dipeptide".
3) The empty first tRNA leaves the ribosome which then shifts to the next mRNA codon, moving the second [tRNA-dipeptide] to the P site.
4) A new [tRNA-amino acid] complex binds to the A site. The ribosome enzyme detaches the dipeptide from the second tRNA (in the P site) and attaches it to the amino acid of the third tRNA (in A site), forming a "tripeptide".
5) The depleted tRNA leaves the P site, the ribosome shifts to the next mRNA codon, the third [tRNA-tripeptide] shifts to the P site, and the process continues as the next [tRNA-amino acid] complex binds to the A site.

E. Termination of protein synthesis.
1) The ribosome reaches a "stop" codon near the end of the mRNA's message.
2) No tRNA recognizes a stop codon.
3) Other molecules ("terminator factors") release the finished protein from the last tRNA and cause the large and small ribosomal subunits to detach.

7. **Gene regulation**.

A. General facts.
1) Most cells in the body have the same DNA but don't use all of it all the time.
2) Individual cells express only a small fraction of their genes, those appropriate to the function of that cell type. Example:
   a. Muscle cells make actin and myosin but not insulin.
   b. Pancreas cells make insulin but not actin and myosin.
3) Gene expression also changes over time, depending on the body's needs at any moment.
4) Basic question: How is gene action regulated in both tissue specific and time specific ways?

B. Regulation may occur at many steps:
1) Regulation of transcription: depends on the cell type, stage of the cell cycle, and metabolic activity of cell and organism.
2) Processing of pre-mRNA: eukaryotic genes are much longer than the final mRNA transcribed from them. So,
   a. Genes are first transcribed into very long "pre-mRNA" molecules.
   b. Differential processing of pre-mRNAs can produce different types of final mRNA that are translated into different proteins.
   c. Before used by cells, mRNAs are also modified by addition or removal of

nucleotides at their ends.

3) <u>Regulation of translation</u>. mRNAs vary:

    a.   In stability: very stable mRNAs are translated many times, while others are rapidly degraded.

    b.   In rate of translation: a cell may block translation of certain mRNAs, depending on metabolic needs.

4) <u>Protein modification</u>.

    a.   Many proteins must be modified to become active.

    b.   Example: protein-digesting enzymes from stomach wall and pancreas cells initially are made in an inactive form; once in the digestive tract, certain amino acids are snipped off to activate their active sites.

5) <u>Regulation of enzyme activity</u>: by competitive or allosteric inhibition (see Chapter 4).

C.   <u>Transcriptional regulation</u> of genes in prokaryotes.

1) Prokaryotic (bacterial) DNA is often organized into <u>operon</u> units in which genes for related functions lie next to each other.

2) An <u>operon</u> <u>system</u> has 4 parts:

    a.   <u>Regulatory gene</u> ("<u>R</u>"), making proteins that may bind to the operator gene.

    b.   <u>Promotor region</u> ("<u>P</u>"), recognized by RNA polymerase as the place where it binds to start transcription.

    c.   <u>Operator gene</u> ("<u>O</u>") that controls whether the RNA polymerase has access to P.

    d.   <u>Structural genes</u> ("<u>S</u>") coding for the amino acid sequences of the enzymes.

3) Whole operons are regulated as units, so that related enzymes are made simultaneously when needed.

    a.   Some operons are active continuously, making enzymes needed to make necessary nutrients.

    b.   Other operons are active only when the enzymes are needed to break down a food molecule that happens to be in the environment. Example: the <u>lactose</u> operon in <u>E</u>. <u>coli</u> bacteria has 3 "S" genes to break down the milk sugar lactose:

        i.   This operon is normally <u>repressed</u> (turned off) unless activated by the presence of lactose.

        ii.   Normally, the "R" gene makes <u>repressor protein</u> that binds to "O", not allowing RNA polymerase to bind at "P". Thus, no "S" genes are transcribed into mRNA.

        iii.   Lactose, however, binds to the repressor protein, inactivating it. RNA polymerase then binds to "P" and moves through "O" to the "S" genes, transcribing mRNA for enzymes that break down lactose, allowing <u>E</u>. <u>coli</u> to use it as an energy source.

        iv.   When the lactose supply is exhausted, new repressor proteins are free to bind to "O", turning the operon off again.

8.   **Transcriptional regulation** of genes in eukaryotes.

A.   Genes for related functions are found scattered in different chromosomes and the individual genes themselves are split up within a chromosome.

B.   Eukaryotic gene structure.
- 1)   Many genes consist of several interspersed regions of 2 types:
  - a.   Exons: DNA segments that code for amino acid sequences in proteins.
  - b.   Introns: DNA segments that lie between exons and code for nothing at all.
- 2)   Each gene has a promotor and an enhancer region. When regulatory protein binds to the enhancer, the binding of RNA polymerase to the promotor is intensified, thus enhancing transcription of the gene.
- 3)   Because the introns and long regions before the first exon and after the last exon are all transcribed along with the exons of a gene, the resulting RNA has many more nucleotides than needed to code for the protein.
- 4)   Two steps occur:
  - a.   RNA nucleotides are added at the front and back to form the mRNA that is translated.
  - b.   Then, enzymes cut this long RNA apart and splice together the exon regions that code for the protein.
  - c.   Functions of fragmented genes:
    - i.   To produce multiple proteins from a single gene.
      - a)   In rat thyroid, one splicing arrangement of a gene's pre-mRNA results in making calcitonin hormone.
      - b)   In rat brain, another splicing arrangement of the same gene's pre-mRNA makes a peptide used for neuro-communication.
    - ii.   To provide a quick and efficient way for eukaryotes to evolve proteins with new functions.

C.   Regulation of transcription in eukaryotes.
- 1)   Can occur at 3 levels:
  - a.   Individual genes.
  - b.   Large parts of chromosomes.
  - c.   Entire chromosomes.
- 2)   Single gene regulation.
  - a.   Best understood example: cellular action of steroid hormones.
  - b.   Estrogen stimulates production of egg white (albumin) protein in female birds.
    - i.   During breeding seasons, birds' ovaries secrete estrogen into the bloodstream.
    - ii.   The estrogen enters the oviduct cells and binds to receptors in the cytoplasm.
    - iii.   The [estrogen-receptor] complexes enter the nucleus and bind to DNA near the enhancer for the albumin gene.
    - iv.   Rapid transcription occurs and albumin is made.
    - v.   The mechanism for timing estrogen release by the ovaries is unknown.
- 3)   Large chunks of DNA, or even entire chromosomes, may be regulated by the intensity of their coiling:
  - a.   Highly condensed parts of chromosomes (i.e., near the centromeres)

prevent RNA polymerase from binding DNA there, thereby blocking transcription of genes in these regions.

b. Some chromosome regions may change from a condensed to a looser configuration at certain stages in the life cycle to allow transcription during those times.

c. <u>Random X-chromosome inactivation</u> in female mammals:

    i. In female body cells, only one X remains genetically active. Any additional Xs present condense and become inactive <u>Barr bodies</u> (densely staining bodies in interphase nuclei) after the first few cell divisions following fertilization.

    ii. In ovaries, both Xs remain active.

    iii. Which X remains active is initially a matter of chance, but all daughter cells maintain the same pattern as the parent cell.

    iv. So, female mammals are <u>mosaics</u>: about half the body cells in an X_1X_2 female have X_1 active and the other half have X_2 active, with the inactive X becoming a Barr body (●). Example: normal tortoiseshell cats are females, being heterozygous for the sex-linked alleles for black (X^B) and orange (X^O) fur color:

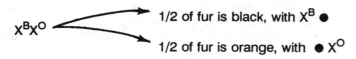

$X^B X^O$
- 1/2 of fur is black, with X^B ●
- 1/2 of fur is orange, with ● X^O

Male offspring of a tortoiseshell female may be either black or orange:

$X^B Y \rightarrow$ black male

$X^O Y \rightarrow$ orange male

9. **Mendelian genetics revisited.**

A. <u>Mutations and the genetic code</u>.

1) A mutation is a change in the sequence of nucleotides in DNA.

    a. Deletions and insertions of bases usually cause severe enzyme malfunction since all mRNA codons beyond the mutated region will be misread.

    b. Point mutations (one base replacing another) have more subtle and varied effects:

        i. If CTC (glutamic acid) → CTT (glutamic acid), the mutation has <u>not</u> <u>changed</u> <u>the</u> <u>amino</u> <u>acid</u> sequence of the encoded protein.

        ii. If CTC → CTA (aspartic acid), the mutation has <u>replaced</u> <u>one</u> <u>amino</u> <u>acid</u> <u>with</u> <u>its</u> <u>functional</u> <u>equivalent</u>, since both are hydrophilic. This is a <u>neutral</u> <u>mutation</u>.

        iii. If CTC → CAC (valine), the mutation has <u>replaced</u> <u>one</u> <u>amino</u> <u>acid</u> <u>with</u> <u>a</u> <u>functionally</u> <u>different</u> <u>one</u>, since valine is hydrophobic. This is the mutation that occurs in sickle cell anemia.

iv.   If CTC → ATC (= stop codon UAG in mRNA), incomplete proteins will be made, causing loss of function.

2)   Usually, mutations are harmful, but if one causes an altered protein that functions better, the mutation will confer a selective advantage on its possessor. This is how most new alleles become established in populations.

B.   Molecular explanation of <u>dominance relationships</u> among alleles.

1)   <u>Complete dominance</u>: one dominant and one recessive allele. Example: body fat color in rabbits.

a.   <u>AA</u> or <u>Aa</u> rabbits have white fat (dominant) while <u>aa</u> rabbits have yellow fat (recessive).

b.   Rabbits eat plants containing yellow pigment called xanthophyll that is fat soluble and will color the fat yellow.

     i.   Allele <u>A</u> makes an enzyme that breaks down the xanthophyll, keeping the fat white.

     ii.   Allele <u>a</u> is mutant, producing a non-functional enzyme.

c.   So, <u>AA</u> and <u>Aa</u> rabbits make functional enzymes in sufficient amounts to degrade all dietary xanthophyll. Thus, <u>A</u> is dominant to <u>a</u>.

d.   In general, dominant alleles make functional enzymes while recessive alleles make non-functional enzymes or no enzymes at all.

2)   <u>Incomplete dominance</u>.

a.   Simplest examples involve one allele making a functional enzyme and the other allele making a nonfunctional enzyme.

b.   Difference between complete and incomplete dominance is the effect of one vs. two copies of the functional allele.

     i.   In incomplete dominance, the amount of enzyme made by one functional allele is not enough to catalyze all the reactions needed to produce a fully dominant phenotype.

     ii.   If a snapdragon has 2 alleles for red pigment (<u>AA</u>), its flowers are dark red.

     iii.   If a snapdragon has one allele for red pigment (<u>Aa</u>), its flowers make less red pigment and appear pink in color.

     iv.   If a snapdragon has no alleles for red pigment (<u>aa</u>), its flowers are white.

3)   <u>Multiple alleles</u> and codominance.

a.   Since mutations can occur anywhere in a gene, different organisms in a population often suffer quite different mutations of the same gene, leading to multiple alleles.

b.   Such multiple alleles can usually be detected due to their different phenotypes.

c.   Example: human ABO blood type system.

     i.   Alleles I^A and I^B make slightly different functional enzymes that produce slightly different red blood cell glycoproteins.

     ii.   Allele <u>i</u> makes a non-functional enzyme incapable of producing glycoprotein. Thus, allele <u>i</u> is recessive.

     iii.   Alleles I^A and I^B are <u>codominant</u> since both are active in heterozygotes.

****

REVIEW QUESTIONS:

1.   Explain what is meant by "Inborn errors of metabolism".
2.   Explain what is meant by the "One-gene, one-enzyme hypothesis".
3.   Name 3 ways that DNA differs from RNA.

4.-13.   **MATCHING TEST**: transcription and translation

         Choices: A.   transcription   B.   translation

4.   DNA → RNA
5.   RNA → protein
6.   Occurs in the nucleus of eukaryotic cells.
7.   Occurs in the cytoplasm of eukaryotic cells.
8.   Involves RNA polymerase.
9.   Involves amino acids.
10.   Involves ribosomes.
11.   Involves codon-anticodon interactions.
12.   Involves copying the genetic code.
13.   Involves deciphering the genetic code.

14.-21. **MATCHING TEST**: types of RNA molecules

         Choices: A.   mRNA   B.   tRNA   C.   rRNA

14.   Has anticodons.
15.   Deciphers the genetic code.
16.   Carries the genetic code to make proteins.
17.   Picks up and transports amino acids.
18.   Part of ribosomes.
19.   Has codons.
20.   Twisted into a cloverleaf shape with a stem.
21.   Fits into the P and A sites.

----

22.   If the DNA strand being copied by RNA polymerase has the base sequence:
          T G C C T T A G A A G,
      What will the base sequence be in the RNA made?
23.   Describe what happens during the initiation, elongation, and termination processes of translation.

24.-28. **MATCHING TEST**: operon systems in bacteria

Choices:	A. Regulatory gene	C. Promotor region
	B. Operator gene	D. Structural gene

24. Where RNA polymerase initially binds.
25. Controls whether RNA polymerase can bind.
26. Makes proteins that may bind to the operator gene.
27. Where transcription begins.
28. Codes for mRNA for enzymes.

----

29. Explain how the lactose operon functions in the absence of lactose and in the presence of lactose.
30. What are "exon" and "intron" DNA regions within a eukaryotic gene?
31. Explain the action of "promotor" and "enhancer" regions in eukaryotic DNA.
32. What is meant by "random X chromosome inactivation" in female mammals?
33. Describe several different effects that point mutations may have on gene function.
34. Explain the molecular basis of dominance and recessiveness of alleles using the alleles for white and yellow body fat in rabbits as an example.
35. Explain the dominance relationships among the alleles controlling the ABO blood group system.

****

ANSWERS TO REVIEW QUESTIONS:

1. see 2.A.1)-2)	23. see 6.C.-E.
2. see 2.B.1)-2)	24.-28. [see 7.C.1)-3)]
3. see 3.B.1)a.-c.	

1. see 2.A.1)-2)

2. see 2.B.1)-2)

3. see 3.B.1)a.-c.

4.-13. [see 5.-6.]

4.	A	9.	B
5.	B	10.	B
6.	A	11.	B
7.	B	12.	A
8.	A	13.	B

14.-21. [see 5.B.1)-3)]

14.	B	18.	C
15.	B	19.	A
16.	A	20.	B
17.	B	21.	B

22. [see 5.A.4)d.]
A C G G A A U C U U C

23. see 6.C.-E.

24.-28. [see 7.C.1)-3)]

24.	C	27.	C.
25.	B	28.	D
26.	A		

29. see 7.C.3)b.i.-iv.

30. see 8.B.1)a.-b.

31. see 8.B.2)

32. see 8.C.3)c.i.-iv.

33. see 9.A.1)b.i.-iv.

34. see 9.B.1)a.-d.

35. see 9.B.3)c.i.-iii.

****

CHAPTER 14:   MOLECULAR GENETICS AND BIOTECHNOLOGY

CHAPTER OUTLINE.

This chapter begins by describing natural examples of genetic recombination and then discusses several techniques used in recombinant DNA technology, including building DNA libraries, identifying and making copies of genes, and using the genes to modify living organisms. Methods to locate and sequence genes are also covered.

1.   **General facts**.

   A.   <u>Biotechnology</u> is the manipulation of the molecular basis of inheritance by using recombinant DNA technology.

   B.   <u>Three goals of biotechnology</u>:
   1)   To understand more about the processes of inheritance and gene expression.
   2)   To provide better understanding and treatment of various genetic diseases.
   3)   To generate economic benefits, including improved plants and animals for agriculture and efficient production of valuable biological molecules.

2.   **DNA recombination in nature**.

   A.   <u>Two processes</u> involved in all DNA recombination:
   1)   Changing the nucleotide composition of cellular DNA.
   2)   Selecting valuable DNA combinations.

   B.   <u>Naturally occurring</u> methods of recombinant DNA.
   1)   <u>Within a species</u>, there are two types of natural recombination:
   a.   <u>Genetic recombination</u>: through crossing over during meiosis I, when genes from a maternal chromosome and a paternal chromosome form a new chromosome with a new sequence of genes.

b. Sexual recombination: chromosomes from two different organisms combine through fertilization to produce genetically diverse offspring.

2) Between-species recombination occurs in nature as well.

a. Transformation: bacteria pick up free DNA, sometimes from another species, from the environment.

b. Transformation often occurs when bacteria pick up plasmids (tiny loops of DNA with 1000-100,000 nucleotides that act as self-replicating cytoplasmic parasites).

i. One bacterium may contain hundreds of copies of a plasmid.

ii. Some plasmids contain genes for antibiotic-digesting enzymes.

iii. When plasmid-containing bacteria die, the plasmids may be liberated into the environment and transform other bacteria.

c. Viruses can transfer genes among eukaryotic organisms.

i. Viral DNA can insert itself into a chromosome of its eukaryotic host cell, and exist there quietly for a time.

ii. When the host is stressed, the viral DNA may leave the chromosome and take a bit of the eukaryotic host DNA with it.

iii. The viral DNA may then replicate itself by taking control of host cell metabolism.

iv. The offspring viruses (with bits of host DNA) may then infect a new host of a different species and inject its DNA (and the piece from the former host) into a chromosome of the new host.

C. Recombination and evolution.

1) Recombination changes the genetic makeup of organisms.

2) Natural recombination is random and undirected and, like random mutation, is tested by natural selection.

3) Most recombinations are harmful or neutral in effect.

4) A few may be beneficial in a particular environment, and organisms with the favored recombinations thrive and pass them on to their offspring.

D. Comparison of recombination in nature and the laboratory.

1) Both involve exchanges of DNA between organisms, either of similar or different species.

2) Naturally occurring DNA recombinations are relatively random and undirected.

3) Lab DNA recombinations are directed by the investigator: specific pieces of DNA are moved between deliberately chosen organisms to achieve a specific goal.

4) The usefulness of naturally occurring DNA recombinations is determined by natural selection, but humans determine whether lab DNA recombinations are useful.

3. **Recombinant DNA technology**: a few examples.

A. One logical sequence of procedures:

1) Production of a "DNA library" for an organism.

2) Identification of individual genes of interest.

3) Producing many copies of the interesting gene.

4) Inserting the gene into the desired organism and regulating the expression of the gene in a useful way.

B.   <u>Building a DNA library</u>.

    1)   <u>DNA library</u>: a readily accessible, easy to duplicate collection of all the DNA of a particular organism.

    2)   Bacterial <u>restriction enzymes</u> sever DNA at particular nucleotide sequences.

        a.   In nature, restriction enzymes defend bacteria against viral infection by cutting apart the viral DNA.

        b.   The bacterial host protects its own DNA from being cut by attaching methyl ($-CH_3$) groups to some of the DNA bases.

        c.   Many restriction enzymes sever <u>palindromic DNA</u> (the sequence of bases in one direction on one strand is the same as the sequence of bases in the other direction on the other strand).

            read →   ...A T T G C A A T...
                        ...T A A C G T T A... ← read

           i.   The DNA is cut between the similar bases on the two strands (example: between the two Ts).

           ii.   This produces two DNA pieces with single-stranded overhanging ends.

             ...A T   T G C A A T...
             ...T A A C G T   T A...

           iii.   Such complimentary single-stranded regions can pair up by hydrogen bonding and become permanently joined by DNA repair enzymes called ligases.

        d.   The specificity and variety of restriction enzymes allows molecular geneticists to identify and isolate desired segments of DNA from many organisms.

    3)   <u>Building a DNA library</u> using restriction enzymes. Steps:

        a.   DNA is isolated from, for example, human white blood cells (wbcs) and cut apart with a restriction enzyme, leaving single-stranded ends protruding from each piece.

        b.   Next, many plasmids are collected, each one having an easily identifiable "marker gene" such as resistance to the antibiotic ampicillin.

        c.   The restriction enzyme is used to cut open the plasmids so that their single-stranded ends are similar to those from the human wbc DNA.

        d.   Mix together the human DNA fragments and the open plasmids so that complimentary single-stranded ends will hydrogen bond to form human/human, plasmid/plasmid, and recombinant human/plasmid DNA molecules.

        e.   DNA ligases are added to covalently bind the DNA fragments and the plasmids together. [Ideally, each recombinant plasmid contains a small piece of human DNA, so that many millions of the recombinant plasmids collectively contain the entire human genome.]

f.   The recombinant plasmids are mixed with bacteria under conditions that allow the bacterium to take up one plasmid each.

g.   Then, the bacteria are plated out in medium containing ampicillin which will kill all bacteria that do not have a plasmid.

h.   The end result: a population of bacteria, all with a plasmid (most being a recombinant plasmid with human DNA). This collection is the human DNA library.

C.   Identifying genes.

    1)   The human genome contains about 6 billion nucleotides and about 1000 nucleotides encode the information for each typical protein.

    2)   It is easy to find a particular gene if the amino acid sequence of its encoded protein is already known.

        a.   From the amino acid sequence, work backwards through the genetic code to determine a likely DNA base sequence.

        b.   Machines are available to make "artificial DNA" chains from the nucleotide sequences fed into them.

    3)   Another way of identifying a gene is to find a cell that makes lots of a particular protein.

        a.   Immature red blood cells make much hemoglobin and have lots of mRNA for hemoglobin.

        b.   This mRNA is complimentary to the DNA of the hemoglobin gene, and can be used to locate the gene.

    4)   One can search the library using synthetic DNA or mRNA labelled with radioactive isotopes.

D.   Making copies of genes located in the library:

    1)   Pick bacteria from the appropriate colonies and grow them under the proper culture conditions. Each time they divide, new copies of the gene will be made.

    2)   A new method for making copies of specific stretches of DNA is the polymerase chain reaction.

E.   Using the gene: several possibilities.

    1)   Protein factories to make medically useful quantities of human proteins.

        a.   Insert the human gene into bacteria for commercial production.

        b.   The recombinant bacteria are grown in large vats.

        c.   The desired product is extracted from the culture medium or the cells themselves, and purified.

        d.   Examples:

            i.   Human growth hormone for short children.

            ii.   Blood clotting factors for hemophiliacs.

    2)   Vaccines.

        a.   Are effective mostly against viral diseases.

        b.   Use recombinant DNA technologies to clone a gene for an appropriate viral protein, insert it into bacteria, and produce large quantities of the protein.

        c.   The vaccine would consist of the pure protein, and would be safe since no living viruses were involved in its production. A useful approach to AIDS.

    3)   Diagnosis of genetic disorders.

      a.   A variety of human diseases are inherited, including sickle-cell anemia, Tay-Sachs disease, cystic fibrosis, and muscular dystrophy.

      b.   There are no cures for these diseases.

      c.   Carriers (heterozygotes for recessive disorders) and/or affected fetuses could be identified using recombinant DNA technology.

4)   Modifying DNA in free-living organisms: bacteria.

      a.   Several labs are working to produce bacteria that selectively metabolize toxic substances (e.g., oil spills).

      b.   "Ice-minus" bacteria have been released into the environment, causing a controversy.

5)   Modifying DNA in free-living organisms: agricultural applications.

      a.   Genes can be inserted into eukaryotic cells which can then grow into multicellular organisms genetically engineered to contain the foreign gene.

           i.   Genes might be introduced into single plant cells via plasmids or viruses.

           ii.   Complete plants are then grown from the transformed cells and used to propagate entirely new crop varieties.

      b.   Example: a gene for resistance to the herbicide glyphosate has been identified and inserted into several types of plants.

      c.   Future goal: to insert genes for nitrogen fixation into common crop plants.

           i.   Since major crop plants cannot use atmospheric nitrogen, they must be given nitrogen fertilizers.

           ii.   If nitrogen fixation genes are inserted into wheat and corn, farmers could grow more food at lower cost in dollars and environmental damage.

6)   Modifying DNA in free-living organisms: human applications.

      a.   Potentially, genetic engineering can correct diseases involving discrete structures (glands or bone marrow) where the gene product normally is made.

      b.   A few children have severe combined immunodeficiency disease (SCID) with no defense against disease organisms.

           i.   25% of these have failed immune systems due to a single defective gene.

           ii.   Recently, normal alleles can be put into white blood cells where they function normally and then the cells are transfused back into the patient.

      c.   These therapies are not really "cures" since the patient's germ line is not "fixed" and the abnormal gene can be passed to offspring.

4.   **Locating and sequencing genes.**

A.   Geneticists do not know much about the molecular nature of most genes or their protein products. Example: Huntington's disease is a lethal brain disorder caused by a single dominant gene but nothing is known about the gene or the protein it encodes.

B.   Locating genes: restriction fragment length polymorphisms (RFLPs).

    1)   Linkage analysis uses the frequency of crossing over to map chromosomes (see

Chapter 11).

    a.   If recombination never occurs between two traits, either the genes lie very close to each other in a chromosome, or the traits are controlled by the same gene.

    b.   If the location of the gene controlling one trait is already known, it should be easy to find the other tightly linked one.

2)   To locate a particular gene in a DNA library, make a probe for a "marker gene" that is closely linked to it.

3)   Problems using linkage analysis to find a gene in DNA library:

    a.   The locations of only a few genes in each chromosome are known.

    b.   Nucleotides coding for proteins make up < 5% of the human genome; the rest are introns, regulatory sequences, or nonfunctional DNA.

    c.   Geneticists need lots more markers a lot closer together.

4)   <u>Using RFLPs to produce markers in chromosomes</u>.

    a.   Restriction enzymes can provide chromosomal markers, just as genes can.

    b.   When a pair of homologous chromosomes is exposed to a restriction enzyme, the DNA is cut up into restriction fragments that can be separated by size using gel electrophoresis.

    c.   The pattern of sizes may or may not differ for both homologous chromosomes, depending on the nucleotide sequences of the chromosomes and the type of enzyme used.

    d.   Differences in the size of the DNA pieces are called RFLPs.

5)   <u>Using RFLPs to locate a gene</u>.

    a.   Gene location by RFLP analysis is conceptually similar to locating a gene by crossing over.

         i.   Look for a restriction enzyme "cut site" that is (nearly) always inherited along with the disorder.

         ii.   Such a cut site is within or very close to the gene causing the disorder.

    b.   A unique cut site might be within a defective allele if the altered nucleotide sequence causing the defect is also cut by the restriction enzyme.

    c.   Since defective alleles for most genetic disorders are rare, all affected persons may be descended from a common distant ancestor.

         i.   If the distant ancestor had a second altered nucleotide sequence close to the one causing the disease, there will have been little crossing over between them.

         ii.   The second nucleotide change can be used to find the locus of the defective allele if either the altered or the normal sequences (but not both) are cut by the restriction enzyme.

6)   <u>Other uses of RFLP analysis</u>.

    a.   Prenatal diagnosis of certain genetic defects.

    b.   "DNA fingerprinting" for forensic medicine to identify the perpetrator of a crime, based on matching a class of RFLPs from blood or semen samples left at a crime with RFLPs from the suspect's tissues.

C.   <u>Sequencing genes</u>.

    1)   Knowing the nucleotide sequence of a gene is useful since:

        a.   The nucleotide sequence determines the amino acid sequence of the encoded protein.

    b.   It is easier to locate and sequence a gene than it is to isolate, purify, and sequence a protein.

    c.   A biochemist can predict the amino acid sequence of a protein, its location in a cell (membrane-bound or dissolved in cytoplasm), and some of its probable functions based solely on the nucleotide sequence of its gene, as was done for cystic fibrosis.

  2)   If one knows a gene's normal nucleotide sequence and the abnormal sequence, and only one sequence is cut by a restriction enzyme, a rapid, definitive test for the genetic disorder can be devised.

    a.   Perhaps the DNA of affected people will not be cut while the DNA of everyone else is cut.

    b.   This type of test is the basis for the prenatal diagnosis of sickle-cell anemia.

  3)   Recently, a huge group of scientists, headed by James Watson, have begun to sequence the entire human genome (the "Human Genome Project").

****

REVIEW QUESTIONS:

1.   What are the two processes involved in all DNA recombination?

2.   Describe two methods of recombination that naturally occur, one involving bacteria and one involving viruses.

3.   Name two ways in which recombination in the laboratory and recombination in nature differ.

4.   Name two ways that molecular geneticists use to identify genes.

5.   Name several agricultural applications of recombinant DNA techniques.

6.   How can recombinant DNA technology produce lots of blood clotting factor to aid hemophiliacs?

7.   Name several uses of RFLP analysis.

8.   Give several reasons why knowing the nucleotide sequence of a gene is useful.

9.-20. **MATCHING TEST:** recombinant DNA

      Choices: A.   DNA library
                      B.   restriction enzymes
                      C.   plasmids
                      D.   DNA ligases
                      E.   RFLPs

9.   Defends bacteria against viral infection by cutting apart the viral DNA.

10.   Self-replicating tiny loops of DNA.

11.   Used to identify marker genes in chromosomes.

12.   Readily accessible, easy to duplicate collection of all the DNA of a particular organism.

13.   Bacteria protect themselves from this by methylating their DNA.

14.   Covalently bind together recombinant DNA molecules.

15.   A bacterium may contain hundreds of these "parasites."

16.   Can cut apart DNA to create single-stranded ends.

17.   Restriction fragment length polymorphisms.
18.   Cut at palindromic DNA sequences.
19.   Often contains genes for antibiotic-digesting enzymes.
20.   Cuts up DNA into fragments of various sizes that can be separated by gel electrophoresis.

****

ANSWERS TO REVIEW QUESTIONS:

1.   see 2.A.1)-2)

2.   see 2.B.2)a.-c.

3.   see 2.D.1)-4)

4.   see 3.C.2)-3)

5.   see 3.E.5)a.-c.

6.   see 3.E.1)a.-d.

7.   see 4.B.5)-6)

8.   see 4.C.1)a.-c.

9.-20. [see entire chapter]

9.	B		15.	C
10.	C		16.	B
11.	E		17.	E
12.	A		18.	B
13.	B		19.	C
14.	D		20.	B

****

CHAPTER 15:   HUMAN GENETICS

CHAPTER OUTLINE.

This chapter focuses on genetic conditions in humans. The natures of recessive, dominant, sex-linked, sex-influenced, and complex inheritance are covered, and the effect of the environment is considered. Chromosomal nondisjunction as a cause of humans with abnormal numbers of chromosomes is discussed, and the more common results of nondisjunction (XO, XXX, XXY, XYY, and Down syndrome) are presented. Finally, the Human Genome Project is discussed.

1.   **General facts.**

    A.   Factors complicating the study of human genetics:
        1)   Humans have long life spans.
        2)   Humans have few children per couple.
        3)   Humans choose their own mates; experimental crosses are not allowed.
        4)   Humans interact extensively with and modify their incredibly diverse environment, obscuring underlying genetic patterns especially for personality traits.

    B.   Factors compensating for the difficulties listed above:
        1)   The enormous numbers of humans on earth.
        2)   Extensive documentation of human families through written records and pedigrees of famous lineages.

2.   **Methods** in human genetics:

    A.   <u>Pedigree analysis</u> (family history records).
        1)   Careful analysis of pedigrees can show the inheritance patterns of certain traits.
        2)   But the trait in question must be clearly defined.
        3)   In a pedigree, one must be certain who possessed which traits.

CHAPTER 15: HUMAN GENETICS

B. Molecular genetics:
1) Geneticists now know the molecular bases of many inherited diseases, like sickle cell anemia.
2) Genetic engineering will increase the ability to predict and cure genetic diseases in the future.

3. **Single gene inheritance** of medically important diseases.

A. Recessive inheritance.
1) General facts.
   a. Normal alleles (A) make functional enzymes and are usually dominant while mutant alleles (a) usually make nonfunctional enzymes and are recessive.
   b. Genetic diseases that are caused by lack of an essential enzyme are mostly recessive conditions.
   c. If the metabolic disease impairs survival, the homozygous recessive (aa) will die.
      i. Allele a is "selected against".
      ii. Heterozygotes (Aa) are carriers, being phenotypically normal but capable of passing allele a to their offspring. Such heterozygotes are relatively rare.
      iii. Related couples (first cousins or closer) have inherited genes from recent common ancestors and are each more likely to be carriers for the same recessive allele (Aa). This increases their risk of having an affected (aa) child.
   d. The only way to prevent genetic disease is to prevent the birth of affected babies. This requires identifying heterozygous carriers and either:
      i. Preventing reproduction by couples who are both carriers, or
      ii. Screening fetuses and aborting those that are homozygous recessive (aa).
      iii. Such actions raise ethical difficulties.
   e. Carriers are detected by relatively expensive medical tests.
      i. Such "screening tests" are useful when the defective alleles are found in high frequencies within a readily identifiable group.
      ii. Example: Tay-Sachs disease and sickle cell anemia are recessive diseases largely restricted to certain segments of the population and for which both carrier detection and prenatal diagnosis are possible.
2) Tay-Sachs disease (TSD).
   a. Brain cells of homozygous children lack an enzyme needed for lipid metabolism. Lipids accumulate, causing progressive mental retardation, blindness, failure of motor control, and death in early childhood.
   b. The carrier (Aa) rate in American Jews of Eastern European ancestry is about 1 in 30 (vs. 1 in 400 in the general population) and the frequency of TSD (aa) among Jewish babies is 1 in 3600.
   c. Heterozygotes for TSD can be detected by a blood enzyme test.
   d. Aa x Aa couples have 3 choices regarding reproduction:
      i. Forego reproduction entirely.
      ii. Take the 25% chance of having a TSD (aa) child.

140

        iii.   Screen fetuses by <u>amniocentesis</u> and <u>chorionic villus sampling</u> and abort the homozygous recessives.

3)   <u>Sickle</u> <u>cell</u> <u>anemia</u> (<u>SCA</u>).

    a.   Nearly all carriers in the US are blacks.

    b.   The SCA mutation changes one amino acid in the hemoglobin molecule:

        i.   This causes the hemoglobin molecules to clump together, forcing the red blood cells into sickle-like shapes.

        ii.   The sickled cells clog capillaries, cutting off circulation and causing fatal strokes and heart attacks in some cases.

    c.   Heterozygote carriers have few sickled cells and show no symptoms.

    d.   Surprisingly, the mutant allele for SCA is quite common in black populations: 8% of US blacks are carriers while in Africa, the frequency is 2 to 3 times higher. The frequency of this harmful allele is so high because carriers are somewhat resistant to malaria which is common in Africa.

    e.   Heterozygote carriers (<u>Aa</u>) can be detected by a blood test, and recombinant DNA analysis of fetal cells collected by amniocentesis or chorionic villus sampling can detect SCA (<u>aa</u>) prenatally.

B.   <u>Dominance inheritance</u>.

1)   The frequency of serious diseases caused by dominant alleles is low because:

    a.   Most mutations cause recessive nonfunctional alleles.

    b.   Dominant diseases cannot be "hidden" in the heterozygous condition, so fewer heterozygous reproduce due to the effects of the disease.

2)   An exception is <u>Huntington's</u> <u>disease</u> (<u>HD</u>), a slow deterioration of the brain, because symptoms normally don't begin until age 30-50, after reproduction has occurred.

    a.   However, well before symptoms begin, scientists can use restriction fragment length polymorphism (RFLP) techniques to detect whether a fetus or child has the gene for HD located on chromosome 4.

    b.   Ethical question: would you want to know that in 20 or 30 years you will develop HD and die? Do you have the right not to know?

C.   <u>Sex-linked</u> and <u>sex-influenced inheritance</u>.

1)   Recessive alleles of X chromosome genes are always expressed in XY males (<u>sex-linked</u> <u>inheritance</u>).

    a.   A male must inherit his X from his mother and can pass them on to all his daughters and none of his sons.

    b.   Thus, human pedigrees show a striking pattern for rare, recessive, sex-linked traits: an affected male has all normal children but half of his grandsons from his daughters are also affected.

    c.   Examples: red-green colorblindness and hemophilia.

2)   Each sex, but especially males, also has <u>sex-influenced</u> <u>traits</u> that occur more strongly or commonly in one sex but are not due to genes on the sex chromosomes. An example is baldness: In males, heterozygotes become bald while in females they do not become bald. This is due to hormonal differences between the sexes.

4. **Complex inheritance**.

    A.   <u>Polygenic inheritance</u>. Most traits are influenced in some way by the interactions of many genes.

        1)   In HD, the age of onset of symptoms varies considerably and this variation is due to the interaction of other genes and as yet unidentified environmental factors.

        2)   In <u>eye</u> <u>color</u>, the iris color varies from very pale blue through green to black.

            a.   All eye colors are caused by the distribution of yellowish-brown pigment called <u>melanin</u>.

            b.   The iris has 2 pigment layers, at the back (usually densely pigmented) and the front:

                i.   Little pigment in front causes "blue" irises due to the scattering of light against the dark background of the rear pigment layer (like the "blue" sky).

                ii.   A bit more pigment in the front appears green due to the mixing of the "blue" with the yellowish-brown melanin.

                iii.   More intense pigment in the front causes brown or black irises

            c.   Probably, more than 2 gene pairs control the amount of pigment in the front layer of the iris, with each gene pair showing incomplete dominance.

        3)   <u>Skin</u> <u>color</u> variation is another case of different amounts of melanin.

            a.   All skin colors are various shades of brown with pinkish tones from blood capillaries.

            b.   At least 3 pairs of genes with incompletely dominant alleles control the intensity of skin pigmentation.

    B.   <u>Environmental effects</u>.

        1)   Genetic and environmental factors interact.

            a.   The sun can darken a white person's skin color but the genotype limits the range of environmental effects on the phenotype.

            b.   How fast we run or how well we sing can be improved by training but without the proper genetic endowment, we will never be "world class".

        2)   At least some mental and personality traits are influenced by both environmental factors and inherited tendencies.

            a.   Schizophrenia is due to both environmental events and genes which produce an altered chemistry in the brain.

            b.   Intelligence has both genetic and environmental components.

                i.   The IQs of identical twins are very similar but become a bit more different if a twin pair is separated at birth and each is raised by a different family (the effect of environmental differences).

                ii.   The IQs of identical twins are considerably more similar than the IQs of non-identical twins or other siblings (the effect of genetic differences).

5. **Chromosomal inheritance**.

    A.   <u>Nondisjunction</u>: Abnormal chromosome movement during meiosis.

        1)   Normally in body cells, human chromosomes exist in homologous pairs, except for the XY pair which is non-homologous.

2)   If chromosomes fail to move properly during meiosis (<u>nondisjunction</u>), gametes with too many or too few chromosomes result.

3)   Most embryos resulting from such gametes spontaneously abort (20-50% of all miscarriages are due to embryos having abnormal chromosome numbers) but some survive and are born.

B.   <u>Abnormal numbers of sex chromosomes.</u>

     1)   Occurs in both males and females.

         a.   In males, abnormal sperm with no sex chromosome ("O") or with XX, YY, or XY can occur.

         b.   In females, abnormal eggs with 0 or XX can occur.

         c.   All human embryos must have at least one X to survive, and many embryos with XO, XXX, XXY, and XYY sex chromosomes do survive.

     2)   <u>Turner's</u> syndrome (<u>XO</u>) <u>females</u>.

         a.   1 in 5000 females is XO.

         b.   At puberty, no menstruation or secondary sexual characteristics develop.

         c.   Sterility, short stature, and webbed skin around the neck occur.

         d.   Barr bodies are absent from cell nuclei.

         e.   Normal mentality is common, but most are weak in math and in spatial perception.

     3)   <u>Trisomy</u> <u>X</u> (<u>XXX</u>) <u>females</u>.

         a.   1 in 1000 females is XXX.

         b.   Normal in phenotype except that some have a subnormal intelligence.

         c.   Have 2 Barr bodies per nucleus.

         d.   Fertile, usually with normal children.

     4)   <u>Klinefelter's</u> syndrome (<u>XXY</u>) <u>males</u>.

         a.   1 in 1000 males is XXY.

         b.   Mixed secondary sexual characteristics at puberty, including partial breast development, broadening of the hips, and small testes.

         c.   Sterile but not impotent.

         d.   Increased incidence of mental deficiency.

         e.   Have a Barr body in cell nuclei.

     5)   <u>XYY</u> <u>males</u>.

         a.   1 in 1000 males is XYY.

         b.   Below average intelligence and above average height.

         c.   Possibly a relatively few might be genetically predisposed to violence.

     6)   <u>Sex</u> <u>determination</u> in humans.

         a.   The Y chromosome determines maleness.

         b.   One X does not determine maleness (XY is male but XO is female).

         c.   In computer terminology, femaleness is the "default" condition for sex: <u>absence</u> of the Y produces a female.

C.   <u>Abnormal numbers of autosomes.</u>

     1)   Autosomal nondisjunction can also occur, producing sperm or eggs with 0 or 2 copies of an autosome instead of the normal one of each type. Numbers of sex chromosomes are normal.

         a.   Fusion of an abnormal with a normal gamete produces an embryo with either one or 3 doses of the affected autosome.

         b.   Embryos with one copy ("<u>monosomies</u>") abort very early in pregnancy.

         c.   Most embryos with 3 copies ("<u>trisomies</u>") spontaneously abort later in

2) <u>Trisomy 21</u> (<u>Down syndrome</u>).
    a. The most common chromosome anomaly, since 1 in 900 births is a trisomy 21 boy or girl.
    b. Common characteristics:
        i. Round face with small nose and small mouth with protruding tongue, and distinctly shaped eyelids.
        ii. Poor muscle tone.
        iii. Low resistance to infectious disease.
        iv. Heart malformations.
        v. Some degree of mental retardation often seen.

3) The <u>frequency of nondisjunction</u> is influenced by the age of the mother: The older the pregnant mother, the greater is the risk that the egg used was the product of nondisjunction.
    a. Meiosis begins in a woman's ovaries while she is still a fetus in her mother's womb, but is suspended before birth during late prophase of meiosis I.
    b. In each egg, meiosis begins again only when the egg is ovulated.
    c. Nondisjunction is more likely in older eggs. Thus, the older the woman when pregnant, the greater the risk of a child with an abnormal chromosome number.
    d. Amniocentesis or chorionic villus sampling should be performed for all pregnant women aged 35 or older.

6. **Human genome project.**

A. "Humankind" is defined by the <u>nucleotide sequences</u> in our genes.
    1) 4000 genetic disorders occur in humans.
    2) Knowing the nucleotide sequence of the entire human genome will allow scientists to:
        a. Devise rapid screening and prenatal diagnosis tests for many genetic disorders.
        b. Design better therapies for many diseases.
        c. Understand much more about brain function.

B. The human genome project, headed by James Watson, began in 1988 with a projected budget of $3 billion. The project is feasible due to developments in:
    1) RFLP (restriction fragment length polymorphism) analysis.
    2) Automated DNA sequencing.
    3) PCR (polymerase chain reaction) technologies.

C. <u>Four major themes</u> of human genome project:
    1) <u>Linkage mapping</u>.
        a. Locations of 3000 or more gene markers will be sought.
        b. Linkage maps will be used in pedigree analysis to locate other genes with phenotypic effects.
    2) <u>Physical mapping</u>.
        a. Each human chromosome's DNA will be cut apart with restriction

enzymes.

    b.   Unique DNA nucleotide sequences, within 100,000 nucleotides of each other, will be identified in each piece.

    c.   These unique sequences will be used as starting points for sequencing the rest of the DNA and identifying previously unmapped genes.

    3)   <u>Sequencing the haploid human genome</u> of about 3 billion nucleotide pairs, and storing the information in a computer.

    4)   <u>Sequencing diversity</u>. Since each person's genome is slightly different, DNA from a large sample of people must be sequenced.

    D.   Critics of the project point out it's hugh cost ($50-100 million per year) will drain needed resources from other important biomedical projects.

    E.   Plans are underway to sequence the genomes of bacteria, fruit flies, nematodes, dogs, and at least one plant.

****

REVIEW QUESTIONS:

1.   How can genetic disease be prevented among the children of parents who are both carriers (<u>Aa</u>) of a recessive condition like Tay-Sachs disease?

2.   What makes Huntington's disease different from most other dominant genetic diseases, and how does this complicate family planning?

3.   Describe the inheritance pattern for sex-linked recessive traits like red-green colorblindness in human pedigrees.

4.   Describe how blue, green, brown, and black eye colors are possible if the irises only produce yellowish-brown melanin pigment?

5.   How does the comparison of IQ scores between identical twins and non-identical twins suggest that both genetic and environmental factors influence a person's level of intelligence?

6.   What is "nondisjunction" and how can it lead to abnormal humans?

7.-21.   **MATCHING TEST:** chromosome anomalies in humans

        Choices: A.   Turner's syndrome
                              B.   Klinefelter's syndrome
                              C.   Trisomy X syndrome
                              D.   XYY syndrome
                              E.   Down syndrome
                              F.   all of the above
                              G.   none of the above

7.   Have an abnormal number of autosomes.

8.   Sterile males with some breast development.

9.   Females with 2 Barr bodies per nucleus.

10.   May be male or female.

11.   Fertile females with normal phenotypes.

12.   Short females with webbed necks.

13. Have a normal number of sex chromosomes.
14. More common among the babies of older mothers.
15. Males with Barr bodies.
16. Sterile females lacking Barr bodies.
17. Have a possible genetic predisposition towards violence.
18. The most common chromosome anomaly among newborns.
19. Body cells have 45 chromosomes.
20. Trisomy 21.
21. Have 46 chromosomes in their body cells.

----

22. In humans, why does the absence of the Y chromosome produce a female while the presence of the Y produces a male?
23. Describe the influence of the mother's age on the frequency of nondisjunction in her egg cells.

****

ANSWERS TO REVIEW QUESTIONS:

1. see 3.A.1)d. and 2)a.-d.

2. see 3.B.1)-2)

3. see 3.C.1)

4. see 4.A.2)a.-c.

5. see 4.B.2)b.i.-ii.

6. see 5.A.1)-3)

7.-21. [see 5.B.1)-5);C.2)

7.	E		15.	B
8.	B		16.	A
9.	C		17.	D
10.	E		18.	E
11.	C		19.	A
12.	A		20.	E
13.	E		21.	G
14.	F			

22. see 5.B.6)a.-c.

23. see 5.C.3)a.-d.

****

UNIT II EXAM:

Chapters 9 - 15. _All questions are dichotomous. Circle the correct choice in each case._

Chapter 9.

1.   Asexually produced organisms are (identical / similar) to their parents.
2.   Prokaryotes have (many / one) chromosome(s) per cell.
3.   During interphase, chromosomes are (coiled and condensed / uncoiled and extended).
4.   Pairs of chromosomes are called (chromatids / homologues).
5.   The somatic or body cells of a human are (diploid / haploid).
6.   A zygote is (diploid / haploid).
7.   Polyploid organisms contain (more than / less than) two homologues of each chromosome type.
8.   DNA replication occurs during the (G_1 / S / G_2) portion of interphase.
9.   Centrioles divide during (interphase / anaphase) of the cell cycle.
10. Cell plates and cell furrows form during (metaphase / cytokinesis) of the cell cycle.

Chapter 10.

1.   Diploid cells produce haploid cells by the process of (meiosis / mitosis).
2.   Crossing over occurs during meiosis (I / II).
3.   In the (advanced / primitive) plants, both the haploid and the diploid stages are independent plants.
4.   In the flowering plants, the (diploid / haploid) stage of the life cycle is greatly reduced.
5.   In plants, gametes are produced by the (diploid / haploid) stage.
6.   Reduction of chromosome number occurs during meiosis (I / II).
7.   DNA (does not replicate / replicates) between meiosis I and meiosis II.
8.   Meiosis (I / II) resembles mitosis.
9.   Sporophyte plants are produced by (gametes / spores).
10. Sister chromatids become daughter chromosomes during (anaphase I / anaphase II) of meiosis.

Chapter 11.

1. Single trait crosses are called (monohybrid / dihybrid) crosses.
2. Each trait is determined by a pair of discrete units called (genes / chromosomes).
3. The members of a gene pair (assort independently / segregate) during meiosis.
4. Alternate forms of a gene are called (chromatids / alleles).
5. In a heterozygote, the gene that is not expressed is (dominant / recessive).
6. PP is a (homozygous / heterozygous) (genotype / phenotype).
7. Genes from different pairs that control different traits show (segregation / independent assortment) during meiosis.
8. An AabbDdEeGg individual will produce (16 / 32) different types of gametes.
9. In an AaBb x aaBB cross, (1/4 / 1/2) of the offspring will be AaBB.
10. Non-sex chromosomes are called (autosomes / Y chromosomes).
11. Linkage modifies Mendel's law of (segregation / independent assortment).
12. New combinations of traits controlled by linked genes occur due to (mutation / crossing over).
13. Changes in the molecular structure of a gene are called (mutations / multiple alleles).
14. When red x white → pink snapdragons, this is an example of (polygenic inheritance / incomplete dominance).
15. A good example of continuous variation is height differences (in pea plants / in humans).

Chapter 12.

1. DNA is (double / single) stranded.
2. Proteins were first thought to be genes because proteins are (more complex / simpler) than nucleic acid.
3. Sugars found in DNA have (five / six) carbons.
4. (Encapsulated / Naked) bacteria cause pneumonia.
5. DNA contains (phosphorus / sulfur).
6. Radioactive sulfur (moves inside / remains outside) bacteria when treated with labeled viruses.
7. The concentration of DNA is (constant / different) for different cells of the same species.
8. Purines pair with (purines / pyrimidines).
9. The duplication of DNA is called (replication / transcription).
10. The building blocks of nucleic acids are (amino acids / nucleotides).

Chapter 13.

1. Mutant Neurospora (would / would not) grow on the minimal medium.
2. Albinism is an example of (the "One gene, one enzyme" hypothesis / an inborn error of metabolism).
3. Protein synthesis occurs in the (nucleus / ribosome).
4. Messenger RNA is (double / single) stranded.
5. Messenger RNA is manufactured in the (nucleus / cytoplasm).
6. The triplets of messenger RNA are called (anticodons / codons).
7. Proteins are made during (transcription / translation).
8. Operons are found in (eukaryotic / prokaryotic) forms.

9. Barr bodies are found in normal mammalian (females / males).
10. Tortoiseshell cats are (heterozygous / homozygous).

Chapter 14.

1. DNA recombination (does / does not) occur in nature.
2. Sexual reproduction between humans (is / is not) an example of DNA recombination in nature.
3. Bacteria pick up free DNA and incorporate it into their chromosomes during (sexual reproduction / transformation).
4. Natural DNA recombination is (nonrandom / random) and (directed / undirected) by humans.
5. A readily accessible, easy to duplicate collection of all the DNA of a particular organism is (palindromic DNA / a DNA library).
6. An example of a palindrome is the word (madam / madman).
7. An enzyme that cuts palindromic DNA open to form "sticky ends" is called (ligase enzyme / restriction enzyme).
8. Bacterial DNA is protected from the action of restriction enzymes by (mutation / methylation).
9. A method for making many copies of a small amount of DNA is the (polymerase chain reaction / restriction fragment length polymorphism reaction).
10. RFLPs are helpful in (gene mapping / making lots of DNA copies).

Chapter 15.

1. Environmental interactions have (little / much) effect on human genetic traits.
2. The more closely related two people are, the (lesser / greater) is their chance of having genetic defects in common.
3. Most genetic defects are controlled by (dominant / recessive) genes.
4. Human (females / males) determine the sex of their children.
5. Males are (diploid / haploid) for sex-linked genes.
6. A son inherits sex-linked traits from his (father / mother).
7. A father passes on sex-linked traits to his (daughters / sons).
8. Sex in humans is dependent on the (presence or absence of a Y chromosome / presence of two Xs).
9. The presence of an extra chromosome in a cell is called (monosomy / trisomy).
10. A person with Turner syndrome is (XXY / XO) and is a (female / male).
11. A person with Klinefelter syndrome is a (XXY / XO) and is a (female / male).
12. People who have Turner syndrome have (no / two) Barr bodies.
13. People with (Turner syndrome / trisomy X) are usually sterile.
14. XYY individuals are (shorter / taller) than normal.
15. An XXXXY individual is (female / male).

****

ANSWER KEY:

Chapter 9:

1. identical
2. one
3. uncoiled and extended
4. homologues
5. diploid
6. diploid
7. more than
8. S
9. anaphase
10. cytokinesis

Chapter 10.

1. meiosis
2. meiosis I
3. primitive
4. haploid
5. haploid
6. meiosis I
7. does not replicate
8. meiosis II
9. gametes
10. anaphase II

Chapter 11:

1. monohybrid
2. genes
3. segregate
4. alleles
5. recessive
6. homozygous, genotype

7. independent assortment
8. 16
9. 1/4
10. autosomes
11. independent assortment
12. crossing over
13. mutations
14. incomplete dominance
15. in humans

Chapter 12:

1. double
2. more complex
3. five
4. encapsulated
5. phosphorus
6. remains outside
7. constant
8. pyrimidines
9. replication
10. nucleotides

Chapter 13:

1. would not
2. inborn errors of metabolism
3. ribosome
4. single
5. nucleus
6. codons
7. translation

8. prokaryotic
9. females
10. heterozygous

Chapter 14.

1. does
2. is
3. transformation
4. random, undirected
5. a DNA library
6. madam
7. restriction library
8. methylation
9. polymerase chain reaction
10. gene mapping

Chapter 15:

1. much
2. greater
3. recessive
4. males
5. haploid
6. mother
7. daughters
8. presence or absence of Y
9. trisomy
10. XO, female
11. XXY, male
12. no
13. Turner syndrome
14. taller
15. male

****

UNIT III. EVOLUTION

CHAPTER 16:   PRINCIPLES OF EVOLUTION

CHAPTER OUTLINE.

This chapter introduces the concept of evolution. An historical account of pre-Darwinian thought is presented, followed by the Darwin-Wallace theory of evolution by natural selection. Finally, various proofs of evolution are given.

1.   **Introduction.**

    A.   <u>Before Darwin</u>, how species originated was a mystery.
        1)   A striking feature about the earth is the wide variety of organisms present.
        2)   Without evidence for evolution, most people explained this diversity by <u>creationism</u>: the separate creation of each type of organism at the beginning of the world by a supernatural being.
        3)   Throughout history, however, scientists have sought natural causes for the origin of species.

    B.   In 1858, Darwin and Wallace published reports on the theory of evolution by descent with modification, driven by natural selection. This is the foundation of our understanding of evolution.

2.   **History** of evolutionary thought.

    A.   Ways to <u>distinguish</u> different "kinds" ("species") of organisms:
        1)   By <u>differences in appearance</u> - an old idea sometimes misleading.
        2)   By <u>reproductive isolation</u> - a newer idea in use today.

      a.   Species: all populations of organisms that are capable of interbreeding under natural conditions and that are reproductively isolated from other populations.

      b.   The members of a species can interbreed among themselves, but usually not with members of other species.

      c.   Occasional interbreeding between different species at best results in hybrid offspring handicapped in some way.

B.   Greek philosophers and the origin of species.

    1)   Plato (427-347 B.C.) thought in terms of "ideal Forms":

      a.   Each object on earth is a "reflection" of its non-material ideal Form.

      b.   The Forms, of unknown origin, are perfect and forever unchangeable.

      c.   This idea led to the belief that all species were created by God at the beginning of time and cannot change substantially, although minor variations may occur due to the imperfect quality of the world.

    2)   Aristotle (384-322 B.C.) developed an orderly scheme for classifying nature.

      a.   He categorized all living things into a Scala Naturae ("ladder of nature"), starting with nonliving matter and ascending, rung by rung, from fungi and mosses to higher plants to primitive animals and finally culminating in humans.

      b.   The Scala Naturae is immutable: each species had its place on the ladder, ordained by God during creation.

C.   Evolutionary thought before Darwin and Wallace.

    1)   Up to the 1700s:

      a.   Creationism went unchallenged, the earth was considered the center of the universe, and man was the center of life on earth.

      b.   Naturalists felt it their duty to catalogue the diversity of organisms, describing the "glory of God's creation".

    2)   Then, naturalists explored lands outside Europe:

      a.   They saw a much greater diversity of living things than existed in Europe.

      b.   Some species closely resembled each other.

      c.   They began to consider that perhaps species could change and evolve from common ancestors.

    3)   Fossils were discovered that partly resembled living organisms.

      a.   Fossils ("dug up") are plants and animals that died long ago and changed into rock by natural processes.

      b.   William Smith discovered that the organization of fossils and rock layers was consistent.

      c.   Fossil remains show a remarkable progression in form:

         i.   Those found in lower (older) rock layers are more primitive looking.

         ii.   There is a gradual advancement to greater complexity and greater resemblance to modern species in younger rocks.

         iii.   Some fossils are the remains of extinct forms.

         iv.   Conclusion: different types of organisms have lived on earth at various times in the past.

    4)   Buffon suggested that:

      a.   The original creation provided a relatively small number of founding species.

        b.   Some modern species had been "conceived by Nature and produced by Time" (i.e., evolved).

5)   Cuvier proposed the theory of <u>Catastrophism</u> to oppose evolution:
        a.   A vast supply of species was created in the beginning.
        b.   Successive catastrophes (like Noah's flood) produced layers of rock and destroyed many species, producing fossils.
        c.   Today's reduced flora and fauna are what remains.
        d.   Agassiz added that there was a new creation after each catastrophe (over 50 in all) and modern species resulted from the most recent creation.

6)   Geologists Hutton and Lyell postulated <u>Uniformitarianism</u>:
        a.   There is no need to invoke catastrophes to explain the geological record.
        b.   The same forces of wind, water, earthquakes, and volcanoes that shape modern geology also shaped the ancient earth.
        c.   This satisfied a scientific notion called "Occam's Razor": the simplest explanation that fits the facts is probably correct.
        d.   If slow natural processes produced all the rock layers, the earth must be many millions of years old ("no Vestige of a Beginning, no Prospect of an End").
            i.   Today, geologists estimate the earth to be 4.5 billion years old.
            ii.   This is enough time for evolution to have occurred.

7)   In 1801, Lamarck hypothesized that organisms evolved through the <u>inheritance of acquired characteristics</u>. He stated:
        a.   Living organisms can modify their bodies through use or disuse (a correct idea to some extent).
        b.   These acquired modifications can be inherited by their offspring (incorrect).
        c.   Lamarck said that:
            i.   Organisms have an innate drive for "perfection", to become more complex and better adapted to the environment.
            ii.   Lamarck's example: ancestral giraffes stretched their necks to feed on leaves high in trees and these longer necks were passed on to their offspring.
        d.   This theory is unsupported scientifically.

8)   In 1858, Charles Darwin and Alfred Russell Wallace independently provided evidence that the driving force behind evolution is natural selection.

3.   Evolution by **natural selection**.

A.   Both Darwin and Wallace:
     1)   Traveled in the tropics and saw the vast diversity of living organisms there.
     2)   Were aware that the fossil record showed a trend of increasing complexity through time.
     3)   Accepted Hutton and Lyell's theory that the earth is quite old.
     4)   Read Thomas Malthus' <u>Essay on Population</u>, which stated that human populations have the capacity to double their size each generation (increase in a geometric ratio).
         a.   They realized this was true for plant and animal populations as well. But natural populations tend to remain constant in size.

           b.   The reason populations don't realize their geometric potential, they theorized, is that countless environmental constraints (lack of food and space, disease, weather) exist in nature causing many organisms to die before they can reproduce.

    5)   Theorized that whether an organism lives or dies depends to some extent on its structures and abilities which vary among members of a population.

           a.   Those with better resistance to cold or disease, for instance, would survive better ("be selected by nature").

           b.   As Darwin said "... favorable variations tend to be preserved, and unfavorable ones to be destroyed."

    6)   Stated that with continual appearance of new variants, subject to further selection, new species could develop from pre-existing ones.

    7)   Darwin "summarized" his theory in the book <u>On</u> <u>The</u> <u>Origin</u> <u>of</u> <u>Species</u> <u>By</u> <u>Means</u> <u>of</u> <u>Natural</u> <u>Selection</u>, forcing the world to consider the new theory.

B.   <u>Summary of the Darwin-Wallace theory</u> in modern terms:

OBSERVATION 1: All natural populations have the potential to increase geometrically in size due to reproductive abilities.

OBSERVATION 2: But most natural populations maintain a relatively constant size.

    CONCLUSION 1: So, many organisms must die young, producing few or no offspring each generation.

OBSERVATION 3: Individuals in a population differ in many abilities that affect survival and reproduction (some are "better adapted").

    CONCLUSION 2: The most well adapted organisms probably reproduce the most, since they survive the best. This differential reproduction is due to <u>natural</u> <u>selection</u>.

OBSERVATION 4: Some of the variation in adaptiveness among individuals is genetic and is passed on to the offspring.

    CONCLUSION 3: Over many generations, differential reproduction among individuals with different genotypes changes the overall frequencies of genes in populations, resulting in evolution.

    1)   Darwin did not know the mechanism of heredity (to explain Observation 4) and could not prove Conclusion 3.

    2)   In desperation, Darwin resorted to a version of Lamarck's inheritance of acquired characteristics and this nearly destroyed his entire theory.

C.   Darwin also conceived of <u>coevolution</u>: two species evolving in reaction to each other. Example: a speedier wolf catches more prey, but the speedier prey escape. So, speedier predators select for speedier prey and vice versa.

4.   **Evidence for evolution.**

A.   <u>Fossils</u>.
   1)   Fossils are the remains of ancestral forms that evolved into modern species; scientists find in the fossil record a gradual series of fossil types from ancient through intermediate stages to modern forms.
   2)   Good examples of this are the fossil records of horses, giraffes, elephants, some molluscs, and to a large extent humans.

B.   <u>Comparative anatomy</u>.
   1)   <u>Homologous structures</u>.
      a.   Species that evolved from recent common ancestors display similar bone arrangements even though they may be adapted to different environments and lifestyles.
      b.   The forelimbs of birds and mammals are variously used for flying, swimming, running, and grasping, but all have remarkably similar internal anatomy.
      c.   All these forelimbs are <u>homologous</u>: similar in internal structure indicating similar evolutionary origin, but evolving different functions in response to different environmental demands (<u>divergent evolution</u>).
      d.   <u>Comparative anatomy</u> is used to determine how closely related species are to one another. The more similar the internal structure of two species are, the more recently they must have diverged from a common ancestor.
   2)   Very distantly related species may undergo <u>convergent</u> evolution if exposed to similar environmental demands.
      a.   Such species independently may evolve superficially similar structures (called <u>analogous</u> <u>structures</u>) by <u>convergent</u> <u>evolution</u>.
      b.   But such outwardly similar analogous structures in distantly related organisms (i.e., wings of birds and butterflies) are completely different in internal anatomy since the parts are not derived from common ancestral structures.
   3)   <u>Vestigial structures</u>.
      a.   Serve no apparent function in the organism (i.e., pelvic bones in whales and snakes, appendix in humans).
      b.   Are homologous to functional structures found in other vertebrates.
      c.   Are "evolutionary baggage," remnants of the functional structures possessed by ancestors but selected out as species evolved in circumstances in which these structures were no longer useful.

C.   <u>Comparative embryological studies</u> support the theory of evolution.
   1)   All vertebrate embryos pass through stages in which they look very similar (i.e., they all possess gill arches and tails during early development).
   2)   This occurs because ancestral vertebrates had the genes for development for tails and gills and all their descendants retain these genes.
      a.   In fish, they remain active throughout development, producing adults with

tails and gills.

b. In reptiles, they act in early development but only the tail genes remain active so that adults have tails but lack gills.

c. In chickens and humans, they act only in the early stages of development so that adults lack tails and gills.

D. <u>Comparative biochemistry studies</u> can be used to determine the degree of relatedness between species.

   1) All living cells are very similar in their basic biochemistry, since all cells:

      a. Have DNA as the genetic information.

      b. Use RNA, ribosomes, and approximately the same genetic code.

      c. Use roughly the same set of 20 amino acids to build proteins.

      d. Use ATP as an intracellular energy molecule.

   2) The degree of similarity in the amino acid sequences of proteins corresponds closely to the evolutionary trees derived by comparing anatomical features.

   3) Relatedness among organisms is also evaluated by comparing chromosome structure and numbers and by analyzing DNA sequences.

      a. Example: the chromosomes of chimpanzees (2n = 48) and humans (2n = 46) are very similar, as are the base sequences of many of their genes.

      b. Conclusion: chimps and humans are very closely related evolutionarily.

   4) Using techniques of molecular biology, DNA rather than secondary features (appearance, behavior, protein structure) can be used to investigate relatedness among organisms.

E. <u>Artificial selection</u>: selective breeding of plants and animals to produce features desirable by humans.

   1) Example: humans have selectively bred dogs (which evolved from wolves) into hundreds of distinct "breeds".

   2) If humans can cause "evolution" to occur in these domestic species, nature can produce even more startling results.

F. <u>Present-day evolution.</u>

   1) Evolution continues today in response to changes in the environment, some caused by humans (such as environmental pollution).

   2) Example: <u>industrial</u> <u>melanism</u> in moths (increase in dark moths as a consequence of industrial pollution).

      a. The peppered moth (<u>Biston</u> <u>betularia</u>) in Great Britain exists in light and dark forms, controlled by a pair of color alleles.

         i. The coloration of light moths matches the lichen-covered tree trunks in unpolluted forests.

         ii. Dark moths match the sooty tree trunks of forests near industrial plants releasing sooty pollution.

      b. In pollution-free areas, light moths are inconspicuous on tree trunks while dark moths are easily seen by bird predators and eaten quickly, reducing their reproductive output. So, light moths are favored by natural selection and are at high frequency.

      c. In polluted areas which developed after the Industrial Revolution, dark moths are inconspicuous on the darkened tree trunks while light moths are easily seen by bird predators and eaten quickly, reducing their reproductive output. So, dark moths are favored by natural selection and are at high frequency.

3)   Points to remember:

a.   The variations upon which natural selection works are produced by <u>chance</u> variations (dark moths were not produced by the polluted environment).

b.   The process of evolution favors organisms that are best adapted to a <u>particular</u> environment based on the genetic variation present (light moths in pollution free and dark moths in polluted environments).

c.   Evolution is easiest to observe in short lived, rapidly reproducing species like insects and bacteria, but occurs consistently in all species with genetic variation.

****

REVIEW QUESTIONS:

1.-16.   Define the following concepts and entities:

1.   Creationism.
2.   Species.
3.   "Ideal Forms".
4.   <u>Scala Naturae</u>.
5.   Fossils.
6.   Catastrophism.
7.   Uniformitarianism.
8.   Inheritance of acquired characteristics.
9.   Natural selection.
10.   Coevolution.
11.   Homologous structures.
12.   Divergent evolution.
13.   Convergent evolution.
14.   Analogous structures.
15.   Artificial selection.
16.   Industrial melanism.

17.-25. **MATCHING TEST:** Theories about life.

Choices:

A.	Lamarck	F.	Agassiz
B.	Cuvier	G.	Aristotle
C.	Wallace	H.	Lyell
D.	Plato	I.	Darwin
E.	Malthus		

17.   Uniformitarianism.
18.   <u>Scala Naturae</u>.
19.   "Ideal Forms".
20.   Inheritance of acquired characteristics.

21.   Catastrophism.
22.   Multiple creations.
23.   Natural selection.
24.   Essay on Population.
25.   Coevolution.

----

26.   Summarize the Darwin-Wallace theory of evolution in modern terms, using their 4 observations
       and 3 conclusions.

27.-33. **Explain** how each of the following supports the concept of evolution.

27.   The fossil record.
28.   Comparative anatomy.
29.   Vestigial structures.
30.   Embryological studies.
31.   Biochemical studies.
32.   Artificial selection.
33.   Industrial melanism.

****

ANSWERS TO REVIEW QUESTIONS:

1.	see 1.A.2)	17.-25.	[see entire chapter]
2.	see 2.A.2)a.	17. H	22. F
3.	see 2.B.1)	18. G	23. C, I
4.	see 2.B.2)	19. D	24. E
5.	see 2.C.3)	20. A	25. I
6.	see 2.C.5)	21. B	
7.	see 2.C.6)		
8.	see 2.C.7)	26.	see 3.B.
9.	see 3.A.1)-7)	27.	see 4.A.
10.	see 3.C.	28.	see 4.B.
11.	see 4.B.1)a.-d.	29.	see 4.B.3)
12.	see 4.B.1)c.	30.	see 4.C.
13.	see 4.B.2)	31.	see 4.D.
14.	see 4.B.2)	32.	see 4.E.
15.	see 4.E.	33.	see 4.F.
16.	see 4.F.		

****

CHAPTER 17:   THE MECHANISMS OF EVOLUTION

CHAPTER OUTLINE.

This chapter covers the basic principles of population genetics, including how to calculate gene and genotype frequencies, and the Hardy-Weinberg Principle. The major forces causing evolution (mutation, migration, small population size, non-random mating, and natural selection) are examined, and the three ways that natural selection acts on populations (stabilizing, disruptive, and directional) are elucidated. The results of natural selection are enumerated. Finally, the ultimate fate for most species (extinction) is discussed.

1.   Evolution and the **genetics of populations**.

    A.   <u>General facts</u>.
        1)   Evolution is an inevitable consequence of the nature of living things.
        2)   Evolution is a direct result of the chemical structure of genes (mutation) and the interaction between organisms and their environments (natural selection).
        3)   Evolution is a genetic change occurring in a population of organisms over a number of generations.
        4)   Inheritance is the link between the lives of individual organisms and the evolution of populations.

    B.   <u>Gene function</u> in individual organisms.
        1)   A gene is a segment of DNA located at a particular place on a chromosome.
        2)   Slightly different nucleotide sequences in genes of similar types produce alleles which generate different forms of the same type of enzyme.
        3)   An organism's phenotype is determined by its specific alleles interacting with the environment.

4) Example: flower colors in pea plants.
    a.   In purple flowers, a chemical reaction converts colorless molecules into purple pigment. These plants have alleles which produce the enzyme catalyzing this reaction.
    b.   In white flowers, the chemical reaction does not occur. These plants have only alleles which are unable to produce the enzyme needed to catalyze the reaction.

C.   Genes in populations.
    1)   The gene pool of a gene is the total of all the alleles of that gene in a population. Each allele has its own allele frequency.
    2)   Example: A population has 130 purple flowered plants (50 AA and 80 Aa) and 70 white flowered plants (aa).
        a.   The allele frequency for allele A is 100 (50 AA x 2) + 80 (80 Aa x 1) = 180/400 (200 individuals x 2 since each has 2 alleles) = 0.45 or 45%.
        b.   The allele frequency for allele a is 140 (70 aa x 2) + 80 (80 Aa x 1) = 220/400 = 0.55 or 55%.
        c.   The total gene pool for all the flower color alleles is 400, of which 45% are A and 55% are a.

D.   Population genetics and evolution.
    1)   If natural selection causes changes in the allele frequencies within the gene pool of a population, evolution is the result.
        a.   Suppose a cow eats all the purple flowered plants before they set seed.
            i.   The frequency of the purple allele (A) will drop to 0.00.
            ii.   The frequency of the white allele (a) will rise to 1.00 since they are the only plants that reproduce.
        b.   Thus, natural selection in the form of foraging by the cow, has caused an evolutionary change in the plant population.
    2)   This example illustrates 4 points about evolution:
        a.   Natural selection does not cause genetic changes in individuals. It does not produce the purple or white alleles or cause them to mutate.
        b.   Natural selection affects individuals but evolution occurs in populations. Plants were eaten or not but the population evolved.
        c.   Evolution is a change in the allele frequencies of a population, due to differential reproduction.
            i.   White flowered plants reproduced while purple flowered plants did not.
            ii.   Consequently, the frequencies of alleles A and a changed.
        d.   Evolutionary changes are not "progressive" or "good" in any absolute sense. Another cow that likes white flowers would have changed the population in a different way.

2.   The equilibrium population.

A.   To understand the forces of change, we first must understand the conditions under which a population will not change or evolve.
    1)   In such an equilibrium population, neither the allele frequencies (A and a) nor the distribution of genotypes (AA, Aa, and aa) change.

2)   In 1908, **Hardy and Weinberg** stated the conditions necessary to keep a population in genetic equilibrium. These <u>conditions</u> are:
    a.   No mutation must occur.
    b.   . No migration must occur.
    c.   The population must be "infinitely" large.
    d.   Matings must be completely at random.
    e.   All individuals must reproduce equally well (no natural selection must occur).

3)   See Figure 17.1 for the mathematics of the Hardy-Weinberg Principle.

B.   Though no real natural population fulfills every condition, the Hardy-Weinberg Principle is a useful starting point in studying the mechanisms of evolution.
    1)   If a population is changing genetically, it may be due to mutation, migration, small population size, non-random mating, or natural selection.
    2)   If a population is not changing genetically, the effects of two or more of these conditions are balancing each other.

3.   Mechanisms of evolution.

A.   <u>Mutation</u>: the raw material of evolution.
    1)   Mutations are chance heritable changes in nucleotide sequences and most usually have harmful effects.
    2)   Since mutations are rare (1 in 10,000 - 1,000,000 genes per generation per individual), mutation by itself is not a major force in evolution.
    3)   But mutations are important since they are the source of <u>new</u> alleles upon which other evolutionary processes can occur.
    4)   A mutation is not goal-directed: it provides "potential" (it might be beneficial under certain circumstances). Acting on that potential, natural selection may favor or disfavor the spread of the mutation through the population.

B.   <u>Migration</u> (gene flow): redistributing genes.
    1)   In evolutionary biology, migration is the flow of genes between populations. When a male baboon leaves one troop and enters another, it carries genes out of one gene pool and into the other.
    2)   Three effects of migration.
        a.   Gene flow helps spread advantageous alleles throughout the species.
        b.   Gene flow helps to maintain the species identity of various populations.
            i.   It prevents populations from becoming very different in allele frequencies.
            ii.   Without gene flow between populations, the isolated populations may become different species.
        c.   Migration and subsequent isolation of small numbers of individuals sometimes lead to significant evolutionary change (see the next section).

C.   <u>Small population size</u>: random changes in gene frequency.
    1)   To remain in equilibrium, a population must be large, since only a small sample of a population actually serves as parents for the next generation.
    2)   Which individuals survive and reproduce depend on both fitness (being favored by natural selection) and chance (having good luck since disaster may befall

even the fittest organism).

3) Genetic drift: Population size greatly influences the potential for chance events to change allele frequencies. Example: 2 populations of ladybugs:

    a. Each population has 50% spotted and 50% unspotted ladybugs but one population has 1000 bugs and the other has only 4.

    b. In each population, assume half the bugs die young and the other half reproduce, each having 2 offspring identical to themselves, keeping the population sizes constant.

        i. In the large population, it is quite likely that about half the reproducers will be spotted and half will be unspotted.

        ii. In the small population, there is only a 50% chance that one reproducer will be spotted and the other unspotted. The other half of the time, both reproducers will be either spotted or unspotted.

        iii. So, each generation, there is a 50% chance of losing either the spotted or the unspotted phenotype from the small population. Eventually, it will happen.

        iv. Such a change in allele frequency in small populations produced by chance is called genetic drift.

4) Genetic drift:

    a. Tends to reduce genetic variability within a small population.

    b. Tends to increase genetic variability between populations.

5) Genetic drift contributes significantly to evolution when population sizes become very small, a phenomenon called a population bottleneck (temporary reduction in size due to an unfavorable environmental condition of short duration).

    a. Population bottlenecks cause:

        i. Changes in allele frequencies.

        ii. Reduction in genetic variability. Example: northern elephant seals and cheetahs.

    b. Due to this reduced genetic variability, elephant seal and cheetah populations are in danger of extinction since without genetic variation they cannot evolve in response to changes in the environment.

6) Founder effect: a special type of population bottleneck occurring when isolated colonies are founded by a small number of organisms.

    a. In humans, this happens when small groups immigrate for religious or political reasons.

    b. By chance, such a small group may have very different allele frequencies than those of the parent population.

    c. If the founders remain isolated as their population expands, a sizeable new population may arise that differs greatly from its neighbors in allele frequencies.

7) Importance of genetic drift in evolution:

    a. Rarely are natural populations extremely small.

    b. However, when small, populations contribute most to major evolutionary changes due to genetic drift.

    c. New species often arise in small populations.

D. Random mating.

1) Organisms seldom mate randomly due to mobility or behavioral features.

2) The 3 common forms of non-random mating among animals are:

    a.   <u>Harem breeding</u>: only a few males fertilize every female (as in elephant seals, baboons, and bighorn sheep).

    b.   <u>Assortative mating</u>: selecting a mate with traits similar to your own (e.g., in humans, height, race, IQ, social status).

    c.   <u>Sexual selection</u>: when mate selection is done by one sex, usually the female in birds and mammals.

        i.   Males display their "virtues" and a female chooses her mate.

        ii.   Due to the selective pressure of female choice (<u>sexual selection</u>), elaborate structures and behaviors found only in males have evolved.

        iii.   Darwin coined the term "<u>sexual selection</u>" to designate the process of evolution through mate choice, though this is really a type of natural selection.

E.   <u>Equivalence of genotypes</u> in relation to natural selection.

    1)   Probably, most alleles provide some advantage or disadvantage to the organisms carrying them.

    2)   Natural selection will act to modify their frequencies through differential survival and/or reproduction.

    3)   Natural selection prunes the growth of a species, molding it to fit its environment.

4.   Natural selection.

A.   <u>General facts</u>.

    1)   Natural selection is more about reproduction than about survival.

    2)   Natural selection acts on the <u>phenotypic</u> traits that organisms display.

    3)   These traits vary within a population due to differences in both the genotypes of individuals and the effects of varying environments.

    4)   Below, the effects of the environment will be ignored.

B.   The 3 major <u>types of natural selection</u>: stabilizing, disruptive, and directional.

    1)   <u>Directional selection</u>.

        a.   If environmental conditions change rapidly in a certain direction (i.e., always colder), a species will evolve rapidly since one extreme phenotype is more adaptive than the rest (i.e., thicker fur).

        b.   <u>Examples</u>: Swifter wolves or larger-necked giraffes got more food than the average or slower/shorter individuals, living longer and leaving more offspring.

        c.   Industrial melanism shows that directional selection can occur rapidly if genetic variation is present and the selective force (differential selection by birds) is strong.

    2)   <u>Stabilizing selection</u>.

        a.   In species already well-adapted to a particular stable environment, selection favors the survival and reproduction of "average" organisms.

        b.   Often occurs when a single trait is under opposing selective pressures.

        c.   <u>Examples</u>:

            i.   Certain small lizards have a hard time defending territories, but large lizards are more likely to be preyed upon by owls. Thus, "average" body size is best.

      ii.   Lengths of necks and legs in giraffes are under stabilizing selection, balancing the demands of feeding (reaching up into trees) and drinking (from ground water).

      iii.   Female mate choice vs. increased predation for sexual displays in birds: if a peacock's tail became too long for flight, it wouldn't live long enough to woo a peahen).

  d.   Sometimes, stabilizing selection will act to maintain variability, not eliminate it. <u>Example</u>: Hemoglobin genes in blacks.

      i.   <u>AA</u> (normal homozygotes) are quite susceptible to severe malaria, often dying.

      ii.   <u>Aa</u> (heterozygotes) are only mildly affected by malaria and do not suffer from anemia under normal conditions.

      iii.   <u>aa</u> (sickle cell homozygotes) usually die young of severe anemia due to abnormal hemoglobin.

      iv.   Thus, stabilizing selection favors heterozygotes (<u>Aa</u> people are healthiest and have the most children). This maintains both alleles (<u>A</u> and <u>a</u>) at high frequency in the populations.

3)   <u>Disruptive</u> <u>selection</u> favors both extremes at the expense of the average or intermediate individuals.

  a.   May occur when a species encounters various microhabitats and different traits best adapt individuals to each microhabitat.

  b.   <u>Example</u>: An island where one bird species exists with different species of plants producing large and small seeds.

      i.   Birds eating large seeds require large beaks, so selection favors large body/ beak size in the bird species.

      ii.   The opposite is true for birds eating small seeds since small birds eat small seeds more efficiently.

      iii.   Birds with intermediate body/beak size are not favored by selection since they are too small to eat large seeds and too large to efficiently get enough energy from small seeds.

      iv.   Thus, disruptive selection favors the survival of both large and small, but not medium-sized birds.

5.   Forces of natural selection.

A.   <u>General facts</u>.

1)   Natural selection acts by eliminating those lacking the traits needed for survival and reproduction in their particular environments.

2)   Thus, natural selection results in better adaptation of a population to its environment.

3)   Adaptation to both abiotic (non-living) and biotic (living) components of the environment occur by natural selection. Example: Buffalo grass is so successful in Wyoming because:

  a.   Its long, deep roots allow it to outcompete other plant life for water in the dry plains.

  b.   Its tough blades, containing silica, discourage buffalo and cattle from eating it.

4)   <u>Coevolution</u>: natural evolution of 2 species that interact extensively, each exerting strong selection pressure on the other. A change in one (greater speed in

rabbits) forces a change in the other (greater speed in wolves) which then forces a change in the first again (still greater speed), etc. in a constant mutual feedback manner.

B. Competition: Competition with members of the same species is a major selective force in the biotic environment. Example:
   1) Robins and cardinals are superficially similar (color and range), but they eat very different diets (don't compete with each other for food).
   2) But robins compete with other robins for worms and insects, while cardinals compete with each other for seeds.
   3) Different species may sometimes compete to a lesser extent for the same resources.

C. Predation: when one organism eats another.

D. Symbiosis: individuals of different species closely interact for much of their lives.
   1) One species usually benefits but the other may benefit, suffer injury, or not be affected at all.
   2) Symbiosis leads to intricate coevolutionary adaptations, since one species must be continually adapting to evolutionary changes in the other.

E. Altruism: behavior that endangers an organism or reduces its reproductive success, but benefits other members of its species.
   1) Cooperation and self-sacrifice (i.e., people helping each other) are important selective forces.
   2) Examples:
      a. A mother killdeer bird feigning a broken wing to lure predators from her young.
      b. Sterile female worker bees caring for the hive queen's offspring.
      c. Young male baboons acting as lookouts for leopards.
   3) Natural selection selects for genes producing altruisms if the altruism benefits relatives who possess the same genes. This is called kin selection.

6. **Extinction**: The usual fate of most species.

   A. Natural selection may lead to extinction.
      1) About 99.9% of all species are extinct, known only by their fossilized remains.
      2) Environmental events are usually the immediate cause of extinction.

   B. Susceptibility to extinction.
      1) Localized distribution: If a species occupies only a very small range, any disturbance of that area could cause extinction.
         a. Example: the Devil's Hole pupfish.
         b. Wide-ranging species will not become extinct due to a local calamity.
      2) Over-specialization: Extremely specialized species are vulnerable to extinction.
         a. Such overspecialization to a particular environment will "imprison" the organism in a very narrow ecological niche.
         b. Example: the specialized diet of the Everglades kite to one species of snail. If the snail is wiped out, the kite must also become extinct.

C. Extinction and the environment.
1) Major environmental changes that drive species to extinction:
   a. Competition among species for limited resources.
     i. A species may become extinct if competitors evolve superior adaptations.
     ii. Example: North American animals displaced South American animals when the Panama land bridge arose and animals migrated southward.
   b. Novel predators or parasites.
     i. Predators other than humans cause few extinctions.
     ii. New parasites, however, can be devastating. Examples: Dutch elm disease and chestnut blight are introduced parasites that nearly destroyed these tree species.
   c. Habitat destruction: The leading cause of extinction today and in the past.
     i. Human activity, destroying many tropical forests (cleared for timber and farming/ grazing land) will probably destroy half the world's species in the next 50 years.
     ii. Prehistoric habitat destruction, due mainly to climatic changes, led to many extinctions. Examples:
       a) Massive volcanic eruptions.
       b) Meteorites hitting the earth and kicking up dust that blocked sunlight, killing plants and the animals that fed on them.

****

REVIEW QUESTIONS:

1. Define evolution.

2.-8. A population of 600 plants contains 294 AA, 252 Aa, and 54 aa individuals. The AA and Aa plants produce purple flowers while the aa plants produce white flowers. What are the frequencies of the following?

2. Allele A.
3. Allele a.
4. Genotype AA.
5. Genotype Aa.
6. Genotype aa.
7. Purple-flowered plants.
8. White-flowered plants.

----

9. Give the 5 conditions that Hardy and Weinberg stated were necessary to keep a population in genetic equilibrium.
10. Explain how mutation by itself is not a major force in evolution but that mutation is necessary for evolution to occur.

11. Name 3 effects of migration on natural populations.
12. Explain why population size greatly influences the potential for chance events to change allele frequencies.
13. Describe the differences between "population bottlenecks" and the "founder effect".
14. Name and describe the 3 common forms of non-random mating.
15. Name, explain, and give an example of each of the 3 ways that natural selection may act in natural populations.
16. How can altruism be evolutionarily advantageous?

17.-34. MATCHING TEST: evolutionary mechanisms

Choices:

A.	Sexual selection
B.	Population bottleneck
C.	Gene flow
D.	Mutation
E.	Altruism
F.	Assortative mating
G.	Founders effect
H.	Species
I.	Genetic drift
J.	Harem breeding
K.	Coevolution
L.	Disruptive selective
M.	Stabilizing selection
N.	Directional selection

17. Only a few males fertilize all females in a population.
18. The frequencies of white and brown guinea pigs in a population changes because a large number of white animals enters the population.
19. A small number of organisms begins a new colony that becomes large and has gene frequencies very different from the parent population and other neighboring populations.
20. Provides "genetic potential" to a population, upon which natural selection acts.
21. Chance loss of genetic variation from a population due to its small size.
22. Temporary restriction in population size due to short term unfavorable environmental conditions.
23. Causes the evolution in males of elaborate structures and behaviors related to reproduction.
24. Average individuals survive and reproduce best since the population is well-adapted to a stable environment.
25. Selection involving sickle cell anemia in African humans.
26. Selecting a mate with traits similar to your own.
27. Selection favoring both extreme phenotypes and not favoring average individuals.
28. A change in a prey species forces a change in a predator species.
29. Selection of one extreme phenotype in populations living in a rapidly changing environment.
30. Selection for flowering times in New England wildflowers.
31. Behavior that endangers an organism or reduces its reproductive success but benefits others in the population.
32. Selection within a bird species encountering only large and small seeds.
33. Selecting causing industrial melanism.
34. The total of all populations of organisms that interbreed under natural conditions.

----

35.   Name 2 situations that increase the susceptibility of a species to extinction.

****

ANSWERS TO REVIEW QUESTIONS:

1.   see 1.A.3); 1.D.2)c.

2.-8.   [see 1.C.2)a.-c.]

2.   588 (294 <u>AA</u> x 2)
            + 252 (252<u>Aa</u> x 1)
        = 840/1200 (600 plants
            x 2 alleles each)
        = 0.70 = 70%.

3.   108 (54<u>aa</u> x 2)
            + 252 (252<u>Aa</u> x 1)
        = 360/1200 = 0.30 = 30%.

4.   294<u>AA</u> /600 total plants
        = 0.49 = 49%.

5.   252/600 = 0.42 = 42%.

6.   54/600 = 0.09 = 9%.

7.   294<u>AA</u> + 252<u>Aa</u> /500 total plants
        = 546/600 = 0.91 = 91%.

8.   54 <u>aa</u> /600 = 0.09 = 9%.

9.   see 2.A.2)a.-e.

10.   see 3.A.2)-4)

11.   see 3.B.2)a.-c.

12.   see 3.C.

13.   see 3.C.5)-6)

14.   see 3.D.2)a.-c.

15.   see 4.B.1)-3)

16.   see 5.E.3)

17.-34. [see entire chapter]

17.	J	26.	F
18.	C	27.	L
19.	G	28.	K
20.	D	29.	N
21.	I	30.	M
22.	B	31.	E
23.	A	32.	L
24.	M	33.	N
25.	M	34.	H

35. see 6.B.1)-2)

****

CHAPTER 18:   THE ORIGIN OF SPECIES

CHAPTER OUTLINE.

In this chapter, the concept of speciation is discussed. After defining biological species, the general mechanisms of allopatric and sympatric speciation are described. Premating and Postmating mechanisms for maintaining reproductive isolation are covered. The distinctions between divergent and phyletic speciation, and gradualism and punctuated equilibrium are elucidated.

1.   **General facts.**

   A.   <u>Species</u>:
      1)   The major qualitative difference between species is the ability to interbreed.
      2)   <u>Definition</u>:   all populations of organisms that are potentially capable of interbreeding under natural conditions and that are reproductively isolated from other populations.

   B.   Questions regarding speciation.
      1)   At what point do enough small differences gradually add up to enough total difference so that a population is a new species?
      2)   Is there a huge difference that suddenly occurs causing speciation distinct from accumulation of small differences?
      3)   Is all evolution the accumulation of small changes over long time periods, or are other mechanisms at work?

2.   **Speciation.**

   A.   To produce a new species, enough changes must occur between populations so that interbreeding cannot occur or hybrids are less fit.

   B.   Speciation depends on:
      1)   <u>Isolation of populations</u>: Little or no gene flow must occur between populations if differences are to accumulate between them.

2) <u>Genetic divergence.</u>
    a.   When formerly isolated populations are reunited, they will become distinct species only if enough genetic differences have accumulated during their time of isolation so that they cannot interbreed to produce vigorous fertile offspring.
    b.   If isolated populations are small, chance events may generate significant genetic differences by genetic drift.
    c.   In both small and large populations, different selective pressures in separate environments may favor evolution of much genetic difference.

C.   <u>Hypotheses</u> for origins of new species:
    1)   <u>Allopatric speciation:</u> diverging populations are geographically separated.
    2)   <u>Sympatric speciation:</u> diverging populations share the same geographical area.

D.   <u>Isolation from gene flow</u> is crucial to both allopatric and sympatric speciation:
    1)   Isolation is obvious if two populations are separated by a physical barrier like a river.
    2)   Even if two populations live in the same area, isolation may occur if:
        a.   They select different habitats within the same area to mate (marshes vs. forest).
        b.   They have different numbers of chromosomes and cannot form fertile hybrids (horses x donkeys → sterile mules).

E.   <u>Allopatric ("having a different fatherland") speciation.</u>
    1)   Occurs when two populations become geographically isolated from one another.
        a.   Migration allows gene flow which reduces genetic differences between populations.
        b.   Presence of a physical barrier allows little or no gene flow.
    2)   Isolated populations could accumulate large genetic differences if:
        a.   Natural selection differs between them.
        b.   They are small enough for genetic drift to occur.
    3)   Founder events may also be important in initiating genetic differences between populations.
    4)   Most biologists think natural selection provides the major impetus to speciation and geographical isolation is involved in most cases of animal speciation.

F.   <u>Sympatric ("having the same fatherland") speciation.</u>
    1)   Occurs within a single population in a single geographical area but still requires no or very limited gene flow.
    2)   Two likely mechanisms:
        a.   Ecological isolation.
        b.   Chromosomal aberrations.
    3)   <u>Ecological isolation.</u>
        a.   One area with two distinct habitats allows different members of the same species to begin specializing in one or the other habitat.
        b.   Natural selection for habitat specialization may lead to speciation.
        c.   Example: The fruit fly <u>Rhagoletis pomonella.</u>
            i.   Is splitting into 2 species based on preference for apples or hawthornes, causing accumulation of genetic differences between the two types.

ii. Flies don't alternate between apples and hawthornes. Each fly chooses only one.

iii. Females usually lay their eggs in the same type of fruit from which they developed, and males rest on the plant type in which they developed.

iv. Apple fruits mature 2-3 weeks later than hawthorne fruits, retarding apple and hawthorne-liking flies from interbreeding.

4) Chromosomal aberrations.

  a. Sometimes, new species arise instantaneously through changes in chromosome changes caused by abnormal meiosis.

  b. Polyploidy (more than 2 doses of each chromosome) in plants often leads to rapid speciation.

    i. If a fertilized egg duplicates its chromosomes but doesn't divide, the resulting cell is tetraploid (4 copies of each chromosome) and will produce a tetraploid plant.

    ii. Tetraploid plants are vigorous and healthy and can produce viable 2n gametes by meiosis.

    iii. If a 2n gamete from a tetraploid fertilizes a normal 1n gamete, the result is a 3n triploid that develops normally but is sterile (odd numbers of chromosomes cannot pair properly during meiosis).

    iv. Thus, a tetraploid plant is reproductively isolated from its diploid parent species.

    v. Nearly half the flowering plants are polyploid, mostly tetraploid.

  c. Speciation by polyploidy is common in plants but rare in animals because many plants can reproduce asexually or can self-fertilize whereas animals cannot.

3. Maintaining **reproductive isolation** between species.

  A. Two categories of reproductive isolation:

    1) Premating mechanisms: members of a species may not mate with members of other species.

    2) Postmating mechanisms: occurs if premating mechanisms fail or have not yet evolved.

      a. Definition: Incompatibilities between species that prevent the formation of vigorous, fertile hybrids.

      b. If members of different species mate, their hybrid offspring may:

        i. Die during development.

        ii. Be born but are less fit or infertile.

  B. Premating isolating mechanisms.

    1) Geographical isolation: members of different species cannot get near each other.

      a. One can't tell if geographically separated populations constitute distinct species (they might mate if given the chance).

      b. This is usually considered a mechanism that creates new species rather than maintains reproductive isolation between them.

    2) Ecological isolation: use of different local habitats for mating by different species with overlapping ranges.

           a.   Example: white-throated sparrows frequent dense thickets while white-crowned sparrows inhabit fields and meadows.

           b.   This does not prevent gene flow entirely; other mechanisms contribute as well.

     3)   Temporal isolation: different breeding seasons by different species living within the same areas.

           a.   In California, bishop pines and Monterey pines coexist. But the former release pollen in summer while the latter release pollen in early spring.

           b.   Hawthorne-liking and apple-liking Rhagoletis fruit flies emerge from their host fruits and breed at somewhat different times of year.

     4)   Behavioral isolation: elaborate courtship colors, patterns and rituals have evolved as recognition and evaluation signals between females and males, especially to determine whether they belong to the same species.

           a.   The colors and calls of male songbirds attract only females of their species.

           b.   Female frogs only approach males of their species because these males emit the correct croaking sounds.

     5)   Mechanical isolation: Failure of the genitalia of males and female animals to fit together properly (preventing sperm transfer) if they belong to different species.

  C.   Postmating isolating mechanisms.

     1)   Gamete incompatibility:

           a.   Animals: fluids of the female reproductive system weaken or kill sperm deposited by males of different species.

           b.   Plants: chemical incompatibility may prevent pollen germination on the stigma of the flower of a different species.

     2)   Hybrid inviability: fertilization occurs but the resulting hybrids are weak and unable to survive.

           a.   The hybrid may abort early in development.

           b.   The hybrid may survive but display non-adaptive, imperfect behaviors that are mixtures of the two parental types.

     3)   Hybrid infertility.

           a.   Animal hybrids, like the mule, are usually sterile since meiosis cannot occur normally.

           b.   Polyploid plants produce sterile triploid offspring if they mate with their diploid parent species.

4.   Phyletic vs. divergent evolution.

  A.   Divergent speciation ("true speciation"): two populations of the same species become different species through isolation from gene flow and genetic divergence, using the mechanisms of reproductive isolation given above. Both species coexist at the same time.

  B.   Complications in the fossil record: Since dead and petrified fossils cannot breed, paleontologists make distinctions among species purely on structural grounds.

  C.   Phyletic speciation ("pseudospeciation"): a whole species, under massive directional selection, changes over time so that the later specimens are very different from the earlier

ones.
1)   If the differences are large enough, we assume that the later organisms are a new species since they could not interbreed with the earlier ones.
2)   However, reproductive isolation between earlier and later forms cannot be tested directly.

D.   Since the fossil record is incomplete, it does not reveal whether divergent or phyletic speciation occurred more frequently.

5.   **The genetics of speciation.**

A.   Two genetic models for speciation:
1)   Gradual accumulation of many small changes.
2)   Sudden appearance of a few major changes.

B.   Gradual accumulation of many small changes.
1)   Two populations, by genetic drift or natural selection, gradually accumulate many small genetic changes.
2)   Over time, these changes may result in reproductive isolation between the two populations, which become separate species.

C.   Sudden appearance of a few major changes.
1)   One or a few regulatory genes may control major developmental processes.
2)   Mutations in just a few regulatory genes might result in significant enough changes in development that the mutants would immediately be reproductively isolated from their parental populations.
3)   Result: instantaneous speciation.

D.   Which model is correct? Biologists aren't sure.
1)   Extreme genetic similarity between certain species (humans and chimps) suggests that mutations of just a few regulatory genes caused the speciation.
2)   The "small changes" model is supported, since many phenotypic traits and behavioral traits are controlled by many genes.
3)   One shot, massive genetic changes are usually harmful.
4)   Molecular genetics may someday resolve this question.

6.   **Rates of speciation.**

A.   Over time, species continually form, exist for a while, and become extinct. But the rate of speciation varies considerably over evolutionary time.

B.   Adaptive radiation: Species give rise to many new species in a relatively short time.
1)   Occurs when populations of one species invade different habitats and change in response to the differing selective pressures in those habitats.
2)   Adaptive radiation usually results from either:
a.   A species encountering a wide variety of unoccupied habitats (i.e., when marsupial mammals first invaded Australia).
i.   With no competitors, all ecological roles are rapidly filled with new species that evolved from the original invaders.

ii.   Unoccupied habitats result from mass extinctions caused by natural disasters (i.e., meteorite impacts).

b.   A species developing a fundamentally new and superior adaptation, enabling it to displace less well adapted species in many habitats. Example: warmblooded mammals displacing reptiles.

7.   **The progress of evolution.**

A.   Biologists aren't sure if natural selection or genetic drift actually has played a greater role in shaping evolutionary history.

1)   Three areas of interest are:
   a.   Rate and timing of evolutionary change.
   b.   Role of speciation in producing morphological change.
   c.   Relative importance of natural selection vs. other forces in determining structures and behaviors.

2)   The opposing views, called gradualism and punctuated equilibrium, are at opposite ends of a spectrum of possibilities. Probably, both types of evolutionary mechanisms occur in nature.

B.   Gradualism.

1)   Speciation occurs gradually (over hundreds of thousands of years or more) due to accumulation of many of small genetic differences between two populations.

2)   Morphological change within a lineage occur continuously, although the rate may change somewhat from time to time.

3)   Morphological change and speciation are not closely linked.

4)   Natural selection drives this process.

C.   Punctuated equilibrium.

1)   Speciation is a geologically rapid ("punctuated") event, driven by a relatively small number of genetic changes in important regulatory genes.

2)   Species remain morphologically the same for long periods of time ("equilibrium").

3)   Morphological change and speciation are tightly linked. Virtually all structural change occurs during the brief period of speciation, and the new species remain unchanged until the next speciation event.

4)   Genetic drift drives this process by "species selection," not individual selection.

D.   Applying the gradualism model to the evolution of the horse.

1)   Fossil record shows progression from Hyracotherium (dawn horse) through many steps to modern Equus.

2)   Gradualist view: fossil horses are the result of changes that occurred gradually throughout the evolution of the horse.

   a.   As environments changed (forests → open woodlands → prairies), natural selection favored larger, faster horses with strong shock absorbing legs, hard hooves, and large grinding teeth.

   b.   Both divergent and phyletic speciation occurred from time to time.

   c.   During phyletic speciation, selection between individuals within a population drove the whole population to change gene frequencies gradually until new species arose.

3)   Problem: every type of fossil horse is quite different from the others; the fossil

record does not show a continuous accumulation of small changes, but rather a series of jumps.

    a.   Gradualists claim the fossilization is a rare event. Perhaps, thousands of years occur when conditions for fossilization are bad.

    b.   Thus, many intermediate stages never show up in the fossil record.

E.   Applying the punctuated equilibrium model to the <u>evolution of the horse</u>.

    1)   Eldredge, Gould, and Stanley claim the "gaps" in the fossil record imply punctuated equilibrium.

        a.   The gaps aren't really gaps because the species don't gradually change.

        b.   Such rapid evolutionary change occurs that fossil remains of intermediate forms would be extremely rare.

    2)   During evolution of the horse:

        a.   <u>Hyracotherium</u> underwent little change for millions of years ("equilibrium").

        b.   Small fringe populations of <u>Hyracotherium</u> split off now and then, and some of them evolved very rapidly by genetic drift to become new species.

        c.   Due to a major environmental change, one of the new species (<u>Mesohippus</u>) quickly replaced <u>Hyracotherium</u> as the dominant form of horse ("punctuation").

    3)   Evolution is driven largely by frequent rapid divergent speciation, followed by selection between species leading to extinction of one species and survival of another.

F.   <u>Synthesis</u>: majority view regarding how speciation occurs.

    1)   <u>Rates of evolution</u>.

        a.   The rate varies greatly, depending on the species studied:

            i.   Some species (like sharks) show long periods of constancy due to an unchanging environment.

            ii.   Rapid changes during times of speciation also occur. Possible mechanisms:

                a)   Genetic drift.

                b)   Directional selection.

                c)   Coevolution.

                d)   Sexual selection.

        b.   Punctuated speciation would seem quite slow to a geneticist (measured over many generations) but quite fast to a geologist (a million years is a small time period to them).

    2)   <u>Selection and morphological change</u>.

        a.   Structural change may occur when a population is isolated from other populations and is subjected to new directional selection pressures.

        b.   Change may occur due to genetic drift and/or founder events if the population is small.

    3)   <u>Species selection vs. individual selection</u>.

        a.   The three possibilities if two species directly compete for resources:

            i.   One will move.

            ii.   One will evolve rapidly to use another resource.

            iii.   One will become extinct.

        b.   Thus, the species is the ultimate unit of selection, although members of the species that survives may have attained superior adaptations through

natural selection acting on individuals.
4) <u>Importance of natural selection</u>.
   a.   Two views:
       i.   "Adaptationists" view: Natural selection is most important.
       ii.   "Punctuationist" view: Non-selective, random events are most
             important.
   b.   Objections to "adaptationist" view:
       i.   It is difficult to prove that a particular trait evolved by natural
             selection.
       ii.   Random forces do operate in nature (i.e., floods, massive forest
             fires, earthquakes, volcanoes, meteorites) and cause extinctions.
5) <u>Consensus view</u>: both natural selection and random forces (genetic drift and
   environmental disasters) shape the evolutionary history of life on Earth. Debate
   continues over which process dominates over what time scales.

****

REVIEW QUESTIONS:

1.-7.   **MATCHING TEST**: speciation

	A.	Migration	D.	Mutation
Choices:	B.	Geographical isolation	E.	Adaptive
	C.	Polyploidy		radiation

1.   Mechanism causing "instant speciation".
2.   Mechanism causing most animal speciation.
3.   Populations become separated by a physical barrier.
4.   Mechanism causing much plant speciation but little animal speciation.
5.   Reduces genetic differences between populations, retarding the process of animal speciation.
6.   One species gives rise to many new species in a short time.
7.   Acquisition of more than 2 copies of each chromosome in the nucleus.
----

8.   Name the two conditions necessary for speciation.
9.   Distinguish between allopatric and sympatric speciation.
10.   What is crucial to both allopatric and sympatric speciation?
11.   Name two models for sympatric speciation.

12.-18. **MATCHING TEST**: maintaining reproductive isolation

	A.	Geographical isolation	E.	Mechanical isolation
Choices:	B.	Ecological isolation	F.	Gamete incompatibility
	C.	Temporal isolation	G.	Hybrid inviability
	D.	Behavioral isolation	H.	Hybrid infertility

12.   Interbreeding does not occur in nature between British peppered moths and Canadian peppered

moths.

13. Interbreeding does not occur between closely related species of fruit flies with slightly different courtship rituals.
14. When horses and donkeys are forced to interbreed, sterile mules are produced.
15. Leopard frogs and pickeral frogs that live in the same area and have similar mating seasons do not interbreed because one species breeds in swamps and the other breeds in clear lakes.
16. Closely related species of katydid insects do not interbreed because the male and female sex organs cannot fit together properly to allow sperm transfer to occur.
17. Wood frogs and green frogs breed in the same lakes but do not interbreed because one species breeds in April while the other species breeds in May.
18. Two closely related species of fruit flies sometimes mate but the female's immune system kills the male's sperm as though it were a foreign invading microbe.

----

19. Name and briefly describe two genetic models for speciation.
20. How do the theories of gradualism and punctuated equilibrium differ?
21. Explain the evolution of the horse in terms of gradualism and punctuated equilibrium.

****

ANSWERS TO REVIEW QUESTIONS:

1.-7.   [see entire chapter]

1.	C	5.	A	
2.	B	6.	E	
3.	B	7.	C	
4.	C			

8.   see 2.B.1)-2)

9.   see 2.E.-F.

10   see 2.D.

11.   see 2.F.3)-4)

12.-18. [see 3.A.-C.]

12.	A	16.	E	
13.	D	17.	C	
14.	H	18.	F	
15.	B			

19.   see 5.A.-C.

20.   see 7.B.-C.

21.   see 7.D.-E.

****

CHAPTER 19:   THE HISTORY OF LIFE ON EARTH

CHAPTER OUTLINE.

This chapter traces the evolution of life from the beginning of the universe through prebiotic evolution, spontaneous generation of the first living prokaryotic cells, metabolic evolution, the rise of eukaryotes and multicellularity, the invasion of land, and finally human evolution.

1.   **Evolution of the Universe** and solar system.

    A.   Before the "Big Bang", only a dense tiny mass of energy and matter existed in a highly unstable form.
        1)   The mass then erupted about 15 billion years ago (the "Big Bang").
        2)   Local accumulations of matter formed as gravity attracted particles.
        3)   Some accumulations grew so large that their centers exploded in thermonuclear reactions. These masses became stars.

    B.   About 5 billion years ago, our solar system started as a cloud of matter, then began to condense.
        1)   The center collapsed to form the sun.
        2)   Further out, local aggregations occurred, forming the planets.
        3)   The first 2 planets near the sun were very hot.
        4)   The 4th planet from the sun was very cold.
        5)   The 3rd planet (Earth) received just enough warmth to permit liquid water to exist. Life evolved here.

2.   **Origin of life**.

    A.   <u>Spontaneous generation</u>: new life appears from non-living matter.
        1)   This theory was accepted as fact into the 1800s.

2)　　It was gradually disproved:
　　　a.　　In 1688, Redi disproved the maggots-from-meat hypothesis of spontaneous generation.
　　　b.　　In the 1860s, Pasteur and Tyndall disproved the broth-to-microbes idea.
3)　　However, in the 1920s and 1930s, Oparin and Haldane proposed <u>prebiotic evolution</u>: billions of years ago, under the right conditions, life could have arisen from non-living matter through ordinary chemical reactions.

B.　　<u>Prebiotic evolution</u>.
1)　　Originally, the earth was very hot - even today, its core is molten metal.
2)　　Gradually, the earth cooled and many compounds formed.
　　　a.　　All oxygen was tied up in CO_2 and water. No O_2 gas existed.
　　　b.　　It rained for thousands of years as the water vapor cooled into liquid water.
　　　c.　　Lightning, volcanic heat, and ultraviolet radiation from the sun poured energy into the young seas.
　　　d.　　The primitive atmosphere of the earth contained CO_2, methane, ammonia, hydrogen, water vapor, and nitrogen. No free O_2 existed.
3)　　<u>Prebiotic synthesis of organic molecules</u>.
　　　a.　　In 1953, Stanley Miller simulated the earth's early atmosphere and found that simple organic molecules formed. So, such molecules could form spontaneously in ancient times.
　　　b.　　Without O_2, these organic molecules did not break down like they do today. They accumulated over time in the seas, making a dilute "nutrient soup".
　　　c.　　In slowly evaporating pools of water near the shores, the "soup" became concentrated.
　　　　　i.　　Such concentrated pools may have provided the molecules for the first living cells.
　　　　　ii.　　The chemical energy in these molecules would be food for these first cells.
4)　　<u>RNA: The first living molecule?</u>
　　　a.　　Recently, Cech and Altman have found RNA molecules that act as enzymes to cut apart RNA and make more RNA.
　　　b.　　These RNAs (called <u>ribozymes</u>) probably arose by chance and had catalyst abilities to make accurate and inaccurate (mutated) copies of itself.
　　　c.　　Perhaps, some ribozymes began to bind amino acids and catalyze the synthesis of short proteins. By mutation, protein enzymes could have thus arisen.
　　　d.　　Further mutation may have caused some ribozymes to copy themselves into DNA to safeguard against attack by other ribozymes.
　　　e.　　According to this theory, RNA was the first "living molecule", with both DNA and proteins evolving later. Eventually, the DNA → RNA → protein scheme evolved.
5)　　The <u>first living cells</u>.
　　　a.　　When proteins and lipids are agitated in water, hollow <u>microspheres</u> form that resemble cells in several respects:
　　　　　i.　　They have well-defined water boundaries.

ii. They form a "membrane" much like a cell membrane.
iii. They absorb materials from solution ("feed"), grow, and divide by splitting.

 b. If a primitive microsphere surrounded the right ribozymes, a primitive cell could have formed. Thus, life would have begun spontaneously.

6) Did all this actually happen?
 a. Enough time was available for it to happen.
 b. Most biologists conclude that the origin of life is an inevitable consequence of the working of natural laws.
 c. However, none of this has been proven and probably never can be.

3. The **age of microbes**.

A. Origin and evolution of <u>Prokaryotes</u>.
 1) The earliest living prokaryotic cells arose about 3.5 BYA (billion years ago).
  a. They obtained nutrients and energy by absorbing organic molecules from the primordial "soup".
  b. With no free O_2 in the atmosphere, the cells <u>metabolized food anaerobically</u>, yielding very little energy.
  c. These were primitive anaerobic bacteria.
 2) Eventually, these cells used up the organic molecules in the "soupy" seas, leaving lots of CO_2 and water (low energy molecules).
  a. <u>Photosynthesis</u> then <u>evolved</u> in some of the bacteria and they evolved into today's cyanobacteria.
  b. Using sunlight energy, they converted CO_2 and H_2O into organic sugar molecules and O_2.
  c. At first, the free O_2 reacted with iron to form iron oxides ("rust").
  d. Then, additional O_2 began accumulating in the atmosphere, reaching high levels about 2 BYA.
 3) Next, some cells <u>evolved aerobic metabolism</u>, using O_2 to completely break down organic molecules into CO_2 and H_2O, releasing lots of energy.
 4) These aerobic cells had a great selective advantage over anaerobic cells because much more energy is released when O_2 is used to metabolize food molecules.

B. The rise of <u>Eukaryotes</u>.
 1) Primitive predatory prokaryotic cells appeared next.
  a. These cells could neither photosynthesize nor undergo aerobic metabolism.
  b. They simply engulfed bacteria and digested them for food.
  c. About 1.4 BYA, these predators evolved into the first eukaryotic cells.
 2) Evolution of eukaryotic organelles: the <u>Endosymbiotic theory</u> of Margulis.
  a. This theory proposes that certain types of bacteria evolved into chloroplasts and mitochondria after they were engulfed by larger predatory bacteria.
  b. If an anaerobic predatory bacteria engulfed an aerobic bacterium but failed to digest it:
   i. The aerobic "prey" would do well since the cytoplasm of the predatory "host" contained many half-digested organic molecules.
   ii. The aerobe would use O_2 to digest these molecules, using the

energy to make lots of ATP.

  iii. Some of the ATP would leak into the cytoplasm to benefit the host cell as well.

  iv. These small aerobic bacteria evolved into mitochondria, and the combination of predatory host with mitochondria soon outcompeted its foes and filled the sea.

  c. Then, such a predatory cell with mitochondria engulfed a photosynthetic cyanobacterium and failed to digest it. The cyanobacterium evolved into the first chloroplast.

  d. The origin of the nucleus is more obscure, perhaps originating by infoldings of the cell membrane to protect the genetic material.

4. **Multicellularity.**

  A. Once predation evolved, increased size became an advantage.
      1) Larger organisms can eat smaller ones, and are harder to be eaten.
      2) Larger organisms can move faster than smaller ones, aiding in food capture or escape.
      3) The best way to grow larger is to become multicellular since large single cells lack sufficient surface area.

  B. Origin of multicellularity.
      1) Little is known except that it occurred.
      2) By about 500 MYA (million years ago), many diverse multicellular animal types had appeared in the fossil record.

5. **Multicellular life first arose in the sea.**

  A. Plants.
      1) Evolved from eukaryotic, chloroplast-containing unicellular organisms.
      2) Advantages of multicellularity to plants:
          a. They became difficult for small unicellular predators to swallow.
          b. They could develop rootlike anchors to remain in a suitable place and leaves to absorb sunlight near the shores. Example: the algae that line our shorelines today.

  B. Animals.
      1) A wide variety of invertebrate animals appeared in the sea about 600 MYA.
      2) One advantage of multicellular animals is the ability to eat larger prey.
          a. The first such animals were jellyfish-like, vase-shaped with a single opening used as both mouth and anus.
          b. Then, worm-like animals evolved with separate mouth and anal openings for more efficient feeding and digestion.
      3) The trend towards greater mobility developed next in both predators and prey by coevolution.
          a. Locomotion occurred by contraction of muscles attached to some sort of skeleton.
          b. Two types of skeletons evolved in invertebrates.

          i.   Internal hydrostatic skeletons (water-filled tubes) in worms and jellyfish.

          ii.   External hard skeletons covering the bodies of arthropods.

4) Rapid mobility necessitated greater sensory capabilities in the head region, and more sophisticated nervous systems.

    a.   Senses for detecting touch, chemicals and light evolved.

    b.   Nervous systems capable of handling the sensory information and directing appropriate responses evolved.

5) About 500 MYA, chordates evolved with internal skeletons.

    a.   Chordates evolved into fish about 400 MYA.

    b.   Fish had better sensory organs, larger brains, and were faster than most invertebrates and became dominant in the seas.

6.   The **invasion of land**.

A.   <u>Sea versus land</u>.

    1) The ocean provides for:

        a.   Buoyant support against gravity.

        b.   Access to water.

        c.   Simple reproduction since sperm and/or eggs can swim to each other.

    2) The land provides for:

        a.   No cushion against gravity.

        b.   No easy access to water.

        c.   More difficult sexual reproduction since the sperm can dry out.

        d.   Much more sunlight for rapid photosynthesis.

        e.   More soil nitrogen and phosphorus than is present in the oceans.

    3) About 500 MYA, the seas swarmed with life, especially plant-eating animals, but the land was devoid of animal life, giving plants that began growing on the land a great selective advantage.

B.   <u>Land plants</u>.

    1) First, a few small green algae began to grow in moist soil near shorelines. They gave rise to multicellular land plants about 400 MYA.

    2) These multicellular plants developed:

        a.   <u>Roots</u> to penetrate the soil to absorb water and minerals.

        b.   <u>Waterproof coatings</u> on above-ground parts to reduce water loss from evaporation.

        c.   <u>Vascular tissue</u> to conduct water from roots to leaves.

        d.   <u>Extra-thick cell walls</u> to enable stems to stand erect.

    3) <u>Reproduction</u> on land.

        a.   The first land plants had swimming sperm.

             i.   They had to live in swamps and marshes or in areas of abundant rainfall so that sex cells could be released into water.

             ii.   This was no problem during the Carboniferous period (360-286 MYA) of warm, moist climates. Club mosses and tree ferns flourished, their fossils becoming coal.

        b.   Meanwhile, primitive conifers evolved that retained eggs in the parent body and produced sperm within drought-resistant pollen grains.

             i.   The pollen was carried by the wind.

ii. If pollen landed near an egg, it grew towards the egg, releasing the sperm directly into living tissue.

iii. Conifers flourished about 250 MYA, when most moist climates dried up.

c. Flowering plants appeared about 130 MYA, having evolved from conifer-like ancestors.

i. An initial advantage was pollination by insects (less wasteful than the wind pollination of conifers).

ii. Other advantages: more rapid reproduction and growth.

iii. Today, flowering plants dominate except in cold regions where conifers still prevail.

C. Land animals.

1) Arthropods evolved shortly after the land plants which served as food for the animals.

a. Aquatic arthropods already had exoskeletons that would be useful on land to withstand the stress of gravity and were waterproof. Thus, arthropods were preadapted to life on land.

b. They developed tracheae within their moist bodies for breathing.

c. Arthropods ruled the land until about 400 MYA, when "lobefin" fish evolved from the sea.

2) Lobefins evolved into bony fish and amphibians.

a. Lobefins had 2 preadaptations to land:

i. Their stout fleshy fins for locomotion.

ii. A primitive lung for breathing, formed from the digestive tract.

b. Lobefins evolved into 2 groups about 350 MYA.

i. Modern bony fish: The lung evolved into a swim bladder and they migrated back to the sea.

ii. Amphibians: The fins evolved into limbs for crawling.

c. The early amphibians had:

i. Plenty of food (arthropods and plants).

ii. No predators against them.

iii. A warm moist climate.

d. Amphibians never fully adapted to land since:

i. Their lungs remained as simple sacs, so that some O_2 had to be obtained through moist skin, restricting amphibians to wet environments.

ii. They shed eggs and sperm into water.

iii. So, when climates became dry, amphibians were in trouble and lost their dominance to the reptiles.

3) Reptiles evolved from amphibians and displaced them as climates became drier and cooler in the Permian period (286-248 MYA).

a. Reptiles are adapted to dry conditions since:

i. They evolved internal fertilization.

ii. They developed waterproof eggs.

iii. They developed scaly, waterproof skin.

iv. They developed improved lungs.

b. When the climate became moist again, dinosaurs evolved from groups of small reptiles.

i. Dinosaurs remained dominant from 165 to 65 MYA.

        ii.   They became extinct for unknown reasons, perhaps due to climatic changes triggered by a gigantic meteorite impact that kicked up enough dust to block sunlight and cause less vegetation to grow, resulting in famine.

    c.   The smaller reptiles needed insulation to retain body heat.
        i.   One group developed feathers and evolved into birds.
        ii.   Another group developed hair and evolved into mammals.

4)   Birds.
    a.   Their initial advantage was that insulating feathers retained body heat, allowing them to be more active than reptiles at night and in cooler climates.
    b.   Later, flight evolved and allowed them to become better predators than the reptiles.

5)   Mammals.
    a.   Some stayed small and nocturnal, eating seeds and insects.
    b.   Some evolved larger sizes, colonizing habitats left empty when the dinosaurs became extinct.
    c.   Most evolved live birth and feed the young secretions from mammary glands.
    d.   One group remained in the trees, giving rise to the primates (including humans).

7.   Human evolution.

A.   General facts.
  1)   Primate (lemurs, monkeys, apes, humans) fossils are relatively rare since:
    a.   They had small populations.
    b.   They were small in size.
    c.   They lived in habitats unsuitable for producing fossils.
  2)   Paleontologists disagree about the interpretation of the scanty human fossil evidence.
  3)   Primates began as insectivorous tree shrews about 80 MYA.
    a.   Over 50 million years, they evolved into the modern tarsiers, lemurs, and monkeys, remaining in trees.
    b.   Adaptations for life in trees:
        i.   Grasping appendages ("hands").
        ii.   Large, forward-facing eyes, allowing for depth perception (3-dimensional or binocular vision).
        iii.   Color vision to see fruit.
    c.   These 3 traits were preadaptations allowing humans to easily evolve tool use.

B.   Fossil humans.
  1)   Between 20 and 30 MYA in tropical Africa, primates called Dryopithecines diverged from monkeys.
    a.   They are ancestral to:
        i.   The hominids (humans and their fossil relatives).
        ii.   The pongids (the great apes).
    b.   About 18 MYA, diversification of habitats and isolation of small

populations led to diversification of the Dryopithecines, including <u>Proconsul africanus</u>, progenitor to humans and pongids.

2) <u>Australopithecines</u> ("southern apes"), the first true hominids, appeared about 4 MYA.
   a.   They lived in the African grasslands.
   b.   They could walk upright.
   c.   Their brains were fairly large, but smaller than modern human brains.
   d.   They split into at least 2 distinct groups:
       i.   <u>Australopithecus</u> <u>afarensis</u> and <u>A</u>. <u>africanus</u>, small omnivorous forms.
       ii.   <u>A</u>. <u>robustus</u> and <u>A</u>. <u>boisei</u>, large herbivorous forms.

3) <u>Homo</u> <u>habilis</u> arose about 2 MYA, probably from <u>A</u>. <u>afarensis</u>. Traits of <u>H</u>. <u>habilis</u>:
   a.   They had a larger brain than their ancestors.
   b.   They were the first hominids to use stone and bone tools and crude weapons.
   c.   They lived in Europe, Asia, and Africa.

4) <u>Homo</u> <u>erectus</u> arose about 1.8 MYA, probably from <u>H</u>. <u>habilis</u>. Traits of <u>H</u>. <u>erectus</u>:
   a.   They had brains as large as the smallest modern adult human brains ($1000\ cm^3$).
   b.   Their faces were slightly protruding, with large brow ridges and no chins.
   c.   They fashioned sophisticated stone tools.
   d.   They ate animals and used fire.
   e.   They spread throughout much of the Old World (e.g., Peking Man and Java Man).

5) <u>Homo</u> <u>sapiens</u>, modern humans, arose about 200,000 years ago.
   a.   The earliest fossils are fragmentary (not much is known).
   b.   We have complete fossils of the <u>Neanderthal</u> <u>Man</u> variety from 100,000 years ago.
   c.   Neanderthals:
       i.   Were heavily muscled.
       ii   Walked erect.
       iii.   Had brains slightly larger than ours.
       iv.   Had heavy brows in Europe but were more like us in the Near East.
       v.   Had ritualistic burial ceremonies that included religious activity.

6) About 90,000 years ago, fully modern humans called <u>Cro-Magnon</u> <u>Man</u> appeared.
   a.   These humans:
       i.   Had domed heads, smooth brows, and prominent chins.
       ii.   Made precision tools.
   b.   Cro-Magnon humans probably evolved from the modern-looking Neanderthals of the Near East and then spread into Europe, though they may have evolved independently from a population of <u>Homo erectus</u>.
       i.   They coexisted with Neanderthals for a time.
       ii.   They survived while the Neanderthals died out.
   c.   Cro-Magnon humans left significant art in caves in France and Spain, indicating their minds were as fully human as ours.

C.   Evolution of <u>human behavior</u>: mainly hypothetical.
   1)   <u>Brain development</u>.

a. Huge brains evolved about 2 MYA.
b. Advantages of a large brain? Perhaps it:
   i. Improved hand-eye coordination.
   ii. Facilitated complex social interactions, especially cooperative scavenging and hunting for large game.
   iii. Allowed for development of language.
c. Early hominids began to walk upright.
   i. Bipedal locomotion allowed them to carry things in their hands as they walked.
   ii. Hominid fossils have powerful shoulder joints and opposable thumbs.
   iii. Brain expansion was related to integration of visual input and control of hand/arm movements.

2) Cultural evolution: learned behaviors passed down from previous generations in a non-genetic way.
a. Our bodies have not changed evolutionarily for thousands of generations, but our behavior has changed greatly.
b. Most human behaviors are strongly influenced by learning.
   i. History influences entire societies.
   ii. We can trace our religious and cultural roots back to the ancient Greeks of 2500 years ago.
c. Most human behavior is the product both of genetic influences and extensive cultural learning.
   i. We learn at least part of nearly every behavior we perform.
   ii. We also have instinctive behaviors such as suckling by infants, smiling, and flashing of eyebrows in greeting.
   iii. Many forms of aberrant behavior are significantly influenced by innate biological functions (i.e., there are genes that predispose a person to develop manic depression or schizophrenia).
   iv. Few if any behaviors are completely controlled by heredity. Any human behavior can be enhanced, suppressed, or modified by culture.

****

REVIEW QUESTIONS:

1. Concerning spontaneous generation:
  A. Define it.
  B. Explain why scientists assert that it cannot happen today.
  C. Explain why scientists assert that it probably did happen billions of years ago.

2. Arrange the following events into the sequence that scientists assert occurred during the history of the earth.
  A. Evolution of terrestrial organisms.
  B. Oxygen gas begins to accumulate in the atmosphere.
  C. Spontaneous formation of simple organic molecules which, in the absence of O_2, accumulated in the seas.
  D. Evolution of anaerobic prokaryotic cells.

E. Evolution of mitochondria and chloroplasts (the Endosymbiotic theory).
F. Chance formation of ribozymes with the ability to make accurate and inaccurate copies of itself.
G. Evolution of aerobic prokaryotic cells.
H. Evolution of multicellular eukaryotic organisms.
I. Evolution of primitive photosynthetic anaerobic cells.
J. By chance, primitive microspheres surround the proper mix of organic molecules and form primitive living cells.

3. Comparing aquatic and terrestrial environments as habitats for plants:
   A. Name the advantages of an aquatic environment.
   B. Name the disadvantages of a terrestrial environment.
   C. Name the advantages of a terrestrial environment.

4. List 4 adaptations that primitive land plants evolved which helped them to cope successfully with life on land.

5.-12. **MATCHING TEST**: terrestrial plants

Choices:
A. Flowering plants
B. Ferns and Mosses
C. Conifers

5. Better adapted to colder climates.
6. Use insects to transport pollen.
7. Need water for sexual reproduction since sperm must swim to the eggs.
8. Primarily use wind to passively transport pollen.
9. Evolved from conifer-like ancestors.
10. The dominant plants today.
11. The dominant plants 250 million years ago.
12. The dominant plants 325 million years ago.

----

13. List 4 adaptations that reptiles evolved which help them cope successfully with a fully terrestrial lifestyle.

14.-24. **MATCHING TEST**: terrestrial animals

Choices:
A. Reptiles   E. Birds
B. Amphibians   F. Bony fish
C. Mammals   G. Jellyfish
D. Arthropods

14. Evolved from reptilian ancestors.
15. Evolved from lobefins whose fins evolved into limbs for crawling.
16. Dinosaurs.
17. Use their lungs and moist skin to exchange gases with the air.
18. First land animals to evolve waterproof eggs.

19. Humans.
20. First land animals.
21. Evolved from lobefins that developed swim bladders from their lungs.
22. Land animals that shed their eggs and sperm into the water.
23. Were preadapted to life on land due to their exoskeletons.
24. Directly evolved from amphibian-like ancestors.

----

25. List 3 adaptations that human ancestors had for living in trees.

----

26. Name the organisms in the human family tree diagrammed below and list several traits for each.

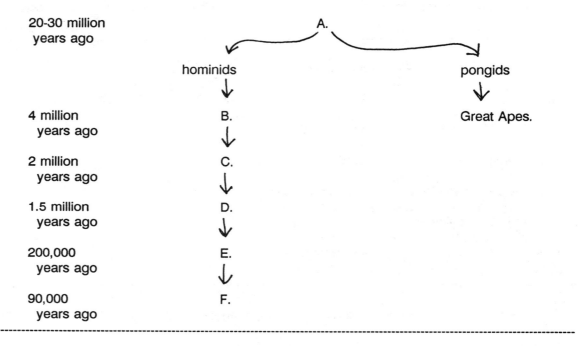

27. List 3 possible advantages favoring the evolution of large brains in our human ancestors.

****

****

ANSWERS TO REVIEW QUESTIONS:

1. A. see 2.A.
    B. see 2.A.2)
    C. see 2.A.3); 2.B.1)-6)

2. C. see 2.B.3)a.-c.
    F. see 2.B.4)a.-e.
    J. see 2.B.5)a.-b.
    D. see 3.A.1)a.-c.
    I. see 3.A.2)a.-b.
    B. see 3.A.2)c.-d.
    G. see 3.A.3)-4)
    E. see 3.B.2)a.-c.
    H. see 4.A.-B.
    A. see 6.A.-C.

3. A. see 6.A.1)a.-c.
    B. see 6.A.2)a.-c.
    C. see 6.A.2)d.-e.

4. see 6.B.2)a-d.

5.-12. [see 6.B.1)-3)]

5.	C	9.	A
6.	A	10.	A
7.	B	11.	C
8.	C	12.	B

13. see 6.C.3)a.i.-iv.

14.-24. [see 6.C.1)-5)]

14.	C and E	20	D
15.	B	21.	F
16.	B	22.	B
17.	B	23.	D
18.	A	24.	A
19.	C		

25. see 7.A.3)b.i.-iii.

26. [see 7.B.1)-6)]
    A. <u>Dryopithecine</u>
    B. <u>Australopithecines</u>
    C. <u>Homo</u> <u>habilis</u>
    D. <u>Homo</u> <u>erectus</u>
    E. <u>Homo</u> <u>sapiens</u> - Neanderthal Man
    F. <u>Homo</u> <u>sapiens</u> - Cro-Magnon Man

27. see 7.C.1)b.i.-iii.

****

CHAPTER 20: TAXONOMY: IMPOSING ORDER ON DIVERSITY

CHAPTER OUTLINE.

The basic principles used to classify organisms into discrete groups for the purpose of systematic study are presented in this chapter.

1. **General facts.**

   A. <u>Taxonomy</u> ("arrangement") is the science by which organisms are classified and placed into categories based on their structural similarities and evolutionary relationships.
      1) These categories form a hierarchy or series of levels each more inclusive than the last.
      2) The 7 major categories:
         a. Kingdom (the most inclusive: many types of organisms are included in a Kingdom.)
         b. Phylum
         c. Class
         d. Order
         e. Family
         f. Genus
         g. species (the least inclusive: only one type of organism is included in a species.)
      3) <u>Facts</u> about these categories:
         a. Each higher category contains several to many of the category just below it (one Kingdom contains several Phyla, one Phylum contains several Classes, etc.)
         b. A phrase to help you remember the first letter of each category in correct sequence: <u>K</u>ing <u>P</u>hilip <u>C</u>alls <u>O</u>ur <u>F</u>riends <u>G</u>ood <u>s</u>ports.
      4) Each category from species to Kingdom is increasingly more general and includes organisms whose common ancestors are increasingly remote (see Table

20-1).

B.   <u>Scientific</u> <u>names</u> are formed from the two smallest taxonomic categories, the Genus and species.

    1)   The first word of each scientific name is the organism's <u>Genus</u> that includes a number of very closely related organisms that do not interbreed.

    2)   The second name is the organism's <u>species</u>, a category limited to naturally interbreeding individuals.

    3)   Example:

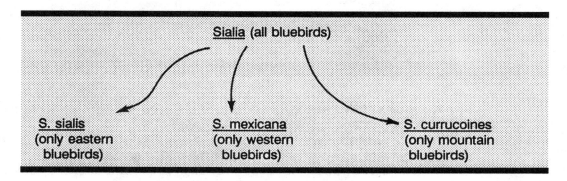

       a.   Interbreeding occurs within a species: eastern bluebirds can breed only with other eastern bluebirds.

       b.   No interbreeding occurs between species: eastern bluebirds cannot breed with western or mountain bluebirds.

    4)   Scientific names are always underlined or italicized and the Genus is always capitalized.

    5)   These names tend to transcend language barriers because most of them are in Latin.

2.   Ancient origins of taxonomy.

A.   <u>Aristotle</u> (384-322 B.C.)

    1)   Found a logical, standardized scheme for naming things based on:

       a.   Structural complexity.

       b.   Behavior.

       c.   Degree of development at birth.

    2)   He classified about 500 different organisms into 11 categories.

B.   <u>Linnaeus</u> (1707-1778) developed the classification system in use today. He:

    1)   Placed organisms into a series of hierarchically arranged categories based on resemblances to other life forms.

    2)   Introduced the <u>Genus-species</u> system for naming organisms scientifically.

C.   <u>Darwin</u> (1822-1884) showed that Linnaeus' categories reflected the evolutionary relatedness of organisms.

3.   Modern criteria for classification.

A.   Four major criteria are used to group organisms.
    1)   Adult <u>anatomical</u> similarities such as skeleton and tooth structures and presence of homologous structures.
    2)   Structures seen during <u>embryonic</u> development.
    3)   <u>Biochemical</u> similarities such as amino acid sequences of protein molecules and the degree of similarity of DNA base sequences, using techniques such as:
        a.   <u>Electrophoresis</u>: separation of molecules according to differences in their charge and molecular weight.
        b.   <u>DNA Hybridization</u>: measures degree of similarity between DNA molecules of different organisms.
        c.   <u>Polymerase Chain Reaction (PCR)</u>: amplifies small amounts of DNA, allowing for sequencing of DNA from museum specimens of extinct organisms.
    4)   Presence of <u>behaviors</u> that prevent matings between organisms of closely related species.

B.   The **Five Kingdoms** of life. As they learn more about living organisms, scientists revise earlier classification categories. Example:
    1)   Originally, all species were placed in one of only 2 Kingdoms: Plantae (including bacteria, fungi, and photosynthetic protists) and Animalia.
    2)   Then a 3 Kingdom model was used, including Plantae, Animalia, and Monera (the bacteria).
    3)   Currently, a 5 Kingdom model (first proposed by Whittaker in 1969) is used, including:
        a.   <u>Monera</u>: unicellular and prokaryotic (mainly bacteria).
        b.   <u>Protista</u>: unicellular and eukaryotic.
        c.   <u>Fungi</u>: eukaryotic, multicellular, non-photosynthetic, and absorbing already digested nutrients.
        d.   <u>Plantae</u> (the plants): eukaryotic, multicellular, and photosynthetic.
        e.   <u>Animalia</u> (the animals): eukaryotic, multicellular, non-photosynthetic, and ingesting food before digesting it internally.
    4)   Today, a sixth kingdom, <u>Archaebacteria</u>, has been proposed by some since these cells are neither strictly prokaryotic nor eukaryotic.

C.   Taxonomy is an inexact science.
    1)   Taxonomic categories are subject to change based on new information, such as breeding behavior or DNA base sequencing.
    2)   In organisms that reproduce asexually, the criterion of interbreeding cannot be used to distinguish species.
    3)   Most unicellular organisms reproduce mostly asexually, making their classification difficult.

D.   Taxonomists do not know how many living species exist.
    1)   At least 1.5 million species have been named: 5% monerans and protistans, 22% plants and fungi, 73% animals.
    2)   As many as 30 million living species may exist.
    3)   About 2/3 of the earth's species live in tropical rain forests, which cover 6% of the earth's surface.

****

REVIEW QUESTIONS:

1. List in order, starting with the most inclusive, the 7 major categories used in classifying organisms.
2. The scientific name for the dog is <u>Canis familiaris</u>. Explain why 2 words are used in the name, and relate it to the names for wolves (<u>Canis lupus</u>) and coyotes (<u>Canis latrans</u>).
3. In what way did Darwin reinterpret the significance of Linnaeus' system of classification?
4. List the 4 major criteria used to group organisms.

5.-10. **MATCHING TEST:** the 5 Kingdoms

Choices:
A. Kingdom Plantae
B. Kingdom Fungi
C. Kingdom Monera
D. Kingdom Animalia
E. Kingdom Protista

5. Unicellular, eukaryotic organisms.
6. Multicellular, non-photosynthetic organisms with extracellular digestion.
7. Prokaryotic organisms.
8. Organisms are unicellular and lack nuclei.
9. All organisms are photosynthetic.
10. Bacteria.
11. Multicellular, non-photosynthetic organisms with digestion within the body.
12. Organisms are unicellular and have nuclei.

****

ANSWERS TO REVIEW QUESTIONS:

1. see 1.A.2)-4)

2. see 1.B.1)-3)

3. see 2.B.-C.

4. see 3.A.1)-4)

5.-12. [see 3.B.3)a.-e.]

5.	E	9.	A
6.	B	10.	C
7.	C	11.	D
8.	C	12.	E

****

CHAPTER 21: THE DIVERSITY OF LIFE: MICROORGANISMS

CHAPTER OUTLINE.

This chapter describes the three major groups of microorganisms: viruses, prokaryotic unicellular microbes (Kingdom Monera), and eukaryotic unicellular microbes (Kingdom Protista).

1.   **General facts** about microorganisms:

    A.   Some are agents of disease.
    B.   Protists are responsible for most of the photosynthetic activity on earth, making food and O_2.
    C.   Some bacteria can convert atmospheric nitrogen into a nutrient used by plants.
    D.   Microorganisms help ruminant animals like cows and sheep to break down cellulose in their diet.
    E.   Bacteria decompose the dead bodies of plants and animals.

2.   **Viruses**.

    A.   <u>General facts</u>. Viruses:
        1)   Fit into no Kingdom comfortably.
        2)   Are not really cellular, lacking membranes, ribosomes, and cytoplasm.
        3)   Cannot move or grow, and reproduce only inside a host cell.

    B.   Viral <u>structure and reproduction</u>.
        1)   Viruses are very small: 0.05-0.2 micrometers (μM) in diameter.
        2)   Viruses have 2 major structural components:
            a.   A protein coat allowing viruses to penetrate and enter host cells.
            b.   A molecule of nucleic acid (DNA or RNA).
        3)   Upon entering a cell, viral genetic material takes over.

a. The host cell is forced to read the viral genes and make components which rapidly assemble into new viral particles.

b. Soon, many new viruses burst out of the host cell and each invades another cell to repeat the process.

C. Viral infections.

1) Each type of virus specializes on a specific host cell and no organism is immune to all viruses.

2) Within a particular organism, viruses specialize on specific cell types. Examples:

a. Common cold virus attacks respiratory tract cells.

b. Measles virus attacks skin cells.

c. Herpes viruses come in 2 types:
   i. One type causes cold sores in the mouth and lips.
   ii. Another type causes sores on or near the genitals.

d. AIDS (acquired immune deficiency syndrome) virus attacks white blood cells that control the immune system.

e. Papilloma viruses cause genital warts and are found in 90% of all cervical cancers.

f. T-cell leukemia viruses attack white blood cells.

3) Viral illnesses are difficult to treat since anti-viral agents often destroy host cells as well. Also, antibiotics don't work against viruses.

D. Viroids and Prions.

1) Viroids:

a. Activity: cause some plant diseases in cucumbers, avocados, and potatoes.

b. Size: about a tenth the size of a typical plant virus.

c. Structure: short intervals of RNA without a protein coat.

d. They enter the host cell nucleus and somehow direct the synthesis of new viroids.

2) Prions (proteinaceous infectious particles):

a. Cause kuru, a human nervous system disease, when victims eat the brains of relatives who died of the disease.

b. Known symptoms resemble those of scrapies, a disease of sheep.

c. Structure: have proteins but no genetic material (method of reproduction is unknown).

d. Possibly related to Alzheimer's disease.

E. The origin of viruses is obscure.

3. **Prokaryotic** and **Eukaryotic** cells.

A. Members of the Kingdom Monera (bacteria) are:

1) Unicellular, with some photosynthetic forms.

2) Prokaryotic, lacking organelles such as nuclei, chloroplasts, and mitochondria.

3) Small in size (0.2-10 μM in diameter).

B. Members of the Kingdom Protista (protists) are:

       1)   Unicellular but complex, with some photosynthetic forms.
       2)   Eukaryotic.
       3)   Large in size (10-100 μM in diameter).

4.   **Kingdom Monera**: over 1700 species.

    A.   Two major Divisions:
       1)   Eubacteria ("true bacteria"): includes the cyanobacteria.
       2)   Archaebacteria ("ancient bacteria").
       3)   Taxonomists classify bacteria according to shape, locomotion, pigments, staining properties, and nutrient requirements.

    B.   Bacterial structure.
       1)   Are the most abundant of living things.
       2)   Almost all have rigid cell walls to protect against osmotic rupture.
          a.   Cell walls give bacteria their characteristic shapes:
             i.   Bacilli - rod shaped.
             ii.   Cocci - spherical shaped.
             iii.   Spirillum - spiral or corkscrew shaped.
          b.   Cell walls contain peptidoglycan (chains of sugars cross-linked by short peptides).
       3)   Some bacteria surround their cell walls with sticky capsules or slime layers (of protein or polysaccharide) which:
          a.   Help bacteria escape detection by the immune systems of their victims.
          b.   Help tooth decay bacteria adhere to teeth (dental plaque).
       4)   Some bacteria are covered with many hair-like protein projections (pili) that attach bacteria to other cells.
       5)   Some have flagella that can rotate rapidly for movement. Some flagellated magnetotactic bacteria have cytoplasmic magnets made of iron crystals that allow the cells to orient to the earth's magnetic field.
       6)   Under harsh conditions, bacteria form spores (protective, resting structures) containing a chromosome within a thick protective coating:
          a.   Metabolic activity ceases within a spore.
          b.   Spores can survive extremely unfavorable conditions (i.e., boiling and desiccation) for long periods of time.
          c.   Spores may be blown to new favorable locations where they germinate.

    C.   Bacterial reproduction.
       1)   **Asexual reproduction**: by simple cell division ("fission") producing genetically identical offspring as quickly as every 20 minutes.
       2)   Such rapid division enhances the effects of mutation to spread rapidly (e.g., drug-resistance mutations), and exploit temporary habitats.
       3)   **Sexual reproduction**: by bacterial conjugation, the transfer (through hollow sex pili) of genetic material from a donor bacterium to a recipient.

    D.   Bacterial habitats: extremely diverse and highly specialized.
       1)   From mountain tops to ocean depths.
       2)   From hot springs and ocean vents to the highly salty Dead Sea to particular parts

of the human body.

E. Bacterial <u>nutrition</u> and <u>community interactions</u>.
   1) Bacteria show dietary versatility:
      a. Cyanobacteria are <u>photosynthetic</u>.
      b. Some bacteria are <u>chemosynthetic</u>, combining O_2 with sulfur, ammonia, or nitrogen to gain energy. These release crucial plant nutrients (sulfates and nitrates) into the soil.
   2) Some are <u>anaerobes</u>, not using O_2 to get energy.
      a. Botulism and tetanus bacteria are killed by O_2.
      b. Some others are aerobic, using O_2 when it is available and anaerobic when O_2 is absent.
   3) Some are <u>symbiotic</u> ("living together") with other organisms.
      a. Bacteria that break down cellulose from plant cell walls live in the intestines of cows, sheep, and goats.
      b. Bacteria normally live in human intestines, where they convert undigested food into vitamins K and B_{12} which we use.
      c. <u>Nitrogen-fixing</u> bacteria live in specialized nodules on the roots of <u>legume</u> plants (beans, clover, alfalfa) and convert N_2 gas from the atmosphere into ammonium usable by the plant.
      d. Bacteria account for the "biodegradable" breakdown of many compounds.
      e. Bacteria are used to make some human foods such as cheese, yogurt, and sauerkraut.
      f. Bacteria are important decomposers in ecosystems, allowing important nutrients like oil to be recycled.

F. <u>Bacteria in human health</u>.
   1) Some bacteria, called <u>pathogens</u> ("disease producers"), make toxins that cause disease. Examples:
      a. Tetanus.
      b. Botulism.
      c. Bacterial pneumonia of the lungs.
      d. Bubonic plague, caused by bacteria carried by fleas on rats.
      e. Tuberculosis.
      f. Leprosy.
      g. Tetanus and botulism bacteria are anaerobes that attack the brain.
      h. The sexually transmitted diseases gonorrhea and syphilis are bacterial in origin and transmitted by direct sexual contact.
      i. Lyme disease.
   2) Many bacteria are beneficial to humans. Examples:
      a. Intestinal bacteria that make vitamin K and some B vitamins.
      b. Vaginal bacteria that retard yeast infections.

G. **Cyanobacteria** (blue-green bacteria).
   1) Found in all moist environments where light and O_2 and available.
   2) Are aerobic and photosynthetic having chlorophyll pigment but lacking chloroplast organelles.
   3) Some form chains of cells unique among prokaryotes due to their "division of labor":

a.   A few cells in the chain capture atmospheric nitrogen.

b.   The rest photosynthesize.

4)   After volcanic eruptions, they are the first to colonize bare rock.

H.   **Archaebacteria** (possibly a new kingdom of organisms).

    1)   They differ from other bacteria in cell wall and cell membrane lipid composition as well as ribosomal RNA subunit sequences.

    2)   Some are <u>methanogens</u>, anaerobic bacteria that convert CO_2 into methane gas ("swamp gas"); found in swamps, sewage-treatment plants, and the stomachs of cows.

    3)   Some are <u>halophiles</u>, thriving in concentrated salt solutions like the Dead Sea.

    4)   Some are <u>thermoacidophiles</u>, thriving in hot acidic environments such as hot sulfur springs.

5.   **Kingdom Protista.**

A.   <u>General facts</u>.

    1)   Over 50,000 species exist.

    2)   Each protist consists of a single eukaryotic cell.

    3)   Most reproduce asexually by <u>mitosis</u> and some can also reproduce sexually by <u>conjugation</u>.

    4)   All 3 major modes of nutrition are represented.

        a.   <u>Photosynthesis</u> by unicellular algae.

        b.   <u>Predation</u> and <u>parasitism</u> by some flagellates.

        c.   <u>Absorption</u> by euglenoids and slime molds.

    5)   Both plant-like and animal-like forms exist and some, like <u>Euglena</u>, fit equally well into either category.

B.   Plant-like protists: **Unicellular Algae** (3 major divisions).

    1)   Often called <u>phytoplankton</u> ("floating plants") since they are found in oceans and lakes. Marine forms account for about 70% of all photosynthesis on earth.

    2)   <u>Dinoflagellates</u> (Division Pyrrophyta).

        a.   Have 2 whip-like flagella that often project through cellulose walls that resemble armor plates.

        b.   Are especially abundant in the oceans and serve as food for larger organisms.

        c.   Many are bioluminescent, glowing blue-green when disturbed.

        d.   Have green chlorophyll and much red pigment to trap light energy.

        e.   Under certain conditions, "red tides" occur due to dinoflagellate population explosions.

            i.   Many fish die due to lack of oxygen.

            ii.   Oysters, mussels, and clams feast, but concentrate a nerve poison produced by the dinoflagellates, causing "shellfish poisoning" in humans eating the tainted molluscs.

    3)   <u>Diatoms</u> (Division Chrysophyta).

        a.   Called "pastures of the sea" since they are an important food source in marine food webs.

        b.   Produce delicate and beautiful glassy protective shells consisting of top and bottom halves that fit together like shoe boxes.

        c.   Accumulation of glassy shells produce "diatomaceous earth" many meters thick.

        d.   Store reserve food as oil which helps diatoms float just below the surface.

4)   <u>Euglenoids</u> (Division Euglenophyta).

        a.   Each has a flagellum for locomotion.

        b.   Have simple light-sensitive organelles (a photoreceptor [eyespot] and a patch of pigment that shades the receptor when light impinges from certain directions).

        c.   Moves towards light levels appropriate for photosynthesis.

        d.   All live in fresh water and lack cellulose cell walls.

        e.   If maintained in darkness, euglenoids lose their chloroplasts and absorb nutrients from their surroundings, resembling animal-like zooflagellates (see below).

C   Fungus-like protists: **Slime Molds**.

    1)   Life cycle has two phases:

        a.   mobile feeding stage.

        b.   stationary reproductive stage (<u>fruiting body</u>).

    2)   Two major Divisions:

        a.   Acellular, plasmodial slime molds (Division Myxomycota).

        b.   Cellular slime molds (Division Acrasiomycota).

    3)   <u>Acellular slime molds</u>:

        a.   A large mass of cytoplasm ("plasmodium") containing thousands of nuclei in common cytoplasm.

        b.   Plasmodia ooze along the forest floor, engulfing food (bacteria and particles of organic material).

        c.   May be brightly colored.

        d.   When dry or starved, plasmodial slime molds reproduce by spores formed in "fruiting bodies" that develop from the mass.

    4)   <u>Cellular slime molds</u>:

        a.   Live in soil as independent amoeboid cells.

        b.   Lack cell walls and feed by ingesting food using pseudopods.

        c.   When food is scarce, the cells swarm together in dense aggregations, forming a slug-like mass called a <u>pseudoplasmodium</u> (false plasmodium).

        d.   After crawling towards light, the pseudoplasmodium develops a reproductive structure called a "fruiting body."

        e.   Fruiting bodies form haploid spores that are wind-dispersed and grow into new amoeboid cells.

D.   Animal-like protists: **Protozoa** ("first animals").

    1)   Three major Phyla:

        a.   Phylum Sarcomastigophora: zooflagellates and sarcodines.

        b.   Phylum Apicomplexa: sporozoans.

        c.   Phylum Ciliophora: ciliates.

    2)   Protozoans:

        a.   all are unicellular.

        b.   all are eukaryotic.

        c.   all are <u>heterotrophic</u>: they obtain energy from the bodies of other organisms.

        d.   differ in their means of locomotion.

3)   Zooflagellates.
    a.   All have at least one flagellum for locomotion and for sensing/grasping food.
    b.   Many are free-living in soil or water.
    c.   Some are symbiotic with other organisms, either benefiting or harming them.
        i.   One type digests cellulose in the gut of termites.
        ii. Another (Trypanosoma) causes African sleeping sickness in humans. This parasite lives parts of its life cycle in tse-tse flies and in humans.
        iii. Another parasite (Giardia) lives in pure mountain streams, released as cysts in the feces of infected animals. Though rarely fatal, Giardia causes diarrhea, nausea, and cramps in humans.

4)   Sarcodines (the amoebae).
    a.   Possess flexible cell membranes and can form pseudopodia ("false feet") used for locomotion and food capture.
    b.   Most are free-living but some are parasites. Example: one type causes amoebic dysentery.
    c.   Heliozoans, fresh water sarcodines, secrete silica shells and produce many needle-like pseudopods.
    d.   Some marine sarcodines produce elaborate and beautiful shells of glass (radiolarians) or calcium carbonate (foraminiferans). Deposits of calcium shells formed the "white cliffs of Dover" in England.

5)   Sporozoans.
    a.   All are parasites, forming infectious spores that are transmitted from victim to victim.
    b.   Adult sporozoans have no means of locomotion.
    c.   Many have complex life cycles. Example: the malarial parasite (Plasmodium) which lives parts of its life cycle in the stomach and salivary glands of mosquitoes, and in the liver and red blood cells of humans. Malaria is transmitted from one human to another through mosquito hosts biting one infected person and later biting an uninfected one.

6)   Ciliates (Paramecium and Didinium).
    a.   Fresh and salt water protozoans with cilia for locomotion.
    b.   Have the most complex unicellular structures of all protozoans.
    c.   Are accomplished predators, some having explosive darts (trichocysts) in their cell coverings.
        i.   Prey is directed by cilia to a mouth-like oral groove.
        ii. Prey is digested in a temporary stomach formed from a food vacuole.
        iii. Wastes are excreted by exocytosis.
    d.   Excess water is collected in contractile vacuoles which pump the fluid out through pores in the cell membrane.

****

REVIEW QUESTIONS:

1.   Name four reasons why microorganisms are important organisms.
2.   Name two Kingdoms of microorganisms and state in what ways they are similar to each other and in what ways they differ.
3.   Why are viruses not considered to be "cellular"?
4.   Describe the structure and reproduction of viruses.
5.   Name four human diseases caused by viruses.
6.   Name a structural difference between viroids and prions.
7.   List the three common shapes of bacteria.
8.   Briefly describe the following modes of bacterial nutrition:
        A.   Photosynthetic.
        B.   Chemosynthetic.
        C.   Anaerobic.
        D.   Symbiotic.
9.   Name five diseases of humans caused by bacteria.
10.   From the following list, pick only those terms that apply to the cyanobacteria.

A.	Dry habitats	G.	Symbiosis
B.	Moist habitats	H.	Have chloroplasts
C.	Aerobic	I.	Lack chloroplasts
D.	Anaerobic	J.	Have chlorophyll
E.	Photosynthetic	K.	Lack chlorophyll
F.	Chemosynthetic		

11.-17.   **MATCHING TEST**: on Arachaebacteria

                Choices: A.   Halophiles
                                 B.   Thermoacidophiles
                                 C.   Methanogens

11.   Found in hot sulfur springs.
12.   Produce "swamp gas".
13.   Thrive in bodies of concentrated salt water.
14.   Thrive in very hot environments.
15.   Found in the stomachs of cows.
16.   Found in very acidic environments.
17.   Found in the Dead Sea.

----

18.   Name and describe the three modes of nutrition seen among protist individuals.
19.   What is "phytoplankton" and why is it important to life on earth?

20.-31.   **MATCHING TEST**: on unicellular algae

                Choices: A.   Euglenoids
                                    B.   Dinoflagellates
                                    C.   Diatoms
                                    D.   All these algae

20.   Will absorb nutrients if maintained in darkness.
21.   Produce glassy shells that fit together like shoe boxes.
22.   Have pairs of flagella.
23.   Lack cellulose cell walls.
24.   Some are bioluminescent.
25.   Have single flagella for locomotion.
26.   All live in fresh water.
27.   Cause "red tides" that kill fish.
28.   Have simple light-sensitive organelles.
29.   Store reserve food as oils that help them stay buoyant.
30.   Produce a toxic nerve poison.
31.   Make food by photosynthesis.

----

32.   Name two phases of the slime mold life cycle.
33.   Name a major structural difference between the two divisions of slime molds.
34.   Name three characteristics possessed by all protozoans.

35.-48.   **MATCHING TEST**: on protozoans

                Choices: A.   Sarcodines
                                    B.   Zooflagellates
                                    C.   Ciliates
                                    D.   Sporozoans
                                    E.   All these protozoans

35.   The "white cliffs of Dover" are made from their calcium shells.
36.   All are parasites.
37.   Move by using flagella.
38.   Use cilia for locomotion.
39.   Some are symbiotic.
40.   Use pseudopods for locomotion.
41.   Have no means of locomotion.
42.   Have trichocysts which aid in predation.
43.   One type digests cellulose in the guts of termites.
44.   One type causes malaria.
45.   One type causes amoebic dysentery.
46.   Some have contractile vacuoles to excrete excess water.
47.   One type causes African sleeping sickness.
48.   Some types form glassy shells.

****

ANSWERS TO REVIEW QUESTIONS:

1.	see 1.A.-E.	18.	see 5.A.4)a.-c.
2.	see 3.A. and B.	19.	see 5.B.1)

3.   see 2.A.1)-3)

4.   see 2.B.1)-3)

5.   see 2.C.2)a.-f.

6.   see 2.D.1) and 2)

7.   see 4.B.2)a.i.-iii.

8.   see 4.E.1)-3)

9.   see 4.F.1)a.-i.

10.   [see 4.G.1)-4)]
       B, C, E, I, J

20.-31. [see 5.A.-B.1)-4)]

20.	A	26.	A
21.	C	27.	B
22.	B	28.	A
23.	A	29.	C
24.	B	30.	B
25.	A	31.	D

32.   see 5.C.1)a.-b.

33.   see 5.C.3) and 4)

34.   see 5.D.2)a.-d.

11.-17. [see 4.H.1)-4)]

11.	B	15.	C
12.	C	16.	B
13.	A	17.	A
14.	B		

35.-48. [see 5.D.3)-6)]

35.	A	42.	C
36.	D	43.	B
37.	B	44.	D
38.	C	45.	A
39.	B	46.	C
40.	A	47.	B
41.	D	48.	A

****

CHAPTER 22: THE DIVERSITY OF LIFE: II. ANIMALS

CHAPTER OUTLINE.

This chapter covers the major Phyla of animals including the invertebrates (sponges, cniderians, flat- round- and segmented worms, arthropods, molluscs, and echinoderms) and vertebrates (chordates including fish, amphibians, reptiles, birds, and mammals). Table 22-1 is an excellent summary of the major characteristics of each Phylum of animals discussed in this chapter. Learn as much of it as you can.

1. **General facts.**

    A.   <u>Characteristics</u> of the Kingdom Animalia. All animals:
        1)   are <u>multicellular</u>.
        2)   have cells <u>lacking cell walls</u>.
        3)   can <u>reproduce sexually</u>.
        4)   are <u>heterotrophic</u> (obtain energy by eating other organisms).
        5)   are <u>motile</u> during some stage in life.
        6)   make <u>rapid responses</u> to external stimuli.

    B.   See Figure 22.1 for a **general classification scheme** for animals.

    C.   <u>Trends in Cellular Organization</u>.
        1)   Evolution of complexity: Cells → tissues → organs → organ-systems
        2)   Sponges: specialized cell types but no tissues.
        3)   Cnidarians: tissues but no organs, and two tissue ("germ") layers called <u>ectoderm</u> (outer layer) and <u>endoderm</u> (inner layer).
        4)   Flatworms and more complex animals:
            a.   have organs and organ-systems.
            b.   have three tissue layers:
                i.   inner layer of <u>endoderm</u>, forming most internal organs.

   ii. middle layer of <u>mesoderm</u>, forming muscles, circulatory and skeletal systems (if present).

   iii. outer layer of <u>endoderm</u>, forming epithelium and nervous systems.

D. <u>Trends in Symmetry</u>.
  1) <u>Free-form</u>: irregular, variable shape, found in many sponges.
  2) <u>Radial symmetry</u>:
   a. Any line through a central axis divides the animal into roughly equal halves.
   b. Earliest true symmetry in animals.
   c. Found in some sponges, the cnidarians, and some adult echinoderms.
  3) <u>Bilateral symmetry</u>.
   a. An animal with bilateral symmetry:
    i. Has roughly mirror-image right and left halves.
    ii. Has upper (dorsal) and lower (ventral) surfaces.
    iii. Has anterior (head) and posterior (tail) regions.
   b. Characteristic of flatworms and all more complex animals, including larval echinoderms.

E. <u>Trends in Cephalization</u>.
  1) Accompanies bilateral symmetry.
  2) Creates anterior head and a posterior tail ends.
  3) Sense organs, aggregations of neurons, and organs for ingesting food are concentrated in the head end.
  4) First seen in flatworms (though food ingestion occurs elsewhere).

F. <u>Trends in Body Cavities</u>.
  1) Simple animals (cnidarians and flatworms) lack any internal space between the body wall and the gut.
  2) Roundworms have a space (<u>pseudocoelom</u>) between the body wall and the gut.
  3) In earthworms and nearly all other complex animals, a true body cavity (<u>coelom</u>) is present, serving several functions:
   a. In earthworms, the fluid-filled coelom acts as a hydrostatic skeleton.
   b. It generally serves as a protective buffer between internal organs and the outer body wall.
   c. It allows internal organs to move independently of the body wall.

G. <u>Trends in Segmentation</u> (presence in the body of similar repeated units).
  1) Segmentation is first seen in annelid worms (earthworms and their relatives).
  2) Segmentation increases body size with a minimum of new genetic information.
  3) In more complex animals, repeated segments have become specialized to perform specific functions.
  4) Segmentation may be obvious only in certain body parts (e.g., the vertebrae).

H. <u>Trends in Digestive Systems</u>.
  1) Sponges lack a specialized digestive tract: digestion occurs completely within cells.
  2) Cnidarians and flatworms have a <u>gastrovascular cavity</u>: a digestive sac with a single opening serving as both mouth and anus.

       3)   Roundworms and all more complex animals have efficient, tubular, one-way digestive systems:

           a.   The <u>mouth</u>, at one end, ingests food.

           b.   Food is processed (physically broken down → enzymatically digested → absorbed) as it passes through a series of specialized regions.

           c.   Wastes are voided through a separate anus, usually near the posterior end of the animal.

I.   Major categories of animals.

     1)   <u>Vertebrates</u>: have vertebral columns (backbones).

         a.   Include fish, amphibians, reptiles, birds, and mammals.

         b.   Include about 1% of all animal species all within a portion of one Phylum (Chordata).

     2)   <u>Invertebrates</u>: lack backbones.

         a.   Include sponges, jellyfish, worms, molluscs, arthropods, and echinoderms.

         b.   Include over 99% of all animal species within 27 Phyla.

     3)   <u>Evolution</u> of animals:

         a.   Primitive invertebrates (sponges) probably evolved from colonial protozoans.

         b.   As animals evolved, the trend has been towards increasing complexity and adaptation to diverse habitats and lifestyles.

         c.   Within the vertebrates, the trend has been towards increasing the size and sophistication of the brain.

2.   **Phylum Porifera**: the sponges (5000 species).

     A.   <u>General facts</u>.

         1)   Sponges are the simplest multicellular animals.

         2)   Sponge cells are relatively independent, resembling colonial cells that live together for mutual benefit. [Wilson mashed a sponge through a piece of fine silk, dissociating it into single cells. After 3 weeks, the cells reaggregated spontaneously into a functional sponge.]

         3)   Sponges lack specialized tissues and a nervous system.

         4)   Sponge cells have intracellular digestion.

         5)   All are aquatic and most are marine.

         6)   Adults are sessile (don't move).

     B.   <u>General body plan</u>.

         1)   <u>Flow of water</u>.

            a.   The body walls are perforated by many pores through which water enters.

            b.   Within the sponge body, water travels through canals and cells extract O_2 and food (microorganisms) and dump wastes and CO_2.

            c.   Water leaves through a few larger openings called <u>oscula</u>.

         2)   Sponges have 3 major cell types:

            a.   Flat <u>epithelial</u> <u>cells</u> cover the inner and outer surfaces. Some surround pores and control their size to regulate flow of water.

           b.   <u>Collar</u> <u>cells</u> have beating flagella that maintain water flow. The collars

filter out microorganisms that are ingested by the cells for food.
- c. Amoeboid cells roam freely between epithelial and collar cells. Functions:
  - i. Digesting and distributing nutrients.
  - ii. Making sex cells.
  - iii. Secreting internal skeletons of spicules that may be chalky, glassy, or proteinaceous in composition.

C. Reproduction: Asexual (budding) or sexual (sperm and eggs).

3. **Phylum Cnidaria**: the hydra, anemones, and jellyfish (9000 species).

A. General facts.
  1) Their cells are organized into distinct tissues (muscles and a nerve net).
  2) They lack organs and have no brains.
  3) All are predatory, using tentacles to capture prey that happen to swim by.
  4) All are aquatic and most are marine.

B. Body plan.
  1) Two basic body plans exist in cnidarians:
     - a. Polyp: adapted to a life attached to rocks, with tentacles that trap prey swimming by.
     - b. Medusa (jellyfish): swim weakly, carried by ocean currents and trailing tentacles used like fishing lines.
  2) Both forms show radial symmetry.
     - a. Body parts are arranged in a circle around a central axis (like a wagon wheel) through the digestive cavity and mouth.
     - b. This allows cnidarians to capture prey or defend themselves from any direction.
  3) Cnidarian tentacles are armed with nematocysts (cells containing poisonous or stinging darts that are injected explosively into prey).
     - a. The tentacles grasp and force prey through the mouth into the digestive sac (gastrovascular cavity).
     - b. Gland cells secrete digestive enzymes into the sac to partially digest the food.
     - c. Cells then absorb the food and finish digestion intracellularly.
     - d. Wastes are expelled through the common mouth/ anus opening. This is inefficient since it prevents continuous feeding.
  4) Coral polyps secrete a hard protective skeleton of limestone that remains after the corals die. These skeletons accumulate, producing coral reefs in warm, clear, tropical waters.

C. Reproduction.
  1) Asexual: some polyps (i.e., hydras and sea anemones) and some medusae bud off miniature replicas of themselves.
  2) Sexual: fusion of eggs and sperm produce free-swimming ciliated larvae that eventually settle down and become tiny polyps.

4.   **Phylum Platyhelminthes**: the flatworms.

    A.   <u>General anatomy</u>.
        1)   Have a gastrovascular cavity and a larval stage similar to cnidarians, but are generally more complex.
        2)   Have <u>bilateral symmetry</u>, an adaptation to active movement.
        3)   Free-living flatworms have light sensitive eyespots and cells responsive to chemical and tactile stimulation in the head region.
            a.   In the head, aggregations of nerve cells form <u>ganglia</u> that act as a simple brain.
            b.   Pairs of nerve cords conduct nervous signals to and from the head region.
        4)   Flatworms are the simplest organisms with true <u>organs</u> (tissues grouped into functional units). Examples:
            a.   A <u>pharynx</u> sucks up food which is then digested in a highly branched intestine that distributes it to all parts of the body. The pharynx also serves as an anus for eliminating solid wastes from the intestines.
            b.   A primitive excretory system consists of tubes with <u>flame cells</u> possessing beating cilia that drive liquids through the system, excreting excess fluids through tiny pores in the skin.
        5)   Flatworms lack both respiratory and circulatory systems. The flat body allows all cells to exchange gases by diffusion through the moist skin.

    B.   <u>Life styles</u>.
        1)   Many flatworms are <u>free-living</u> (the <u>planarians</u>).
        2)   Some are <u>parasites</u> of humans.
            a.   Example: <u>tapeworms</u>.
                i.   Infection occurs by eating improperly cooked pork, beef, or fish infected by the worms whose larvae form <u>cysts</u> in the meat muscle.
                ii.   Cysts hatch in the human digestive tract and attach to the intestines where they grow and mature.
                iii.   They grow up to 7 meters long absorbing food from the intestines and releasing egg packets that are shed in the host's feces.
                iv.   If an animal eats grass contaminated with the feces, the eggs hatch in the digestive tract, releasing larvae that burrow into muscles and form cysts.
            b.   Example: the blood fluke <u>Schistosoma</u> that causes schistosomiasis common in Africa and parts of South America.
                i.   Symptoms include dysentery, anemia, and brain damage.
                ii.   Have complex life cycles including humans and snails as hosts.

    C.   <u>Reproduction</u>.
        1)   <u>Asexually</u> by splitting in half, each half then regenerating its missing parts.
        2)   <u>Sexually</u> through self-fertilization since many are <u>hermaphrodites</u> (possessing both male and female sexual organs).

5. **Phylum Nematoda**: the roundworms (10,000 named species).

    A. <u>General anatomy</u>.
        1) Nematodes have <u>tubular</u> <u>bodies</u> with a gut running from mouth to anus.
        2) Nematodes have a tough dead <u>cuticle</u> protecting the delicate skin.
        3) Nematodes have brain ganglia like flatworms.
        4) Reproduction is always sexual and the sexes are separate with the smaller males fertilizing the larger females internally.
        5) Most species are microscopic in size.
        6) Nematodes lack circulatory and respiratory systems; due to their this shape, gases diffuse directly through the skin.

    B. <u>Life styles</u>.
        1) <u>Free-living</u> forms are important decomposers and are harmless to other organisms.
        2) Some are important <u>parasites</u> of humans.
            a. Example: <u>hookworm</u>.
                i.   Larvae live in the soil.
                ii. They bore into human feet (walking barefoot in the summer is a danger), enter the bloodstream, and burrow into the intestines where they cause continuous bleeding.
            b. Example: <u>trichinella</u> worms causing trichinosis.
                i.   Ingested by eating improperly cooked pork infected with larval cysts (up to 15,000 per gram of meat).
                ii. Cysts hatch in the human digestive tract and invade blood vessels and muscles, causing damage and pain.
        3) <u>Heartworms</u> in dogs is caused by a nematode parasite.

6. **Phylum Annelida** ("little rings"): the segmented worms (9000 species).

    A. <u>General anatomy</u>.
        1) The body is divided into a series of <u>repeating</u> <u>segments</u> containing identical copies of various structures. Segmentation allows some regions to specialize for specific functions and aids in locomotion.
        2) Annelids have a fluid-filled space, the <u>coelom</u>, that separates the body wall from the digestive tract. The coelom allows for efficient muscular movement since the muscles push against the fluid-filled coelom which acts as a <u>hydrostatic</u> <u>skeleton</u>.
        3) Annelids have a well-developed circulatory system to distribute gases and nutrients throughout the body. Five pairs of "hearts" pump blood throughout the body.
        4) Annelids have pairs of <u>nephridia</u> (excretory organs) in many of their segments.
        5) Annelids have simple ganglionic brains, and series of repeating ganglia connected by nerve cords travel the length of their bodies.
        6) Digestion occurs in a series of specialized compartments. In the earthworm, food passes through the following parts of the system:
            a. Muscular pharynx (mouth).
            b. Esophagus.
            c. Crop (storage organ).

      d.   Muscular gizzard (grinds up the food).

      e.   Intestines (food is digested and absorbed).

      f.   Anus (solid wastes are excreted).

B. The three Classes of annelids:

    1) <u>Class</u> <u>Oligochaeta</u> (the earthworms): gas exchange occurs through the moist skin.

    2) <u>Class</u> <u>Polychaeta</u> (the tube worms): mostly marine forms that use gills to exchange gases.

    3) <u>Class</u> <u>Hirudinea</u> (the leeches): parasitic or carnivorous forms that suck the blood of vertebrate hosts or eat smaller invertebrates.

7. **Phylum Arthropoda** ("jointed foot"): includes the insects, arachnids, and crustaceans.

A. In terms of numbers of individuals and species, arthropods are the dominant animals on earth (over 1,000,000 species have been discovered).

B. <u>General anatomy</u>.

    1) Arthropods have <u>exoskeletons</u>.

        a.   External skeletons with thin, flexible regions to allow movement of the paired, jointed appendages.

        b.   Secreted by the epidermis (skin).

        c.   Composed chiefly of protein and chitin (a polysaccharide).

        d.   Provides defense against small predators.

        e.   Allows for increased agility and locomotion.

        f.   Is waterproof, allowing arthropods to invade and inhabit dry terrestrial habitats.

        g.   Must be molted (shed) periodically as the arthropod grows, leaving the animal vulnerable during molting (i.e., "soft-shell" crabs).

        h.   Is relatively heavy, so large terrestrial arthropods can't evolve.

    2) <u>Segmentation</u> (evidence of ancestry from annelids). The segments are highly specialized for locomotion, feeding, flight, and sensing the environment.

    3) Arthropods have <u>well-developed sensory systems</u> including complex compound eyes and nervous systems that allow for complex behavior.

    4) Efficient <u>gas exchange</u> occurs due to:

        a.   <u>Gills</u> in aquatic forms (crustaceans).

        b.   <u>Book</u> <u>lungs</u> in arachnids (spiders).

        c.   <u>Tracheae</u> in insects.

    5) Arthropods have a well-developed but <u>open circulatory system</u>.

        a.   Blood travels out of the heart through arteries.

        b.   The arteries empty into a blood-cavity, the <u>hemocoel</u>, where internal organs are bathed in blood.

        c.   Blood then flows into veins and ultimately back to the heart.

    6) Arthropod sexes are separate and fertilization is internal.

C. <u>Class Insecta</u>: the largest (800,000 species) and most diverse class of arthropods.

    1) Insects have 3 pairs of legs and usually 2 pairs of wings (the only invertebrates that can fly).

    2) During development, insects undergo <u>metamorphosis</u> (radical changes in body form as juveniles become adults).

    a.   The immature form is usually a worm-like <u>larva</u> (i.e., fly maggots or butterfly caterpillars).

    b.   Larva and adult often have very different diets to reduce competition between them (i.e., caterpillars eat leaves while adult butterflies drink nectar).

D.   <u>Class Arachnida</u>: 50,000 species of terrestrial spiders, mites, ticks, and scorpions.
1)   Arachnids have 4 pairs of walking legs.
2)   Most are carnivorous and predatory.
3)   Arachnids breathe by tracheae and/or book lungs.
4)   They have simple eyes able to detect movement.

E.   <u>Class Crustacea</u>: 30,000 species of aquatic crabs, crayfish, lobsters, shrimps, and barnacles.
1)   All have 2 pairs of antennae, but the other appendages vary in form and number, depending on the lifestyle of the species.
2)   Most have compound eyes.
3)   Nearly all respire using gills.

8.   **Phylum Mollusca** ("soft"): 100,000 species of snails, clams, octopuses, and squids, evolved from annelid ancestors.

A.   <u>General anatomy</u>.
1)   Have soft, moist, muscular segmented bodies without skeletons.
2)   Some protect their bodies with a calcium carbonate (limestone) shell.
3)   Have an open circulatory system (like arthropods).
4)   Their brains have more ganglia than annelids or arthropods.
5)   Reproduction is always sexual; both separate sexes and hermaphrodites are known.

B.   <u>Class Gastropoda</u> ("stomach foot"): 35,000 species of snails and slugs.
1)   Gastropods crawl using a muscular foot near the stomach.
2)   They feed with a <u>radula</u>, a type of spiny tongue that scrapes algae from rocks or grasps prey.
3)   Aquatic forms breathe using gills or by diffusion through moist skin.
4)   A few species live in moist terrestrial environments and breathe using simple lungs.

C.   <u>Class Pelecypoda</u> ("hatchet foot"): marine scallops, oysters, mussels, and clams.
1)   Possess 2 shells ("<u>bivalves</u>") connected by a flexible hinge and held shut by powerful muscles.
2)   Most are sessile and lack heads.
3)   Are filter-feeders, driving water over their gills for gas exchange and trapping food particles which are conveyed by beating cilia to the mouth.
4)   Clams use a muscular foot to burrow into sand and mud while scallops flap their shells together, squirting water to move by jet propulsion.

D.   <u>Class Cephalopoda</u> ("head foot"): the marine octopuses, squids, nautiluses, and

cuttlefish.
1) Are the largest, swiftest, and smartest of all invertebrates.
2) All are predatory carnivores.
3) The foot has evolved into tentacles with chemosensory abilities and suction discs to detect and grasp prey. Salivary venom poisons the prey and beaklike jaws rip them apart.
4) The cephalopod eye is more complex than our own.
5) Cephalopods move by jet propulsion and by using the tentacles as legs.
6) The brain is quite complex and is encased in a skull-like case of cartilage. They can learn and remember.

9. **Phylum Echinodermata** ("spiny" or "hedgehog" skin): the sea stars (starfish), sea urchins, sea cucumbers, sand dollars, and sea lilies.

    A. Have bilaterally symmetrical free-swimming larvae and radially symmetrical slow moving or sessile adults.
    B. Most lack heads.
    C. Most feed on algae, but sea stars are predatory.
    D. Have <u>water-vascular</u> <u>systems</u> that function in locomotion, respiration, and food capture. In this system:
        1) Water enters through a bony <u>sieve</u> <u>plate</u> opening.
        2) It is conducted through a bony <u>ring</u> <u>canal</u> from which branch a number of <u>radial</u> <u>canals</u>.
        3) The radial canals conduct water to the <u>tube</u> <u>feet</u>, each with a muscular squeeze bulb (<u>ampulla</u>).
        4) Contractions of the ampullae force water into the tube feet, causing them to extend.
        5) Suction cups at the end of each tube foot can then adhere to a substrate or food object.
    E. Have relatively simple nervous systems without brains.
    F. Lack circulatory systems. Instead, they circulate fluid in the coelom.
    G. Gas exchange occurs through the tube feet and tiny skin gills in some species.
    H. Eggs and sperm are shed into water and free-swimming larvae develop after fertilization. Sexes are usually separate.
    I. Sea stars can regenerate lost parts, as long as a piece of the central body is present.
    J. Possess an <u>endoskeleton</u> (internal skeleton) of calcium carbonate (limestone) plates just beneath the skin.

10. **Phylum Chordata**: the tunicates, lancelets, and vertebrates.

    A. <u>Common features</u> of all chordates (including humans during embryonic development):
        1) A <u>notochord</u>: stiff flexible rod providing attachment sites for muscles.
        2) A <u>dorsal</u> <u>hollow</u> <u>nerve</u> <u>cord</u> producing a brain at the anterior end.
        3) <u>Pharyngeal</u> <u>gill</u> <u>grooves</u> just behind the mouth.
        4) A <u>tail</u>: extension of the body past the anus.

    B. The <u>invertebrate chordates</u>: all are marine.
        1) Lack heads and backbones.
        2) Adult lancelets (<u>Amphioxus</u>) have all the typical chordate features.

3) Adult tunicates ("sea squirts") look like sessile filter-feeding vases.
    a. They move only by contracting their sac-like bodies to squirt a stream of water.
    b. However, their actively swimming tadpole-like larvae possess all the chordate features.

4) The evolutionary link between the invertebrate and vertebrate chordates remains a mystery.

C. <u>Subphylum Vertebrata</u>: the vertebrates.
    1) During development, vertebrate embryos replace the notochord with a <u>vertebral column</u> (backbone) of either cartilage or bone.
        a. The backbone provides:
            i. Support for the body.
            ii. Attachment sites for muscles.
            iii. Protection for the delicate nerve cord and brain.
        b. The backbone is capable of growth and self-repair.
        c. The backbone has allowed vertebrates to achieve great size and mobility due to its ability to support without adding great weight.
    2) The earliest vertebrates were jawless fish that arose in the seas.
    3) Classes of fishes:
        a. <u>Class Agnatha</u>: two groups of jawless fishes (hagfishes and lampreys).
            i. Have skeletons of cartilage.
            ii. Are eel-like in shape.
            iii. Have unpaired fins along the midline of the body.
            iv. Have smooth slimy skin without scales.
            v. Have circular gill slits.
            vi. <u>Hagfishes</u>:
                a) Pink or purplish and exclusively marine.
                b) Live in colonial mud burrows.
                c) Feed mainly on polychaete worms, but also attack dying fish.
                d) Produce slime as protection against predators.
            vii. <u>Lampreys</u>:
                a) Found in both salt and fresh water, although the salt water species return to fresh water to spawn.
                b) Include both parasitic and non-parasitic species.
                c) Have multiplied greatly in the Great Lakes.
            viii. About 425 million years ago, jawless fish gave rise to fish with jaws that evolved into the two major classes of modern fish.
        b. <u>Class Chondrichthyes</u>: 625 species of marine cartilage fish including the sharks, skates, and rays.
            i. Skeleton is made entirely of cartilage.
            ii. Skin is leathery with many tiny scales.
            iii. Respire using gills; lack swim bladders.
            iv. Have a 2-chambered heart and several rows of razor sharp teeth.
            v. Sharks are the largest fish.
        c. <u>Class Osteichthyes</u>: 17,000 species of marine and fresh water bony fish.
            i. Skeleton is bony.
            ii. Respire using gills although a few fresh water species have evolved lungs.

         iii.   Have swim bladders for flotation at any depth, probably evolved from the lungs of fresh water ancestors.

         iv.   Have 2-chambered hearts.

         v.   A few species have fleshy fins that can be used as legs if the pool dries up. The ancestors of these fish evolved into the amphibians and invaded land.

4)   Terrestrial vertebrate Classes:

    a.   Evolving from the sea.

        i.   Life on land offered the advantages of abundant food and shelter, and no predation.

        ii.   Successful colonization of land depended on development of several adaptations:

            a)   Supporting the body.

            b)   Waterproofing the skin and eggs.

            c)   Protecting respiratory membranes.

            d)   Controlling body temperature.

            e)   Efficient circulation.

    b.   Class Amphibia ("double life"): 2500 species of frogs, toads, and salamanders.

        i.   Amphibia are on the boundary between aquatic and terrestrial existence.

        ii.   Have legs for crawling and leaping.

        iii.   Lungs replace gills in adult forms.

        iv.   Have a 3-chambered heart.

        v.   Some gas exchange continues to occur through the skin which must remain moist.

        vi.   External fertilization occurs in water.

    c.   Class Reptilia: 7000 species of lizards, snakes, turtles, alligators, and crocodiles.

        i.   Evolved from amphibian ancestors about 250 million years ago (MYA); for about 150 million years, dinosaur lizards ruled the land.

        ii.   Became fully adapted to dry habitats due to:

            a)   Having dry, tough, scaly, waterproof skin.

            b)   Having internal fertilization.

            c)   Having waterproof shelled amniotic eggs with an inner membrane (the amnion) enclosing the embryo in a watery environment during development.

        iii.   Have very efficient lungs for gas exchange, but are coldblooded.

        iv.   Have a very efficient 3-chambered heart for better separation of oxygenated and deoxygenated blood.

        v.   Limbs give good support for efficient movement on land.

    d.   Class Aves: 8600 species of birds.

        i.   Evolved about 150 MYA from flying reptiles when scales became modified into feathers.

        ii.   Are warmblooded which allows the rapid metabolism necessary for flight.

        iii.   Have a 4-chambered heart that completely separates the flow of deoxygenated blood to the lungs from oxygenated blood to the body. Air sacs increase the efficiency of respiration.

iv. Feathers insulate and protect the body.

v. Hollow bones decrease body weight.

vi. Female birds have only one ovary (keeps body weight down).

vii. Lay shelled amniotic eggs.

viii. Have excellent coordination and acute eyesight due to complex nervous systems.

e. Class Mammalia: 4500 species of mammals.

i. Evolved from reptiles, modifying scales into hair; became dominant about 70 MYA.

ii. Are warmblooded with rapid metabolism.

iii. Fur protects and insulates the body.

iv. Like birds, have a 4-chambered heart.

v. Have evolved a remarkable diversity of body forms by adaptive radiation.

vi. Have mammary glands for suckling their young. Also have sweat, scent, and sebaceous (oil producing) glands not found in reptiles.

vii. Mammals are grouped according to how their young are born, although all are amniotes:

a) The monotremes (spiny anteater and duck-billed platypus) lay eggs.

b) The marsupials (like opossums, koalas, and kangaroos) produce young that develop briefly in the uterus, then crawl to a protective pouch where they firmly grasp a nipple and complete development nurtured by milk.

c. The placental mammals (like humans, whales, and elephants) retain their young in the uterus for extended development before birth.

viii. Have highly developed brains and nervous systems that allow behavioral adaptations to varied environments as well as unparalleled curiosity and learning behavior.

11. **Major evolutionary advances** seen in each group of animals:

A. Porifera: multicellular.

B. Cnidaria: two tissue layers
gastrovascular cavity.

C. Platyhelminthes: true organs
bilateral symmetry
three tissue layers.

D. Nematodes: separate mouth and anal openings to the digestive tract.

E. Annelids: segmentation
circulatory system
internal body cavity.

F.	Arthropods:	exoskeleton jointed appendages.	
G.	Mollusca:	enlarged brain.	
H.	Echinoderms:	internal skeleton.	
I.	Chordates:	internal cartilaginous or bony skeleton complex nervous system.	

****

REVIEW QUESTIONS:

1. List six characteristics of members of the animal Kingdom.
2. Which are more common, invertebrates or vertebrates? Name 5 groups of each.

3.-17. **MATCHING TEST:** sponges and jellyfish

Choices:
A.	Poriferans	C.	Both of these
B.	Cnidarians	D.	Neither of these

3. Coral reefs.
4. Have some extracellular digestion.
5. Resemble colonies of cells.
6. Have polyp and medusae stages.
7. Hydras.
8. Simplest multicellular animals.
9. Have separate mouth and anal openings.
10. Lack true tissues.
11. Have radial symmetry.
12. Predatory.
13. Have nematocysts.
14. Body wall has many pores.
15. Have collar cells.
16. Have bilateral symmetry.
17. Have internal skeletons of spicules.

---

18. Name the different ways a bilaterally symmetrical animal can be divided.
19. What are the advantages of having a tubular digestive system with separate mouth and anal openings?
20. What are the advantages of having a coelomic cavity?
21. Name, in sequence, the various parts of an earthworm's digestive system.
22. Name three advantages and one disadvantage of exoskeletons.

23.-37. **MATCHING TEST**: worms

Choices: A. Roundworms D. All worms
B. Flatworms E. No worms
C. Annelid worms

23. Have bilateral symmetry.
24. Planarians.
25. Have nerve ganglia.
26. Segmented.
27. Hookworms.
28. Lacks a separate anus.
29. Have coelomic cavities.
30. Have true organs.
31. Heartworms.
32. Have flame cells.
33. Have circulatory systems.
34. Trichinosis.
35. Have nephridia.
36. Tapeworms.
37. Leeches.

38.-53. **MATCHING TEST**: arthropods and molluscs

Choices: A. Arthropods C. Both of these
B. Molluscs D. Neither of these

38. Octopuses.
39. Have internal skeletons.
40. Largest Phylum of animals.
41. Have limestone shells.
42. Have open circulatory systems.
43. Have exoskeletons.
44. Clams.
45. Have tracheae.
46. Are segmented.
47. Evolved from annelid worms.
48. Show metamorphosis during development.
49. Snails.
50. Crabs.
51. Can learn and remember.
52. Have water-vascular systems.
53. Barnacles.

----

54. Describe the water-vascular system of echinoderms.
55. List the four features common to all chordates.
56. List three functions of the vertebrate backbone.
57. Name three traits that allowed reptiles to become truly terrestrial animals.

58.   Based on how their young are born, list and describe the 3 general groups of mammals.

59.-78. **MATCHING TEST:** chordates

Choices: A.   Cartilage fish   E.   Birds
            B.   Bony fish   F.   Mammals
            C.   Amphibians   G.   All of these
            D.   Reptiles   H.   None of these

59.   Produce amniotic eggs.
60.   Are segmented.
61.   Produce placentas.
62.   Frogs.
63.   Have non-bony skeletons.
64.   Have 2-chambered hearts.
65.   Dinosaurs.
66.   Have notochords at some stage of development.
67.   Have 4-chambered hearts.
68.   On the boundary between aquatic and terrestrial existence.
69.   Have swim bladders.
70.   Feed milk to their young.
71.   Have open circulatory systems.
72.   Have 3-chambered hearts.
73.   Have backbones.
74.   Are warmblooded.
75.   Evolved into amphibians.
76.   Have hollow bones and air sacs.
77.   Whales.
78.   Sharks.

****

****

ANSWERS TO REVIEW QUESTIONS:

1. see 1.A.1)-6)

2. see 1.l.1) and 2)

3.-17. [see 2. and 3.]
3.	B	11.	B
4.	B	12.	B
5.	A	13.	B
6.	B	14.	A
7.	B	15.	A
8.	A	16.	D
9.	D	17.	A
10.	A		

18. see 1.D.3)a.i.-iii.

19. see 1.h.3)a.-c.

20. see 1.F.3)a.-c.

21. see 6.A.6)a.-f.

22. see 7.B.1)d.-h.

23.-37. [see 4.-6.]
23.	D	31.	A
24.	B	32.	B
25.	D	33.	B
26.	C	34.	A
27.	A	35.	C
28.	B	36.	B
29.	C	37.	C
30.	D		

38.-53. [see 7. and 8.]
38.	B	46.	C
39.	D	47.	C
40.	A	48.	A
41.	B	49.	B
42.	C	50.	A
43.	A	51.	B
44.	B	52.	D
45.	A	53.	A

54. see 9.D.1)-5)

55. see 10.A.1)-4)

56. see 10.C.a.i.-iii.

57. see 10.C.4)c.ii.a)-c)

58. see 10.C.4)e.vii.a)-c)

59.-78. [see 10.C.3) and 4)]
59.	DEF	69.	B
60.	G	70.	F
61.	F	71.	H
62.	C	72.	CD
63.	A	73.	G
64.	AB	74.	EF
65.	D	75.	B
66.	G	76.	E
67.	EF	77.	F
68.	C	78.	A

****

CHAPTER 23: THE DIVERSITY OF LIFE: III. FUNGI.

CHAPTER OUTLINE.

This chapter describes the general features and ecology of fungi, then provides a brief description of the major fungal divisions.

1. **The Fungal Body.**

   A.   Consists of <u>hyphae</u> (microscopically thin, threadlike filaments) that grow into an interwoven <u>mycelium</u>.
   1)   Hyphae of some fungi are single elongated cells with many nuclei.
   2)   Hyphae of other fungi have many cells (each with one or more nuclei) separated by <u>septa</u> with pores to allow cytoplasm to flow, distributing nutrients among the cells.

   B.   Unlike animals, fungal nuclei are <u>haploid</u> (have one set of chromosomes).

   C.   Like plants, fungal cells have <u>cell walls</u>, a few containing cellulose but most others containing chitin.

   D.   The fungal mycelium:
   1)   Infiltrates into dead tissues or cell products.
   2)   Periodically condenses and differentiates into visible reproductive structures like mushrooms and puffballs.

2. **Fungal Nutrition.**

   A.   Like animals, fungi are <u>heterotrophic</u>, using nutrients stored in the bodies or wastes of other organisms.

B.   Types of heterotrophic nutrition seen in fungi:
  1)   Some are <u>saprobes</u>, digesting bodies of dead organisms.
  2)   Some are <u>parasites</u>, feeding on living organisms.
  3)   Some are <u>symbionts</u>, living in harmony with other organisms (examples: lichens and mycorrhizae).

C.   Fungi absorb nutrients, as do bacteria, since cell walls prevent direct ingestion of food. Steps:
  1)   Fungi secrete enzymes outside their bodies to break nutrients down into small molecules.
  2)   Hyphae have much surface area for nutrient absorption.
  3)   The small nutrient molecules diffuse through the cell walls and into the fungal cells.

3.   **Fungal Reproduction**.

A.   During simple <u>asexual reproduction</u>, a mycelium breaks into pieces, each piece growing into a new individual.

B.   Many fungi reproduce both asexually and sexually, using different types of spores:
  1)   Spores form on or in structures that project above the mycelium to allow dispersal of spores by wind or water.
  2)   Haploid asexual spores are produced by mitotic divisions of haploid fungal cells.
  3)   Sexual spore formation begins with fusion of two haploid nuclei to produce a diploid zygote.
    a.   A zygote undergoes meiosis to produce haploid spores.
    b.   These spores, once dispersed, divide by mitosis to form new haploid mycelia.

4.   **Economic and Ecological Impacts**.

A.   Microscopic fungal filaments are found throughout moist rich soils, penetrating decaying vegetation and animal bodies.

B.   As decomposers, fungi aid ecosystems since:
  1)   Digestive activities liberate nutrients used by plants.
  2)   Fungal bodies are eaten by small insects and worms.
  3)   If fungi and bacteria disappeared, nutrients would remain locked in dead plant and animal bodies and ecosystems would collapse.

C.   Parasitic fungi:
  1)   Cause the majority of plant diseases (e.g., Chestnut blight and Dutch-elm disease).
  2)   Insect parasites are important in pest control as "fungal pesticides" (e.g., against citrus mites).
  3)   Can attack humans, causing:
    a)   Skin diseases: Ringworm and athlete's foot.
    b)   Lung diseases: Valley fever and histoplasmosis.
    c)   Vaginal diseases: Candidiasis.

D. Some types of fungi are edible, including mushrooms and truffles, and others are important in making bread, cheese, wine, and beer.

5. **Fungal Symbiotic Relationships.**

   A. <u>Lichens</u>: associations between ascomycete fungi and either unicellular algae or cyanobacteria.
      1) Usually among the first to colonize newly formed volcanic islands.
      2) The algal partner provides food from photosynthesis.
      3) The fungal partner provides support and protection.
      4) About 20,000 types are known.

   B. <u>Mycorrhizae</u>: fungi (5000 species of either basidiomycetes or ascomycetes) found growing beneficially with the roots of about 80% of all vascular plants.
      1) Mycorrhizal hyphae surround plant roots and often invade root cells.
      2) The fungal partner digests organic compounds in the soil and absorbs water for the plant.
      3) The plant partner passes photosynthetic food to the fungus.
      4) Mycorrhizal associations may have been important in the invasion of land by plants over 400 million years ago.

6. **Fungal Taxonomy and Diversity**: Nearly 100,000 species are known, with about 100 new species described each year.

   A. <u>Division Zygomycota</u> ("zygote fungi"): 600 species.
      1) Hyphae have the ability to mate, fusing their nuclei to produce diploid <u>zygospores</u>.
      2) Zygospores are dispersed through the air and remain dormant until conditions are favorable for growth. Then, they undergo meiosis, producing haploid cells that become new hyphae.
      3) Hyphae can also reproduce asexually, forming haploid spores in black spore cases called <u>sporangia</u>.
      4) Examples: the black bread mold (<u>Rhizopus</u>) and the dung fungus (<u>Philobolus</u>).

   B. <u>Division Oomycota</u> ("egg fungi"): 475 species.
      1) Very different from other fungi since:
         a) Cell walls often contain cellulose.
         b) Sexual reproduction involves fertilization of a large egg cell.
         c) They reproduce asexually using swimming flagellated cells (<u>zoospores</u>) unique to this fungal group.
         d) They are diploid for most of their life cycle.
      2) Causes various plant diseases, including:
         a) Downy mildew of grapes in France.
         b) "Late blight" of potatoes in Ireland.
      3) Also includes the water molds.

   C. <u>Division Ascomycota</u> ("sac fungi"): 30,000 species.
      1) Named after the saclike case (<u>ascus</u>) in which spores form during sexual reproduction.

2) Secrete the enzymes cellulase and protease that damage textiles like cotton and wool in warm humid climates.

3) Examples:
   a) Dutch-elm and Chestnut blight disease fungi.
   b) <u>Claviceps purpurea</u> attacks rye plants and produces toxins like the active ingredient in LSD, and ergot (used in labor-inducing and hemorrhage-controlling drugs).
   c) Yeasts used in making bread and wine.
   d) Edible morels and truffles.

D. <u>Division Basidiomycota</u> ("club fungi"): 25,000 species.
   1) The mushrooms, puffballs, shelf fungi ("monkey stools"), rusts and smuts (attack wheat and corn plants).
   2) Mushroom and puffball mycelia are reproductive structures that emerge from an underground network of hyphae to undergo sexual reproduction and release spores.
      a) Leaflike gills on the underside of mushrooms produce club-shaped diploid cells (<u>basidia</u>).
      b) Basidia form haploid <u>basidiospores</u> by meiosis.
      c) Basidiospores are dispersed by wind and water to germinate into new mycelia.

E. <u>Division Deuteromycota</u> (the "imperfect fungi"): 25,000 species.
   1) Sexual reproduction has not been observed.
   2) Examples:
      a) The fungus producing penicillin (<u>Penicillium</u>).
      b) Those used in making Roquefort and Camembert cheeses.
      c) Some human parasites (ringworm and athlete's foot).
      d) Some predatory forms that trap nematodes.

****

REVIEW QUESTIONS:

1.-18. **MATCHING TEST:** fungi

Choices:	A.	Zygomycota	D.	Basidiomycota
	B.	Ascomycota	E.	Oomycota
	C.	Deuteromycota		

1. Mushrooms.
2. Sac fungi.
3. Egg fungi.
4. Causes ringworm.
5. Produce zygospores.
6. Sexual reproduction has not been observed.
7. Yeasts.
8. Black bread mold.

9. Late blight of Irish potatoes.
10. Makes penicillin.
11. Causes Dutch-elm disease.
12. Puffballs.
13. Truffles.
14. Have cellulose in their cell walls.
15. Used in making Roquefort cheese.
16. Water molds.
17. Club fungi.
18. Causes athlete's foot.

---

19. In what ways do fungi differ from plants and from animals? In what ways are fungi similar to plants and to animals?

20.-26.   **MATCHING TEST:** symbiotic fungi.

Choices: A.   Lichens   C.   Both of these
             B.   Mycorrhizae   D.   Neither of these

20. Hyphae often invade plant root cells.
21. Association between ascomycete fungi and cyanobacteria.
22. Harm the non-fungal partner.
23. Early colonizers of bare rocky volcanic islands.
24. Association between basidiomycetes fungi and the roots of vascular plants.
25. Association between ascomycete fungi and animals.
26. Digest organic compounds in the soil.

****

ANSWERS TO REVIEW QUESTIONS:

1.-18. [see 6.A.-E.]   19.   see 1. and 2.

1.	D	10.	C
2.	B	11.	B
3.	E	12.	D
4.	C	13.	B
5.	A	14.	E
6.	C	15.	C
7.	B	16.	E
8.	A	17.	D
9.	E	18.	C

20.-26. [see 2.C.1) and 2)]

20.	B	24.	B
21.	A	25.	D
22.	D	26.	B
23.	A		

****

CHAPTER 24: THE DIVERSITY OF LIFE. IV. PLANTS.

CHAPTER OUTLINE.

This chapter describes the plants. After a discussion of evolutionary trends, the multicellular algae, bryophytes, ferns, conifers, and flowering plants are described.

1. **General facts.**

    A. Plants are eukaryotic, multicellular, photosynthetic organisms.

    B. Plants capture sunlight and make sugars from water and carbon dioxide, as do the unicellular algae and the blue-green bacteria.

    C. The energy that these organisms harvest from sunlight powers nearly all life on earth.

2. **Evolution of multicellular plants.**

    A. Origins.
       1) Definite algae first appear in deposits from the Cambrian era (500-600 million years ago).
       2) Terrestrial plants first appeared about 430 million years ago.
       3) Two groups of plants arose from ancestral green algae:
          a. Vascular plants, with specialized vessels that transport water and nutrients and provide support for the plant body.
          b. Bryophytes (mosses and liverworts), lacking specialized tissues and straddling the boundary between aquatic and terrestrial existence.

    B. Nearly 300,000 species of multicellular plants have been identified, including:
       1) Three groups of aquatic algae (the reds, browns, and greens).
       2) The bryophytes (Division Bryophyta): non-vascular terrestrial plants that require

wet conditions for reproduction.
3)   The tracheophytes (vascular plants):
     a.   Seedless plants, such as ferns (Division Pteridophyta).
     b.   Seed plants:
         i.   Non-flowering gymnosperms (Division Coniferophyta).
         ii.   Flowering angiosperms (Division Anthophyta).

C.   <u>Evolutionary trends in plant structure</u>.
1)   Plant life arose in the seas, which:
     a.   Support the plant body.
     b.   Provide a relatively constant temperature.
     c.   Bathe the entire plant in nutrients.
2)   <u>Adaptations</u> to terrestrial environments:
     a.   <u>Roots</u> and rootlike structures: anchor the plant and absorb water and nutrients from soil.
     b.   <u>Conducting vessels</u>:
         i.   Transport water and minerals upward from roots.
         ii.   Transport photosynthetic products from leaves to other body parts.
     c.   Stiffening <u>lignin</u>: impregnates water and mineral conducting vessels and supports the plant body.
     d.   <u>Waxy cuticle</u> covering stem and leaf surfaces: limits evaporation of water.
     e.   <u>Stomata</u> pores in leaves and stems: allow gas exchange but close when water is scarce.

D.   <u>Evolutionary trends in plant reproduction</u>.
1)   For algae, water is the medium for reproduction:
     a.   Gametes of some algae are carried passively by water currents.
     b.   Other algae produce flagellated, actively swimming gametes and spores ("zoospores").
     c.   Algal gametes meet in the water, and zygotes and spores are dispersed by water.
2)   When plants began living on land:
     a.   They developed new methods for transporting gametes (pollen grains, and later flowers to attract animals to carry pollen).
     b.   Their embryos required:
         i.   Protection from drying (seeds).
         ii.   A means of dispersal independent of water (seed coats and fruits).

E.   <u>Evolutionary trends in plant life cycles</u>.
1)   Plant life cycles exhibit <u>alternation of generation</u>:
     a.   A haploid plant (<u>gametophyte</u>) produces sex cells by mitosis. Gametes unite to form a diploid zygote.
     b.   The diploid zygote develops into a diploid plant (<u>sporophyte</u>) that develops haploid spores by meiosis.
     c.   Each haploid spore can develop into a gametophyte plant.
2)   <u>General trend</u> seen as ancient algae evolved into bryophytes, ferns, gymnosperms and angiosperms: decreased size, duration, and prominence of the gametophyte generation relative to the sporophyte generation.

        a.   In some green algae, the sporophyte is absent and only the zygote is diploid.

        b.   In mosses, the sporophyte is present but it is smaller than the gametophyte and remains attached to it.

        c.   In ferns, the sporophyte is dominant, while the gametophyte is smaller, though independent.

        d.   In seed plants, male and female gametophytes are microscopic and barely recognizable, but still produce gametes that unite to form the sporophyte embryo.

3.   Watery origins: algae.

A.   General facts.
1)   Plant life arose in the sea and early plants were relatively simple algae:
      a.   Lack true roots, stems, and leaves.
      b.   Lack complex reproductive structures such as flowers and cones.
      c.   Gametes are shed directly into water.
2)   Algal life cycles are varied and complex.
3)   Are classified into 3 Divisions (= Phyla) based on their characteristic colors caused by red, brown, and green pigments that absorb light energy for photosynthesis.

B.   Division Rhodophyta ("red algae"): 4000 species.
1)   Mostly marine forms in deep, clear, tropical waters.
2)   Some accumulate calcium carbonate (limestone) and contribute to coral reefs.
3)   Others are harvested as food in the Orient.
4)   Others form gelatinous carrageenan used to emulsify paints, cosmetics, and ice cream.
5)   They form the food-producing photosynthetic foundation of marine ecosystems.

C.   Division Phaeophyta ("brown algae"): 1500 species.
1)   Entirely marine "seaweeds" and "kelps".
2)   Found in cool, temperate oceans, often along rocky shores.
3)   Some use gas-filled "floats" to support their bodies.
4)   Some giant kelps grow 100 meters tall, forming "undersea forests" and providing food, shelter, and breeding areas for marine animals.

D.   Division Chlorophyta ("green algae"): 7000 species.
1)   Extremely diverse group that gave rise to terrestrial plants.
2)   Are like land plants in three ways:
      a.   They use the same pigments for photosynthesis.
      b.   They store food as starch and have cell wall composition similar to plants.
      c.   Most live in fresh water where they have evolved adaptations to temperature fluctuations and periodic dryness.
3)   Are mainly multicellular, with a few unicellular and colonial forms.
4)   Examples:
      a.   Volvox, a colonial form with some division of labor.
      b.   Spirogyra, forms long chains of cells.

        c.   <u>Ulva</u>, the "sea lettuce".

4.   **Land**: the new frontier.

    A.   <u>Advantages</u> for plants to move onto land:
        1) CO_2 and sunlight for photosynthesis are present in abundance in the air.
        2)   There was abundant space and resources on the land.

    B.   <u>Problems</u> with living on land: land plants need:
        1)   Support against gravity.
        2)   Conducting tissues to move substances around the plant body.
        3)   Special adaptations to reproduce since sex cells cannot continuously be in water.
        4)   Protection against desiccation.

    C.   Two major groups of land plants evolved from algal ancestors.
        1)   <u>Bryophytes</u>, on the boundary between aquatic and terrestrial life.
        2)   <u>Tracheophytes</u> (vascular plants), completely adapted to land.

5.   **Division Bryophyta**: 16,000 species of mosses and liverworts.

    A.   Lack true roots, stems, and leaves.
        1)   They lack well-developed vessels for conducting water and nutrients.
        2)   Rootlike structures (<u>rhizoids</u>) bring water and nutrients into the plant body.
        3)   Most are quite small due to lack of conducting tissue.

    B.   They vary in their ability to withstand desiccation and most are confined to moist habitats.

    C.   Have evolved enclosed reproductive structures that protect the gametes from drying out.
        1)   Eggs develop within <u>archegonia</u>.
        2)   Sperm develop within <u>antheridia</u>.
        3)   Sperm must swim to the eggs through a film of water.

    D.   <u>Life cycle</u>.
        1)   The larger "leafy" plant bodies are the haploid gametophytes that form gametes by mitosis.
        2)   Diploid sporophytes develop from fertilized eggs that remain in the archegonia.
            a.   The mature sporophytes produce haploid spores by meiosis and release them with explosive force from capsules that burst open.
            b.   Spores that land on fertile soil develop into new gametophytes.

6.   **Tracheophytes: the vascular plants**.

    A.   Small bryophytes were successful at colonizing moist terrestrial habitats only, leaving vast dry areas unoccupied. Other plants evolved adaptations to exploit these habitats:
        1)   Taller plants could outcompete bryophytes for sunlight. In these plants, <u>vascular tissue</u> evolved for support and for conducting materials between roots and leaves.
        2)   Also, other plants evolved methods to avoid desiccation and not depend on moisture for reproduction.

B.   Seedless vascular plants: club mosses, horsetails, and ferns.
1)   These reached tree-like proportions and dominated during the Carboniferous era (355-265 million years ago), their fossils becoming the coal we use today.
2)   Today, the club mosses (such as Lycopodium) and horsetails (Equisetum) are of minor importance.
3)   The ferns (12,000 species of the Division Pterophyta) are somewhat successful today.
     a.   Their broad leaves capture much energy from sunlight.
     b.   Unlike bryophytes, the diploid fern sporophyte is dominant, producing haploid spores by meiosis in structures on special leaves.
     c.   The spores are blown by wind and develop into tiny haploid gametophyte plants.
     d.   The gametophytes produce gametes by mitosis, but water is still needed for sexual reproduction since:
          i.   Sperm must swim to the eggs.
          ii.   The small gametophytes lack conducting tissue.

C.   Seed plants (gymnosperms and angiosperms). Have dominated the land for the past 350 million years since they no longer need a watery environment for reproduction due to evolution of pollen grains and seeds.
1)   The haploid gametophyte is reduced so much that it acts as a reproductive organ within the diploid sporophyte.
2)   Female gametophytes make eggs.
3)   Male gametophytes are pollen grains containing sperm.
     a.   The waterproof pollen grains are dispersed by wind or by animal pollinators.
     b.   So, sexual reproduction does not depend on water.
4)   Seeds consist of:
     a.   Embryonic plants.
     b.   A supply of food for the embryo as it develops roots and leaves.
     c.   A protective coat, allowing the embryo to remain dormant until conditions are favorable for growth.

D.   Gymnosperms ("naked seeds"): nonflowering seed plants.
1)   The first seed plants to evolve.
2)   Include the cycads, ginkgos, and conifers (Division Coniferophyta with 500 species).
3)   Are most abundant in colder latitudes and higher dry elevations.
4)   Conifers (pines, firs, spruce, hemlocks, and cypresses) are adapted to withstand dry, cold, conditions since:
     a.   Their leaves are thin needles covered with a thicker cuticle (little evaporation possible).
     b.   They retain their leaves year-round ("evergreens"), extending the short northern growing season.
     c.   They produce an "antifreeze" resin in their sap so that nutrient transport can continue in subfreezing temperatures.
5)   Reproduction in conifers.
     a.   The diploid sporophyte ("tree") develops male and female cones.
     b.   Male cones are smaller and delicate, disintegrating after releasing clouds

of pollen grains (the males' waterproof gametophytes).

    c.   Female cones are larger with woody scales. Each scale has 2 haploid female gametophytes, each producing an egg.

    d.   After a pollen grain lands on a female cone, it sends out a slowly growing pollen tube that takes 14 months to reach the eggs.

    e.   After fertilization, seeds develop and are released when the female cone matures and the scales open.

E.   <u>Angiosperms</u> ("covered seeds"): the flowering seed plants (<u>Division Anthophyta</u> with 250,000 species).

    1)   Evolved about 125-130 million years ago from gymnosperm ancestors that used insects to carry pollen from plant to plant, benefitting both the insects (they eat some of the pollen) and the plant (good efficiency in pollen transfer).

    2)   Plants evolved structures (flowers) to attract pollinator animals. These flowering plants have been dominant for the last 100 million years.

    3)   Angiosperms show incredible diversity of size, form, and habitat.

    4)   <u>Life cycle</u>.

        a.   Angiosperms have dominant sporophytes that produce and nurture tiny male (pollen grains) and female (embryo sacs) gametophytes within their flowers.

        b.   Gametophytes produce sex cells and fertilization occurs within flower ovaries.

        c.   Zygotes develop into embryos. Embryos with food cells are enclosed within protective seeds that are then enclosed within fruits (the ripened ovaries of the flowers).

        d.   Just as flowers encourage animals to transport pollen, fruits entice them to disperse seeds.

        e.   Some seeds pass through animal digestive tracts unharmed when fruit is eaten. Also, animals may pick up "burr" fruits in their fur.

        f.   Some trees (i.e., maples) have fruits with "wings" for wind-dispersal.

    5)   Angiosperms have broad leaves that are shed under adverse conditions, allowing them to exist in warmer wetter climates than gymnosperms.

    6)   So, the <u>three major adaptations</u> allowing angiosperms to become the dominant terrestrial plants are:

        a.   Flowers.

        b.   Fruits.

        c.   Broad leaves and deciduous lifestyle.

    7)   <u>Two classes</u> of flowering plants:

        a.   <u>Monocots</u> (Class Monocotyledoneae): 65,000 species including the grasses, corn and other grains, irises, lilies, and palms.

        b.   <u>Dicots</u> (Class Dicotyledoneae): 170,000 species including nearly all the trees, shrubs, and herbs.

****

REVIEW QUESTIONS:

1. Name three characteristics of all plants.
2. Name the two major groups of seed plants.
3. Name five adaptations of plants to terrestrial conditions.
4. Describe the general trend seen in life cycles during plant evolution.
5. Name the three major components of a seed.
6. List the three major adaptations that allowed angiosperms to become the dominant type of land plants.

7.-14.   **MATCHING TEST**: alternations of generations.

Choices: A.   Sporophyte   C.   Both of these
              B.   Gametophyte   D.   Neither of these

7. Haploid.
8. Begins as a zygote.
9. Multicellular.
10. Produces sex cells by meiosis.
11. Begins as a spore.
12. Produces spores by meiosis.
13. Grows by mitosis.
14. Dominant form seen in flowering plants.

15.-29.   **MATCHING TEST**: algae.

Choices: A.   Division Rhodophyta   D.   All of these
              B.   Division Phaeophyta   E.   None of these
              C.   Division Chlorophyta

15. Shed gametes directly into water.
16. Brown algae.
17. Contribute limestone to coral reefs.
18. Red algae.
19. Much like land plants.
20. Live in clear, tropical, marine habitats.
21. Stores food as starch.
22. Spirogyra.
23. Seaweeds and kelps.
24. Photosynthesize.
25. Green algae.
26. Produce food emulsifiers.
27. Live in cool temperate oceans.
28. Live in fresh water environments.

29.   Have vascular tissue.

30.-43.   **MATCHING TEST:**   land plants.

Choices: A.   Angiosperms   D.   Bryophytes
             B.   Gymnosperms   E.   All of these
             C.   Ferns   F.   None of these

30.   Do not produce seeds.
31.   Make pollen grains.
32.   Lack well-developed vascular tissues.
33.   Some rely on insects for pollination.
34.   Sperm must swim to eggs.
35.   Produce fruits.
36.   Seedless vascular plants.
37.   Gametophytes are larger than sporophytes.
38.   Monocots and dicots.
39.   Make cones.
40.   Most successful land plants today.
41.   Vascular plants.
42.   Mosses.
43.   Seed plants.

****

ANSWERS TO REVIEW QUESTIONS:

1.   see 1.A.

2.   see 2.B.3)b.i.-ii.

3.   see 2.C.2)a.-e.

4.   see 2.E.2)a.-d.

5.   see 6.C.4)a.-c.

6.   see 6.E.1)-6)

7.-14.   [see 2.E.1)-2)]

7.	B		11.	B
8.	A		12.	A
9.	C		13.	C
10.	D		14.	A

15.-29.   [see 3.A.-D.]

15.	D		23.	B
16.	B		24.	D
17.	A		25.	C
18.	A		26.	A
19.	C		27.	B
20.	A		28.	C
21.	C		29.	E
22.	C			

30.-43.   [see 5.-6.A.-E.]

30.	CD		37.	D
31.	AB		38.	A
32.	D		39.	B
33.	A		40.	A
34.	CD		41.	ABC
35.	A		42.	D
36.	C		43.	AB

****

UNIT III EXAM:

Chapters 16 - 24. All questions are dichotomous. Circle the correct choice in each case.

Chapter 16.

1. Before Darwin, most people thought that species were (capable of change / incapable of change).
2. The idea that God created some new species after every catastrophe was proposed by (Agassiz / Cuvier).
3. The mutual evolution of predator and prey species is an example of (coevolution / convergent evolution).
4. (Darwin / Lamarck) proposed that an innate or internal drive toward change within cells is the driving force in evolution.
5. The similarity in the bones making up a bird's wing and a horse's foot are due to (common ancestry / convergent evolution).
6. Amino acid sequences in proteins of different animals tend to (deny / support) evolution.
7. Aristotle's <u>Scala Naturae</u> was considered (flexible / immutable).
8. In convergent evolution, the two forms being modified are (closely related / unrelated).
9. The many different varieties of dogs are the result of (artificial / natural) selection.
10. Analogous structures arise due to (divergent / convergent) evolution.

Chapter 17.

1. Individual plants or animals (change / do not change) in response to selection.
2. For a population to remain at equilibrium, it must be (large / small).
3. (Mutation / Selection) is the factor that controls the direction of evolution.
4. Genetic drift is a characteristic of (large / small) populations.
5. Natural selection acts on (genotypes / phenotypes) and on (individuals / populations).
6. Stabilizing selection results in (change / lack of change).
7. Gene flow tends to (decrease / increase) differences between populations.
8. Both large-beaked and small-beaked varieties of a single species of birds can be maintained in an area by (stabilizing / disruptive) selection.
9. Behavior that endangers an organism but benefits its close relatives is (symbiosis / altruism).
10. Most species eventually (give rise to new species / become extinct).

Chapter <u>18</u>.

1.   Phyletic evolution (does / does not) increase the total number of species existing at any one time.
2.   Polyploidy is most common in (animals / plants).
3.   Mechanical incompatibility is a (premating / postmating) isolating mechanism.
4.   The most valid way to determine whether two organisms belong to different species is to look (for physical differences / at mating behavior).
5.   Speciation depends on (isolation / lack of isolation) between populations.
6.   Populations that are geographically separated are (sympatric / allopatric).
7.   Sympatric speciation can occur if (chromosome aberrations / random mating) occurs.
8.   Geographic isolation is a (postmating / premating) isolating mechanism.
9.   Hybrid infertility is a (postmating / premating) isolating mechanism.
10.   Adaptive radiation causes (convergent / divergent) evolution.

Chapter <u>19</u>.

1.   The universe is (15 million / 15 billion) years old.
2.   Primitive earth was characterized by an (absence / abundance) of free oxygen.
3.   The first living organisms were most likely (aerobes / anaerobes) and (eukaryotic / prokaryotic).
4.   The first living organisms were (autotrophic / heterotrophic).
5.   Photosynthesis arose first in the (cyanobacteria / green algae).
6.   Multicellular plants evolved from unicellular (eukaryotic / prokaryotic) forms.
7.   A separate mouth and anus is (less / more) efficient than a single common opening.
8.   The first skeletons were (external / internal).
9.   In plants, spores are (asexual / sexual) reproducing agents.
10.   Spores germinate and grow into (gametophytes / sporophytes).
11.   Gymnosperms, as a rule, produce (less / more) pollen than angiosperms.
12.   The dinosaurs (gradually / suddenly) became extinct.
13.   Feathers probably evolved first for (flight / insulation).
14.   (Cro-Magnon / Neanderthal) man came first.

Chapter <u>20</u>.

1.   The science which places organisms into categories based on their evolutionary relationships is (classification / taxonomy).
2.   A (Class / Order) contains several (Classes / Orders).
3.   A (Kingdom / Phylum) is a more general category.
4.   The two part name for a species is its (Kingdom and Phylum / Genus and species).
5.   Organisms within the same (Genus / species) interbreed in nature.
6.   Eastern and mountain bluebirds have the same (Genus / species) name.
7.   (Linnaeus / Darwin) developed the classification system we use today.
8.   One important criterion used to classify organisms is (anatomical similarity / geographical location).

9.   All moneran and protistan organisms are (unicellular / prokaryotic).
10.   Plants are, but fungi are not (multicellular / photosynthetic).

Chapter 21.

1.   Eukaryotic organisms are (monerans / protistans).
2.   Viruses are (cellular / non-cellular).
3.   The better defense against viral infection is (antibiotics / vaccines).
4.   Bacilli bacteria are (rod-shaped / spherical).
5.   Bacteria reproduce asexually by (fission / mitosis).
6.   Cyanobacteria are (heterotrophic / autotrophic).
7.   Unicellular algae are often called (zooplankton / phytoplankton).
8.   (Protozoa / Dinoflagellates) are unicellular, eukaryotic, heterotrophic protists.
9.   The organism causing African sleeping sickness is a (sporozoan / zooflagellate).
10.   The organism causing malaria is a (sporozoan / zooflagellate).

Chapter 22.

1.   Large animals are likely to have (less / more) stable internal environments.
2.   All adult sponges are (motile / sessile).
3.   Water (enters / leaves) a sponge through the osculum.
4.   Free-swimming cnideria are called (medusae / polyps).
5.   Flatworms have (radial / bilateral) symmetry.
6.   Terrestrial arthropods have (open / closed) circulatory systems.
7.   Molluscs evolved from (annelids / arthropods).
8.   Echinoderms have (endoskeletons / exoskeletons).
9.   The osteichthyes are (cartilaginous / bony) fishes.
10.   Mammals evolved from (reptiles / birds).

Chapter 23.

1.   The single thread-like filaments that make up the body of a fungus are the (hyphae / mycelia).
2.   The cell walls of most fungi are composed of (cellulose / chitin).
3.   Fungi are (autotrophic / heterotrophic).
4.   Most fungal nuclei are (diploid / haploid).
5.   Fungi digest food particles (outside their bodies / inside their bodies).
6.   In fungi, haploid sexual spores are produced from (diploid zygotes / haploid hyphal cells).
7.   Fungi and (protozoans / bacteria) are decomposers.
8.   A fungus causes (malaria / Dutch-elm disease).
9.   A fungus found growing beneficially with plant roots is called a (mycorrhiza / lichen).
10.   Yeast belong to the fungal division (Zygomycota / Ascomycota).

Chapter 24.

1. The gametophyte plant is (diploid / haploid).
2. Spores develop into the (gametophyte / sporophyte) generation.
3. The gametophyte produces (gametes / spores).
4. The brown algae are commonly found in the (temperate / tropical) zone.
5. The archegonium is a (female / male) reproductive organ.
6. The dominant generation of the bryophytes is the (gametophyte / sporophyte).
7. A liverwort is an example of a (nonvascular / vascular) plant.
8. The (gametophyte / sporophyte) generation of ferns is the large, visible generation.
9. The (angiosperms / gymnosperms) are the flowering plants.
10. The green protective parts of a flower are the (petals / sepals).
11. The (anther / carpel) is the male part of a flower.
12. Pollination takes place on the (stigma / style).

****

ANSWER KEY:

Chapter 16:

1. incapable of change
2. Agassiz
3. coevolution
4. Lamarck
5. common ancestry
6. support
7. immutable
8. unrelated
9. artificial
10. convergent

Chapter 17:

1. do not change
2. large
3. selection
4. small
5. phenotypes, populations
6. lack of change
7. decrease
8. disruptive
9. altruism
10. become extinct

Chapter 18.

1. does not
2. plants
3. premating
4. at mating behavior
5. isolation
6. allopatric
7. chromosome aberrations
8. premating
9. postmating
10. divergent

Chapter 19:

1. 20 billion
2. absence
3. anaerobes, prokaryotic
4. heterotrophic
5. cyanobacteria
6. eukaryotic
7. more
8. external
9. sexual

10. gametophytes
11. more
12. suddenly
13. insulation
14. Neanderthal

Chapter 20:

1. taxonomy
2. class, orders
3. Kingdom
4. Genus and species
5. species
6. Genus
7. Linnaeus
8. anatomical similarities
9. unicellular
10. photosynthetic

Chapter 21:

1. protistans
2. non-cellular
3. vaccines
4. rod-shaped
5. fission
6. autotrophic
7. phytoplankton
8. protozoa
9. zooflagellate
10. sporozoan

Chapter 22:

1. more
2. sessile
3. leaves
4. medusae
5. bilateral
6. open
7. annelids
8. endoskeletons
9. bony
10. reptiles

Chapter 23:

1. hyphae
2. chitin
3. heterotrophic
4. haploid
5. outside their bodies
6. diploid zygote
7. bacteria
8. Dutch-elm disease
9. mycorrhiza
10. Ascomycota

Chapter 24.

1. haploid
2. gametophyte
3. gametes
4. temperate
5. female
6. gametophyte
7. nonvascular
8. sporophyte
9. angiosperms
10. sepals
11. anther
12. stigma

****

UNIT IV. PLANT ANATOMY AND PHYSIOLOGY

CHAPTER 25: *UNIFYING CONCEPTS IN PHYSIOLOGY: HOMEOSTASIS AND THE ORGANIZATION OF LIFE*

CHAPTER OUTLINE.

This chapter enumerates the life processes common to all organisms, and the adaptations required for life on land.

1.   **Homeostasis and the challenges of life**:

   A.   <u>Homeostasis</u>: For physiological processes, maintenance of internal constancy required for the maintenance of life.
   1)   The internal condition maintains a "dynamic equilibrium."
   2)   Change occurs, but within a narrow range required for cellular function.
   3)   The narrow range is maintained by use of negative feedback systems.

   B.   <u>Negative feedback</u>: change initiates a series of events to counteract the change.
   1)   The most important principle governing maintenance of homeostasis.
   2)   The home thermostat is a familiar example. Negative feedback requires:
      a.   <u>Set point</u>: the thermostat setting.
      b.   <u>Sensor</u>: the thermometer.
      c.   <u>Effector</u> that accomplishes the change: the heater.
   3)   Biological example: maintenance of body temperature at 98.6° F. (37° C.)

      a.   Set point is established by neurons in the hypothalamus.

      b.   Neurons in hypothalamus, abdomen, spinal cord, and large veins are temperature sensors and transmit the information to the hypothalamus.

      c.   When body temperature drops, hypothalamus activates various effector mechanisms:

          i.   Shivering (heat through muscle contractions).

          ii.   Constriction of blood supply to skin (lessens heat loss).

          iii.   Elevation of metabolic rate.

C.   Positive feedback and homeostasis: a series of events that amplifies a change.

    1)   Tends to create explosive events that must be carefully restricted.

    2)   Example: nuclear fission.

    3)   Biological example: during childbirth.

        a.   Early contractions of labor force baby's head against the cervix, dilating it.

        b.   Stretch receptors neurons in cervix signal hypothalamus, which responds by triggering release of hormones that stimulate more and stronger uterine contractions.

        c.   End result: expulsion of baby and placenta.

2. Challenges of life.

A.   The environment places stresses on the organism which it must overcome if it is to survive and reproduce.

B.   Common challenges that all organisms must overcome if they are to sustain life:

    1)   Obtaining materials.

        a.   Animals typically ingest food substances made of proteins, fats, and carbohydrates.

           i.   These large molecules are broken down into single amino acids, fatty acids, and sugars in the digestive tract.

           ii.   Various body parts use these subunits to make their own larger molecules.

        b.   Plants:

           i.   Obtain simple inorganic molecules of CO_2, H_2O, and minerals from the environment.

           ii.   Make their own amino acids, fatty acids, and sugars from scratch.

    2)   Obtaining energy.

        a.   Plants: use energy from sunlight to make sugars from minerals, CO_2, and H_2O (photosynthesis).

        b.   Animals: eat food which contains both nutrients and chemical energy.

    3)   Gas exchange.

        a.   Plants absorb CO_2 and release O_2 during photosynthesis.

        b.   Plants and animals use O_2 to help break down sugars and release the maximum amount of chemical energy (cellular respiration).

        c.   Small organisms exchange gases directly through the body surface.

        d.   Plants have slow metabolism and many internal tissues are dead so that gas exchange with internal body parts are minimal.

4) Regulating body composition.
   a. Organisms must eliminate surplus material and toxic wastes while retaining desirable molecules.
   b. Many types of excretory organs exist (central vacuoles of plants, contractile vacuoles of protozoans, kidneys of animals, etc.).
5) Distributing materials.
   a. Animals have circulatory systems with a heart to pump fluid around the body.
   b. Plants have rigid tubes through which they move sugar from leaves and water from roots to other body parts by a process involving osmosis and evaporation of water.
6) Coordinating body activities.
   a. Plant and animal cells produce hormones (that trigger responses in other cells) that coordinate growth, development, and functioning of the body.
   b. Animals also have nervous systems for rapid information exchange between body parts. Nerve cells release neurotransmitter hormones which affect nearby "target cells".
7) Defense.
   a. Every organism is food for another.
   b. Some plants and animals have bad taste, poisons, or armor as defense mechanisms.
   c. Animals can run away, hide, or fight back.
   d. Animals have "immune systems" to defend against invading parasites.
8) Growth and development.
   a. All adult structures develop from a fertilized egg.
   b. During development, cells differentiate into different tissues.
9) Reproduction.
   a. Most plants and animals reproduce sexually.
   b. Animals actively seek out mates.
   c. Plants passively use H_2O, wind, and animals to transport pollen to the eggs.

C. Processes required for life on land. Life arose in the sea. When plants and animals invaded the land, they evolved adaptations to compensate for the loss of their watery environment:
   1) Obtaining water.
      a. Plants absorb H_2O from the soil with their roots, or from fog and dew.
      b. Animals acquire H_2O:
         i. By drinking it.
         ii. From the foods they eat.
         iii. As "metabolic water" from decomposition of food (especially needed in desert animals).
   2) Preventing water loss by evaporation.
      a. Most advanced land plants and animals have waterproof body coatings.
      b. Desert-dwellers have special adaptations to minimize H_2O loss during gas exchange (lower metabolic rates during daytime, excretion of wastes with minimal H_2O content).
   3) Protection against temperature extremes. Some mechanisms are:
      a. Heat-resistant proteins.
      b. Antifreeze molecules.

      c.   Heat or cold-resistant stages of the life cycle.
      d.   Hibernation of arctic animals.
      e.   Migration, especially in birds.
      f.   Evaporating water during hot weather ("sweating").

4)   <u>Supporting the body against gravity</u> (skeletal support).
      a.   Plant cells have rigid cellulose cell walls.
      b.   Larger plants have "woody" tissue made of cells with extra thick cell walls.
      c.   Animal skeletons may be:
            i.   Internal and bony (in the vertebrates).
            ii.   External and chitinous (in insects and other arthropods).
            iii.   Hydrostatic: tubes filled with H_2O (in earthworms).

5)   <u>Reproduction on land</u>.
      a.   Primitive plants (mosses and ferns) and animals (amphibians) still need H_2O for reproduction since sperm must swim to the eggs.
      b.   Some advanced animals (insects, birds, reptiles) evolved:
            i.   Internal fertilization.
            ii.   Waterproof eggs.
      c.   Mammals have internal embryo development but the embryo is enclosed within the watery amniotic sac.
      d.   In some advanced land plants, non-swimming sperm are enclosed in pollen grains that need no water for transport or survival.
      e.   Seed plants protect their embryos from desiccation with seeds (conifers and flowering plants) and fruits (flowering plants).

3.   Organization of the animal body.

A.   The animal body is made up of a number of <u>organ systems</u>, each made up of <u>organs</u> composed of a number of <u>tissues</u> constructed of many structurally similar <u>cells</u> that act together to perform a similar function.

B.   <u>Tissue</u>.
    1)   Composed of:
      a.   Cells with similar structure and function.
      b.   Non-cellular components made by the cells (i.e., bone and cartilage).
    2)   Four general <u>types</u> of tissue:
      a.   Epithelial.
      b.   Connective.
      c.   Muscle.
      d.   Nerve.
    3)   <u>Epithelial tissue</u>.
      a.   Cells form continuous sheets or <u>membranes</u> with glands.
      b.   Membranes cover the body and line all body cavities.
      c.   Forms a barrier that either:
            i.   Resists movement of substances across it (skin).
            ii.   Only allows specific substances to cross (in small intestine lining).
      d.   Has no blood vessels; epithelium is nourished by diffusion from capillaries beneath.
      e.   Are continuously lost and replaced by cell division.

       f.   Are classified according to shapes of their cells and number of cell layers present (Table 25.1).

4)   <u>Connective tissue</u>.

    a.   Includes dermis (beneath the skin), tendons, ligaments, cartilage, bone, fat, and blood.

    b.   <u>Common feature</u>: all secrete large quantities of extracellular substances between the living cells.

    c.   Underlie all epithelial tissues, and nourish them via capillaries and fluid-filled spaces.

    d.   All (except blood) is interwoven with extracellular fibrous strands of <u>collagen</u> protein secreted by the cells.

    e.   <u>Tendons</u> and <u>ligaments</u> attach muscles to bones and bones to bones, respectively. They contain densely packed collagen fibers in a parallel arrangement.

    f.   <u>Cartilage</u> has widely spaced <u>chondrocyte</u> cells surrounded by collagen.

    g.   <u>Cartilage</u>:

        i.   Covers ends of bones at joints.

        ii.   Supports respiratory passages.

        iii.   Supports ear and nose.

        iv.   Forms shock-absorbing pads between vertebrae.

    h.   <u>Bone</u>: resembles cartilage hardened by deposits of calcium phosphate.

    i.   Bone cells (<u>osteocytes</u>) become embedded in concentric layers of hardened collagen surrounding a central canal containing a capillary that nourishes the osteocytes.

    j.   <u>Haversian system</u>: concentric circles of bone each surrounding a canal.

    k.   <u>Fat cells</u> (adipose tissue): specially modified to act as storage sacs for triglycerides.

    l.   <u>Blood</u>: largely composed of extracellular fluid called <u>plasma</u>.

5)   <u>Muscle tissue</u>: specialized for contraction.

    a.   Contract via orderly arrangement of actin and myosin proteins which use energy to move past each other.

    b.   Three types of muscle: skeletal, smooth, and cardiac.

    c.   <u>Skeletal (striated) muscle</u>: has striped appearance due to rows of actin and myosin.

        i.   Generally under voluntary or conscious control.

        ii.   Main function: move the skeleton.

    d.   <u>Cardiac muscle</u>: striated looking but found only in the heart.

        i.   Is spontaneously active.

        ii.   Cells are interconnected by gap junctions to allow contraction signals to spread rapidly.

    e.   <u>Smooth muscle</u>: not striated.

        i.   Embedded in walls of digestive tract, uterus, bladder, and large blood vessels.

        ii.   Produces involuntary slow, sustained contractions.

6)   <u>Nerve tissue</u>.

    a.   Composed of neuron cells specialized to generate and conduct electrical signals.

    b.   A <u>neuron</u> has four major parts:

        i.   <u>Dendrites</u>: receive signals.

        ii.   <u>Cell body</u>: maintains and repairs the cell.

iii.   <u>Axon</u>: conducts electrical signal to target cell.

iv.   <u>Synaptic terminals</u>: transmit signal to target cell at a region called the <u>synapse</u>.

C.   <u>Animal organs</u>.
1)   Formed from two or more different tissue types that function together.
2)   <u>The skin</u>: a representative organ.
    a.   <u>Epidermis</u>: outer layer of epithelial tissue.
        i.   Covered by protective layer of dead epidermal cells.
        ii.   Packed with <u>keratin</u> protein to keep it airtight and waterproof.
    b.   <u>Dermis</u>: beneath epidermis: loosely packed cells with arterioles and capillaries.
        i.   Lymph vessels carry off extracellular fluid within dermis.
        ii.   Packed with variety of epidermal glands.
        iii.   Hair follicles produce proteinaceous hair.
        iv.   Sweat glands: watery secretions that cool the skin and excrete salts and urea.
        v.   <u>Sebaceous glands</u>: secrete oily sebum that lubricates the epidermis.
    c.   <u>Nerves</u>:
        i.   Regulates loss of heat through skin via neurons controlling degree of arteriole constriction.
        ii.   Variety of sensory nerve endings are scattered throughout dermis and epidermis
    d.   <u>Muscle tissue</u>: attached to hair follicles.

****

REVIEW QUESTIONS/ANSWERS:

1.   Describe a home thermostat in terms of negative feedback.   [see 1.B.2)]
2.   Explain how body temperature is regulated. [see 1.B.3)]
3.   How do positive and negative feedback differ? [see 1.C.]
4.   List nine life processes common to all organisms. [see 2.B.1)-9)]
5.   Name three ways that animals obtain water. [see 2.C.1)a.-b.]
6.   Name five ways that organisms react against temperature extremes. [see 2.C.3)a.-f.]
7.   Name and describe three types of animal skeletons. [see 2.C.4)c.i.-iii.]
8.   In what way are mosses, ferns, and amphibians similar? [see 2.C.5)a.]
9.   Name and briefly describe four types of animal tissues. [see 3.B.3)-6)]
10.   In what ways do skeletal, cardiac, and smooth muscles differ? [see 3.B.5)c.-e.]
11.   Describe a neuron. [see 3.B.6)b.]
12.   Describe the skin as a representative organ. [see 3.C.2)a.-d.]

****

CHAPTER 26: THE STRUCTURE OF LAND PLANTS

CHAPTER OUTLINE.

This chapter deals with the structures and functions of the major plant parts (roots, stems, and leaves). It also discusses plant tissues and cell types, and special adaptations of roots, stems, and leaves.

1. **General facts.**

    A.   Plants:
        1)   Cannot move to escape enemies, to find food or water, to avoid winter weather, or to locate a mate.
        2)   Have evolved many successful adaptations enabling them to thrive in a variety of habitats.

    B.   The most abundant land plants are:
        1)   <u>Angiosperms</u> (Division Anthophyta): flowering plants.
        2)   <u>Gymnosperms</u> (Division Coniferophyta): conifers.

2. **Overview of flowering plant structure.**

    A.   <u>Roots</u> (below ground); have 5 functions:
        1)   Anchor plant to the ground.
        2)   Absorb water and minerals.
        3)   Store surplus sugars.
        4)   Transport water, minerals, and other materials.
        5)   Produce hormones.

    B.   <u>Shoots</u> (stems, leaves, and reproductive structures, all above ground); have 4 functions:
        1)   Photosynthesis (leaves, young green stems).

2) Transport of materials.
3) Reproduction.
4) Synthesis of hormones.

C. Two <u>types of flowering plants</u>:
1) <u>Monocots</u> (Class Monocotyledonae): grasses, lilies, orchids.
2) <u>Dicots</u> (Class Dicotyledonae): deciduous trees and bushes, most garden flowers.

3. **Plant development.**

A. Animals and plants develop differently.
1) As a human infant grows to become an adult:
    a. All body parts become larger.
    b. At a certain height, growing upwards ceases.
2) Flowering plants:
    a. Grow throughout their lives, never reaching a stable "adult" form.
    b. Grow longer only at the tips of branches and roots.

B. Plants have two different <u>categories of cells</u>.
1) <u>Meristem cells</u>: embryonic, undifferentiated, capable of cell division.
2) <u>Differentiated cells</u>: specialized in structure and function, cannot divide.
3) Each time a meristem cell divides, one daughter cell becomes differentiated and the other remains meristematic.
4) Continued division of meristem cells keeps a plant growing throughout life.

C. <u>Plant growth</u>.
1) Occurs by division and differentiation of meristem cells in 2 regions.
    a. <u>Apical meristems</u>: tips of roots and shoots (main stems and branches).
    b. <u>Lateral meristems</u> or <u>cambium</u>: tracts of cells within roots and shoots.
2) Types of growth:
    a. <u>Primary growth</u>:
        i. Mitotic division of apical meristems and differentiation of resulting daughter cells.
        ii. Growth in length of shoots and roots.
    b. <u>Secondary growth</u>:
        i. Mitotic division of lateral meristems and differentiation of resulting daughter cells.
        ii. Growth in width and thickness of stems and roots which become woody.

4. **Plant tissues and cell types.**

A. Land plants have <u>three tissue systems</u>:
1) <u>Dermal tissue</u>: covers the outer surface of the plant.
2) <u>Vascular tissue</u>: transports water, minerals, sugars, and hormones throughout the plant.
3) <u>Ground tissue</u>:
    a. All non-dermal and non-vascular tissues.
    b. Makes up most of the body of young plants.

      c.   Functions include:
            i.   Photosynthesis.
            ii.   Support.
            iii.   Storage.

    4)   Each tissue system arises from division of meristem cells and differentiation of daughter cells.

B.   <u>Dermal tissue system</u>: two tissue types.
    1)   <u>Epidermal tissue</u>.
        a.   Covers leaves, stems, and roots of young plants.
        b.   Covers flowers, seeds, and fruits.
        c.   Shoot epidermis composed of thin-walled cells packed close together and made waterproof with a waxy <u>cuticle</u>.
        d.   Root epidermis not covered with cuticle since roots must absorb water and minerals.
        e.   Some epidermal cells produce fine hairs.
            i.   Root hairs greatly increase the surface area of root in contact with soil.
            ii.   Stem and leaf hairs in desert plants reflect sunlight and produce an unstirred layer of air close to the epidermis to minimize water loss.
    2)   <u>Periderm tissue</u>.
        a.   Replaces epidermis on roots and stems of woody plants as they age.
        b.   Composed of thick walled cork cells that die at maturity.
        c.   Cork cells form the bark of trees and woody shrubs, and the woody covering of their roots.

C.   <u>Ground tissue system</u>: three tissue types.
    1)   <u>Parenchyma tissue</u>:
        a.   Most abundant ground tissue.
        b.   Thin-walled cells alive at maturity.
        c.   Carries out plant metabolic activities.
        d.   Serves diverse functions:
            i.   Photosynthesis.
            ii.   Storage of sugars and starches.
            iii.   Secretion of hormones.
        e.   Capable of cell division.
    2)   <u>Collenchyma tissue</u>:
        a.   Elongated, many-sided cells with irregularly thickened but flexible cell walls.
        b.   Alive at maturity
        c.   Incapable of cell division.
        d.   Provides support in herbs, and in leaf petioles and young growing stems.
    3)   <u>Sclerenchyma tissue</u>:
        a.   Cells with thick, inflexible, secondary cell walls.
        b.   Supports and strengthens plant bodies.
        c.   Die at the last stage of differentiation.
        d.   Found in xylem, phloem, and elsewhere.
        e.   Comprises fibrous portions of hemp and jute: used to make rope.
        f.   Forms nut shells, outer portion of peach pits, and gritty texture of pears.

D.   <u>Vascular tissue system</u>: two complex tissues (xylem and phloem).

    1)   <u>Xylem</u>:
        a.   Conducts water and minerals in thick-walled dead tubes made from two types of cells, tracheids and vessel elements.
        b.   <u>Tracheids</u>:
            i.   Narrow cells with slanted ends like tips of hypodermic needles.
            ii.   Are stacked atop each other with slanted ends overlapping.
            iii.   Slanted ends have <u>pits</u> where secondary walls don't form, so that water and minerals can pass from tracheid to tracheid by crossing only the thin water-permeable primary cell wall.
        c.   <u>Vessel elements</u>:
            i.   Resemble soup cans: large in diameter with blunt bends.
            ii.   Complete perforations (no cell walls) occur at ends of adjoining vessel elements.
            iii.   If ends completely disintegrate, an open pipeline results from root to leaves.

    2)   <u>Phloem</u>:
        a.   Carries concentrated sugar solutions through tubes made of <u>sieve tube element</u> cells containing only a thin layer of cytoplasm lining the plasma membrane at maturity.
        b.   Where adjacent cells meet, holes form in the cell walls, resulting in <u>sieve plates</u> connecting the interiors of the two cells and creating a continuous conducting system.
        c.   Sieve tube elements are alive but lack ribosomes, Golgi apparatus, and nuclei.
        d.   Each sieve tube element is nourished and maintained by smaller, adjacent <u>companion cells</u>.
        e.   Companion cells also regulate movement of sugars into and out of the sieve tubes.

5.   Roots: Anchorage, absorption, and storage.

A.   <u>Types of root systems</u>:
    1)   In dicots (carrots, dandelions), the primary root becomes larger and thicker, forming a <u>taproot</u> with smaller side branches.
    2)   In monocots (grasses), the primary root dies off and is replaced by many new roots of nearly equal size, forming a <u>fibrous root system</u>.

B.   <u>Primary growth in roots</u>: the apical meristem cells divide and give rise to 4 types of tissue:
    1)   <u>Root cap</u>: protects apical meristem from mechanical damage
    2)   <u>Epidermis</u>: along the outer edge of the root.
    3)   <u>Vascular cylinder</u>: at the core of the root, containing tubular cells.
    4)   <u>Cortex</u>: storage cells between 2) and 3).

C.   <u>Epidermis</u>: outermost layer of cells in a root.
    1)   <u>Cell walls</u> are highly water permeable as are the spaces between cells.
    2)   Many epidermal cells have <u>root hairs</u> projecting into the soil. These increase surface area so roots can absorb more water and minerals.

D. <u>Cortex</u>: beneath the root epidermis. Made of 2 very different cell types.
   1) <u>Outer</u> <u>mass</u> of loosely packed larger cells; store starches converted from sugars made during photosynthesis.
   2) Inner ring of smaller close-fitting cells called <u>endodermis</u>:
      a. Cell walls are waterproof where endodermal cells touch each other (like mortar in a brick wall). The waxy waterproofing material is called the <u>Casparian strip</u>.
      b. Water and minerals from the outside can penetrate a root through and between epidermal and outer cortex cells but can't pass <u>between</u> endodermal cells, only <u>through</u> their cytoplasm.
      c. Thus, the Casparian strip plays an important role in water and mineral absorption in roots.

E. <u>Vascular cylinder</u>.
   1) Contains conducting tissue of 2 types:
      a. <u>Xylem</u>: conducts water and minerals up from the roots to other plant parts.
      b. <u>Phloem</u>: conducts concentrated sugar solution from leaves to other plant parts.
   2) <u>Pericycle</u>: outermost cell layer; is meristematic (can divide) and produces <u>branch</u> <u>roots</u> that break through the cortex to the outside as they grow.

6. Stems: Reaching for the light.

A. <u>Primary growth and structure of stems</u>.
   1) Stems develop from <u>apical meristem cells</u> that lie at the tip of the young shoot.
   2) Daughter cells of the apical meristem differentiate into specialized stem, bud, leaf, and flower cell types.

B. <u>Surface structures of stems</u>.
   1) As shoots grow, small clusters of meristem cells are left at spots on the stem surface, forming <u>leaf primordia</u> and <u>lateral buds</u> that appear at <u>nodes</u> on the stem.
   2) Regions between nodes are <u>internodes</u>.
   3) Leaf primordia develop into mature leaves.
   4) lateral buds can grow into branches.

C. <u>Internal organization of primary dicot stems</u>.
   1) Moving inward from the surface, stems have 4 tissue types: outer <u>epidermis</u>, <u>cortex</u>, <u>vascular</u> <u>tissues</u>, and inner <u>pith</u>.
   2) <u>Epidermis</u>.
      a. Secretes a waxy coating (<u>cuticle</u>) to reduce evaporation.
      b. Perforated with adjustable pores (<u>stomata</u>) to regulate H_2O, O_2, and CO_2 diffusion.
   3) <u>Cortex</u> and <u>pith</u>. Functions:
      a. <u>Support</u>: by turgor pressure (due to H_2O filling the central vacuole) and collenchyma and sclerenchyma cells with thick cell walls.
      b. <u>Storage</u>: parenchyma cells convert sugar into starch stored as a food reserve.

      c.   <u>Photosynthesis</u>: the outer layers of cortex cells contain chloroplasts in many plants (e.g., in cactus stems).

   4)   <u>Vascular tissues</u>: transport materials through stems, roots, and leaves.

      a.   In growing shoot tips, <u>primary xylem</u> and <u>phloem</u> arise from apical meristem cells.

      b.   In young stems, primary xylem and primary phloem are arranged in concentric cylinders.

D.   <u>Stem branching</u>. Upon stimulation by hormones, a lateral bud containing meristem cells grows into a branch containing the same cell types as the main stem.

E.   <u>Secondary growth</u> in dicot stems: they become thicker due to cell divisions of the lateral meristem cells in the <u>vascular cambium</u> and the <u>cork cambium</u>.

   1)   <u>Vascular tissues</u>.

      a.   <u>Vascular cambium</u> is a cylinder of meristem cells between the primary xylem and primary phloem. As the vascular cambium divides:

         i.   Daughter cells towards the center of the stem become <u>secondary xylem</u>.

         ii.   Daughter cells towards the outside of the stem become <u>secondary phloem</u>.

      b.   The secondary xylem pushes the vascular cambium and all outer tissues further out, increasing the diameter of the stem.

      c.   The secondary xylem forms the <u>wood</u> of a tree trunk.

      d.   Phloem cells are much weaker than xylem cells.

         i.   As they die with age, sieve tube elements and companion cells are crushed between the hard xylem and the tough cork.

         ii.   Only a thin strip of recently formed phloem remains alive and functional.

      e.   <u>Annual rings</u> occur in wood because spring xylem cells grow larger and are pale while summer xylem cells are smaller and darker in color.

   2)   <u>Surface tissues</u>.

      a.   As the stem expands in diameter due to secondary growth:

         i.   The epidermis splits off and dies.

         ii.   Some cortex parenchyma cells become a new lateral meristem (the <u>cork cambium</u>), dividing to form tough waterproof cork cells.

      b.   <u>Bark</u>: all tissues outside the vascular cambium (phloem, cork cambium, and cork). Removal of a strip of bark around a tree's circumference (called "girdling") kills the tree since the phloem is severed and the roots can't get sugar.

7.   **Leaves: nature's solar collectors**.

A.   Leaves are the major photosynthetic structures of most plants:

   1)   Have a large surface area for gathering light.

   2)   Are porous to permit CO_2 to enter.

   3)   Are waterproof to prevent excessive evaporation.

B.   <u>Leaf structure in a flowering plant</u>.

   1)   <u>Leaf parts</u>.

        a.   A flat <u>blade</u>.

        b.   <u>Petiole</u>: stalk connecting leaf to stem, containing vessels of xylem and phloem.

        c.   <u>Veins</u>: vascular bundles of xylem and phloem branched through the blade.

2)   <u>Leaf cells</u>.

        a.   <u>Epidermis</u>: layer of non-photosynthetic transparent cells that secrete a waxy waterproof cuticle.

        b.   <u>Stomata</u>: adjustable pores that regulate diffusion of H_2O and CO_2. Each stoma has 2 photosynthetic <u>guard</u> <u>cells</u> that surround and regulate the size of the pore.

        c.   Photosynthetic <u>mesophyll</u> ("middle of the leaf") <u>cells</u>, just below the epidermis, are of 2 types:

             i.   Columnar <u>palisade</u> <u>cells</u> below the upper epidermis.

             ii.   Irregularly shaped <u>spongy</u> <u>cells</u> above the lower epidermis.

3)   <u>The photosynthetic process</u>.

        a.   Xylem brings water to the mesophyll cells.

        b. CO_2 diffuses into the leaf through stomata.

        c.   Sunlight energy is transmitted through the epidermis to the mesophyll.

        d.   Sugars from photosynthesis are transported to other plant parts through the phloem.

8. Special adaptations of roots, stems, and leaves.

A.   <u>Root adaptations</u>: Fewer than seen in stems and leaves.

    1)   Storage: beets, carrots, sweet potatoes.

    2)   Photosynthesis: certain orchids that grow on trees have green roots.

B.   <u>Stem adaptations</u>.

    1)   Runners: strawberries; horizontal stems, growing along the ground, sprout roots and develop into new plants .

    2)   Water storage in above-ground stems: saguaro cactus, baobab tree.

    3)   Carbohydrate storage in underground stems: white potatoes (each "eye" is a lateral bud) and irises (have plump underground stems called <u>rhizomes</u>).

    4)   Thorns: roses; modified branches for protection.

    5)   Tendrils: grapes, Boston ivy; modified branches for wrapping around vertical objects for support.

C.   <u>Leaf adaptations</u>.

    1)   Large size in shady watery environments.

    2)   Thick leaves to store water in desert "succulent" plants.

    3)   Thin spines for protection against predators and to reduce water loss in "cactus" plants; water storage and photosynthesis occurs in the fleshy stems.

    4)   Tendrils for climbing in peas.

    5)   Storage organs in onions, daffodils, tulips, etc. The "bulb" is a very short stem containing many fleshy overlapping leaves that store nutrients and energy.

    6)   Predatory leaves of Venus flytrap and sundew plants capture insects as a source of nitrogen.

    7)   Flowers: reproductive leaves of angiosperms.

****

REVIEW QUESTIONS:

1. Name five functions of roots.
2. Name four functions of stems.
3. Explain the differences between meristem cells and differentiated cells in land plants.
4. Explain the differences between primary growth and secondary growth in plants.
5. How are the roots of carrots and crabgrass different?
6. Name the three tissue systems in plants.
7. Name the three types of ground tissue and tell how they differ from each other.

8.-15.   **MATCHING TEST:** roots

	Choices:	A.	Root cap	D.	Vascular cylinder
		B.	Epidermis	E.	Pith
		C.	Cortex		

8. Contains xylem and phloem cells.
9. Along the outer edge of the root.
10. Contains pericycle cells.
11. Protects the apical meristem from mechanical damage.
12. Produces branch roots.
13. Stores starches.
14. Produces root hairs.
15. Contains endodermal cells.

---

16. Explain the important role played by endodermal cells and the Casparian strip in roots.
17. In what ways do xylem and phloem differ in function?
18. Starting from the outside and working inward, name the four types of tissue encountered in a primary dicot stem.

19.-24.   **MATCHING TEST:** dicot stems

	Choices:	A.	Nodes	D.	Vascular tissues
		B.	Epidermis	E.	Pith
		C.	Cortex		

19. Stores starches.
20. Have stomata.
21. Forms leaves and lateral buds.
22. Capable of some photosynthesis.
23. Transport materials up and down stems.
24. Secretes the waxy cuticle.

25.-33. **MATCHING TEST**: vascular tissues in stems

Choices: A. Xylem   C. Both of these
   B. Phloem   D. None of these

25. Conducts water and minerals.
26. Some cells retain their nuclei when mature.
27. Primarily arises from apical meristem cells.
28. Transport sugars.
29. Contains tracheids and vessels.
30. Secondarily arises from vascular cambium.
31. Cells are dead when mature.
32. Arise from cork cambium.
33. Contains sieve tubes and companion cells.

----

34. Explain the relationship between sieve tube elements and companion cells.

35.-40. **MATCHING TEST**: secondary growth in dicot stems

Choices: A. Vascular cambium   D. Wood
   B. Cork cambium   E. Annual rings
   C. Bark

35. Girdling removes this, causing death of the tree.
36. Daughter cells become secondary xylem and secondary phloem.
37. Produces waterproof cells at the surface of a stem.
38. Forms from secondary xylem.
39. All tissues outside the vascular cambium.
40. Forms because spring xylem cells are larger and paler than summer xylem cells.

41.-46. **MATCHING TEST**: leaves

Choices: A. Petioles   D. Mesophyll
   B. Veins   E. Cortex
   C. Stomata

41. Adjustable pores that regulate diffusion of H_2O and CO_2.
42. Photosynthetic cells.
43. Stalk connecting leaf to stem.
44. Contains palisade and spongy cells.
45. Surrounded by guard cells.
46. Vascular bundles of xylem and phloem tubes.

****

ANSWERS TO REVIEW QUESTIONS:

1.	see 2.A.1)-5)			19.-24.	[see 6.B.-C.]			
					19.	CE	22.	C
2.	see 2.A.1)-4)				20.	B	23.	D
					21.	A	24.	B
3.	see 3.B.1)-4)							
				25.-33.	[see 4.D.1)-2); 6.C.4)]			
4.	see 3.C.1)-2)				25.	A	30.	C
					26.	B	31.	A
5.	see 2.C.1)-2); 5.A.1)-2)				27.	C	32.	D
					28.	B	33.	B
6.	see 4.A.1)-3)				29.	A		
7.	see 4.C.1)-3)			34.	see 4.D.2)d-e.; 6.E.1)-2)			

8.-15.	[see 5.B.-E.]			
8.	D	12.	D	
9.	B	13.	C	
10.	D	14.	B	
11.	A	15.	C	

35.-40.	[see 6.E.1)-2)]			
35.	C	38.	D	
36.	A	39.	C	
37.	B	40.	E	

16. see 5.D.2)a.-c.

17. see 4.D.1)-2)

41.-46.	[see 7.B.1)-2)]			
41.	C	44.	D	
42.	D	45.	C	
43.	A	46.	B	

18. see 6.C.1)-4)

****

CHAPTER 27: NUTRITION AND TRANSPORT IN LAND PLANTS

CHAPTER OUTLINE.

In this chapter, the ways plants acquire minerals and water are elucidated. In addition, the transport of water and minerals from roots to leaves is discussed in light of the "cohesion-tension" theory, and the movement of sugars from leaves to other plant parts is explained in terms of the "pressure-flow" theory.

1. **Comparison of plant and animal nutrition.**

    A. Nutrition: the activities by which organisms acquire nutrients from the environment and process them into various molecules of their own bodies.

    B. Nutrition activities:
       1) Acquisition of nutrients.
       2) Digestion, if required.
       3) Distribution of nutrients throughout the body.
       4) Synthesis of molecules for the organism's body.

    C. Similarities between plant and animal nutrition: utilization of inorganic ions not incorporated into organic molecules.
       1) Usually acquired from the environment as ions dissolved in water.
       2) Become important cellular or extracellular components without further modification.

    D. Differences between plant and animal nutrition: use of organic molecules and minerals (like nitrogen, phosphorus, and sulfur) incorporated into organic molecules.
       1) Plants and animals have large carbohydrates, lipids, and proteins composed of smaller subunits.

      2)   Animals make most of the subunits, but lack enzymes to make "essential" nutrients that must be eaten.

      3)   Plants acquire <u>all</u> their nutrients from soil, water, or air and have enzymes to make all necessary organic molecules from these nutrients.

E.   <u>Energy</u> production and use in plants and animals: ATP is the main energy molecule in plants and animals.
      1)   Animals make ATP by breaking down organic molecules acquired in the diet.
      2)   Plants get energy from sunlight (via photosynthesis) and use it to make ATP and carbohydrates.

F.   <u>Plant nutrition</u>.
      1)   Plants get nutrients as simple inorganic molecules.
      2)   Table 27-2 described the sources and functions of:
          a.   <u>Macronutrients</u>: used by plants in large amounts.
          b.   <u>Micronutrients</u>: used by plants in small amounts.
      3) CO_2 and O_2 enter a plant by diffusion from air into leaves, stems, and roots.
      4)   Roots extract water and nutrients (called "minerals") from the soil.

2.   **Acquisition of minerals**.

A.   Only minerals dissolved in soil water are accessible to roots.
      1)   Concentration of minerals in soil water is lower than concentration within plant cells.
      2)   Thus, most minerals are moved into a root against their concentration gradient by <u>active transport</u>.

B.   <u>Four step process</u> of mineral absorption by roots:
      1)   Active transport into root hairs.
      2)   Diffusion through root hair cytoplasm to pericycle cells via pores called <u>plasmodesmata</u>.
      3)   Active transport from pericycle cytoplasm into the extracellular space of the vascular cylinder.
      4)   Diffusion into the xylem.

C.   <u>Role of Casparian strip</u>.
      1)   If water and minerals could pass between endodermal cells, the minerals would leak out of the extracellular space as fast as they were pumped in, causing:
          a.   Energy wastage.
          b.   Reduction of concentration gradient needed for diffusion of minerals into xylem cells.
      2)   Casparian strip "leakproofs" the vascular cylinder, retaining the concentrated mineral solution within the extracellular space of the vascular cylinder.

D.   <u>Symbiotic relationships in plant nutrition</u>: many plants have mutually beneficial relationships with other organisms that aid in acquiring scarce nutrients.
      1)   <u>Mineral acquisition and Mycorrhizae</u>.
          a.   Most plants form root-fungal complexes called <u>mycorrhizae</u> that facilitate mineral extraction and absorption from rocks.

      b.   The fungus convert rock-bound nutrients into simpler water-soluble compounds that root hairs can absorb and transport.

      c.   The fungus gets sugars and amino acids from the plant.

2)   <u>Nitrogen acquisition</u>.

      a.   80% of the atmosphere is N_2 gas, but plants can only take up N from their roots, usually in the form of ammonium (NH_4^+) or nitrate (NO_3^-) ions.

      b.   <u>Nitrogen-fixing</u> <u>bacteria</u> have the enzymes necessary to convert N_2 into NH_4^+ and NO_3^-.

      c.   <u>Legume</u> plants allow nitrogen-fixing bacteria to live in <u>root</u> <u>nodules</u> to the benefit of both:

            i.   The plant supplies the bacteria with sugar for energy.

            ii.   The bacteria produce excess NH_4^+ or NO_3^- for the plant.

3.   **Acquisition of water**.

A.   <u>Osmosis</u>: causes water movement into roots since mineral concentration is greater (and water concentration is lower) inside root cells than in the soil.

B.   <u>Water movement</u>:

1)   The low mineral solution in the soil has a high water concentration relative to the root cells.

2)   At the endodermis, the waterproof Casparian strip blocks movement of water between cells.

3)   The cell membranes of the endodermis act as semipermeable membranes separating the outer low mineral (high water) solutions from the inner high mineral (low water) solutions.

4)   Thus, water moves from the extracellular space outside the Casparian strip, through the endodermal cells, and into the extracellular space within the vascular cylinder by <u>osmosis</u>.

5)   Water then moves from the extracellular space within the vascular cylinder into tracheids and vessel elements of xylem through their porous cell wall pits.

6)   Water is then pulled up the xylem, powered by evaporation of water from the leaves.

4.   **Transport of water and minerals**.

A.   Land plants move water and dissolved minerals up from roots to stems and leaves <u>en masse</u> by <u>bulk</u> <u>flow</u>.

B.   "<u>Cohesion-tension theory</u>": water is pulled up the xylem powered by the force of evaporation of water from the leaves.

1)   The 2 essential parts of the theory:

      a.   <u>Cohesion</u>: water within xylem tubes holds together like a solid rope or wire. The network of hydrogen bonds within water is quite strong, giving water high cohesion.

      b.   <u>Tension</u>: the "water rope" is pulled up the xylem with evaporation providing the necessary energy.

            i.   As leaves transpire, water concentration in the mesophyll falls, causing osmosis of water from xylem in nearby veins into the dehydrating mesophyll cells.

    ii.   When water molecules leave the xylem, they pull adjacent waters up the xylem. Thus, water moves by bulk flow up the xylem.

    iii.   In roots, the upward movement of water in the xylem causes water to move into the vascular cylinder by osmosis through the endodermal cells.

  2)   The flow of water in xylem is unidirectional, from root to shoot, because only shoots can transpire.

C.   <u>Control of transpiration</u>.

  1)   Transpiration has positive and negative effects:

    a.   Positive: provides the force that transports water and minerals to the leaves.

    b.   Negative: transpiration through leaf stomata is the largest source of water loss in plants, and must be closely controlled to avoid desiccation.

  2)   Leaves open and close <u>stomata</u> according to a delicate balancing of CO_2 and water loss.

    a.   Generally, stomata open during the day (for photosynthesis) and close at night.

    b.   Stomata will close if the leaf begins to dehydrate.

  3)   <u>Stoma structure and function</u>.

    a.   A stoma is a central opening surrounded by two kidney-shaped <u>guard cells</u> that regulate the stoma size.

    b.   Guard cells change the size of the opening between them by changing their own shape.

  4)   <u>Relationship between guard cell shape and opening a stoma</u>.

    a.   Stomata open when guard cells take up water and swell, and close when guard cells lose water and shrink.

    b.   Cellulose fibers in the wall encircle the guard cells like inelastic belts.

      i.   When water enters, guard cells cannot become fatter but do become longer.

      ii.   Each pair of guard cells is attached at both ends, and as they become longer, they must bow outward, opening the hole between them.

  5)   <u>Osmosis of water into/out of guard cells</u>.

    a.   When potassium (K) enters guard cells, water follows by osmosis and the stomata open.

    b.   When K leaves, water leaves by osmosis and the stomata close.

    c.   Factors regulating K concentration inside guard cells:

      i.   <u>Light reception</u>. Guard cells have pigments that, when they absorb light, trigger reactions that cause K to be actively transported into the cells.

      ii.   <u>CO_2 concentration</u>: K concentration inside guard cells is regulated by the balance between photosynthesis and aerobic cellular respiration.

        a)   During the day, photosynthesis uses CO_2 faster than respiration makes it. Low CO_2 levels trigger active transport of K into guard cells. Water follows by osmosis, guard cells swell, and stomata open.

        b)   During the night, photosynthesis stops but respiration continues, building up CO_2 levels. This stops active transport of K into guard cells, the K diffuses out, water leaves by osmosis, guard cells shrink, and stomata close.

    iii.   <u>Water</u>. If a leaf loses water too fast, mesophyll cells release a hormone (<u>abscissic acid</u>) which causes guard cells to lose K. Water is lost by osmosis, guard cells shrink, and the stomata close.

5.   **Transport of sugars** from leaves to other plant parts is the function of phloem.

    A.   Using aphids to study phloem, botanists discovered that phloem fluid:
        1)   Is under positive pressure (is pushed through the sieve tubes).
        2)   Contains a high sugar concentration.

    B.   <u>"Pressure-flow theory"</u> for sugar movement in phloem.
        1)   Sucrose movement from leaf into developing fruit.
            a.   Concentrated sugar from photosynthesis enters leaf companion cells in leaf vein phloem via active transport.
               i.   Sucrose then diffuses into adjacent sieve tube elements.
               ii.   Water enters the sieve tubes by osmosis from nearby xylem cells.
            b.   The developing fruit is a "sucrose sink."
               i.   Sucrose is actively transported out of the sieve-tube elements into the cells of a fruit, lowering the sugar level in that end of the system.
               ii.   Water leaves these sieve tubes due to osmosis, and follows the sugar into the fruit.
            c.   <u>Bulk flow</u>, driven by hydrostatic pressure gradient (like in a garden hose).
               i.   If water enters the leaf end and leaves the fruit end of the same sieve-tube element, the water will flow <u>in bulk</u> from leaf to fruit, driven by the difference in water pressure between the two ends.
               ii.   The bulk flow of water carries the dissolved sugar along with it.
        2)   <u>Sinks and sources of sucrose</u>.
            a.   Phloem flow is directed by sugar production and use.
            b.   Any structure that makes sugar will be a source of phloem flow.
            c.   Any structure that uses up sugar will be a "sink" towards which phloem will flow.
            d.   Fruits, roots, flowers, and new leaves are all "sinks." Thus phloem can move either up or down the plant.

****

REVIEW QUESTIONS:

1. Name the four activities involved in nutrition.
2. In what ways are plant and animal nutrition similar and different?
3. What is meant by micronutrients and macronutrients?
4. Name the four steps involved in mineral absorption by roots.
5. Explain the role of mycorrhizae and nitrogen-fixing bacteria in plant nutrition.
6. Explain the "cohesion-tension" theory for water movement up the xylem.
7. Explain how the movement of potassium (K) into and out of guard cells is related to the balance between photosynthesis and respiration and how this regulates the opening and closing of stomata.
8. Explain the "pressure-flow" hypothesis for sugar flow in phloem.
9. Why is the flow of dissolved minerals unidirectional in a plant while the flow of sugars is not unidirectional?

****

ANSWERS TO REVIEW QUESTIONS:

1.	see 1.B.1)-4)		6.	see 4.B.1)-2)
2.	see 1.C.-D.		7.	see 4.C.5)c.ii.
3.	see 1.F.2)a.-b.		8.	see 5.B.1)-2)
4.	see s.B.1)-4)		9.	see 4.B.;5.B.
5.	see 2.D.1)-2)			

****

CHAPTER 28: PLANT REPRODUCTION AND DEVELOPMENT

CHAPTER OUTLINE.

This chapter discusses plant life cycles, and the structure and function of flowers, including pollination and pollinators. Also, gametophyte, seed, and fruit development in flowering plants is covered, as well as the germination and growth of seedlings and the coevolution of flowers and their pollinators.

1.   **Reproduction** and the plant **life cycle.**

    A.   Most plants can reproduce either <u>asexually</u> or <u>sexually</u>.
        1)   <u>Asexually</u> produced offspring arise by mitosis from cells of the parent plant, and thus are genetically identical to the parent. Asexual reproduction is highly efficient but results in a lack of genetic variation among plants.
        2)   <u>Sexually</u> produced offspring usually combine genes from two parents. The new combination of traits they inherit may help them adapt to changes in the environment. Due to this advantage, most plants have evolved sexual reproduction.

    B.   **Plant life cycles.**
        1)   Plants have <u>alternation</u> <u>of</u> <u>generations</u> (2 distinct multicellular "adult" forms):
            a.   A diploid <u>sporophyte</u> ("spore plant") <u>generation</u> with cells that undergo meiosis to produce haploid <u>spores</u> (not gametes).
            b.   A haploid <u>gametophyte</u> ("gamete plant") <u>generation</u> that grows from spores and produces gametes (sperm and eggs) by mitosis.
            c.   The gametes fuse (<u>fertilization</u>), producing a diploid <u>zygote</u> that develops by mitosis into a sporophyte.
        2)   Alternation of generations occurs in all plants:
            a.   In primitive land plants (mosses and ferns), the gametophyte is small and independent, producing mobile sperm that must swim to adjacent

gametophytes to find eggs. Water is essential to sexual reproduction here.

b. Seed plants (gymnosperms and angiosperms) package the sperm in waterproof cases that don't need water for transport. In angiosperms:

    i. Male and female spores (the start of the gametophyte generation) are formed by meiosis in the flowers.

        a) <u>Megaspores</u> (large spores) develop into female gametophytes.

        b) <u>Microspores</u> (small spores) develop into male gametophytes.

    ii. The female gametophyte remains in the flower to produce eggs.

    iii. The male gametophyte becomes a <u>pollen grain</u> (waterproof sperm-transport packet) carried or blown from one flower to another. It germinates and grows down to the female gametophyte, releasing its sperm near an egg.

    iv. After fertilization, the zygote (start of the new sporophyte generation) is enclosed in a drought-resistant <u>seed</u> from which the seedling will sprout.

2. Evolution of flowers.

A. The earliest seed plants were gymnosperms (conifers):

1) In spring, small male cones release many windborne pollen grains.

2) Some grains land on large female cones, germinate and release sperm to fertilize eggs there.

3) Since most pollen grains are lost, the windborne process is inefficient.

4) About 150 million years ago, insects began carrying pollen grains from protein-rich male cones to sugar-rich female cones as they fed, a more efficient process of pollen transfer favored by natural selection.

B. <u>Pollination</u> by insects, especially beetles, increased in efficiency due to:

1) Plants producing enough pollen and sugary secretions (<u>nectar</u>) within the cones to assure regular visits from the hungry insects.

2) Insects visiting a series of plants of the same species, pollinating them along the way.

3) About 130 million years ago, plants developed flowers to attract insect pollinators and allow them to choose which species to visit.

3. Flower structure: flowers evolved from leaves.

A. Complete <u>flowers</u> possess 4 parts, from the outside in:

1) <u>Sepals</u> which may resemble leaves (in dicots) or petals (in monocots). Sepals surround and protect the flower bud.

2) <u>Petals</u>, usually brightly colored and fragrant, advertising the location of the flower to insects.

3) <u>Stamens</u>, or male reproductive structures. Each has 2 parts:

    a. A long, slender stalk or <u>filament</u>.

    b. An <u>anther</u> at the tip containing chambers where pollen is made.

4) <u>Carpels</u>, or female reproductive structures, each having 3 parts:

a. Sticky <u>stigma</u> at the tip for trapping pollen.
b. Elongated <u>style</u> below the stigma.
c. Bulbous <u>ovary</u> with areas (<u>ovules</u>) that contain the haploid female gametophytes with eggs. When mature:
    i.   Ovules become <u>seeds</u>.
    ii. Ovaries become <u>fruits</u>.

B. <u>Incomplete</u> <u>flowers</u> lack one or more of the 4 floral parts, such as stamens (female flowers) or carpels (male flowers).

C. <u>Diagram</u> of a typical flower.

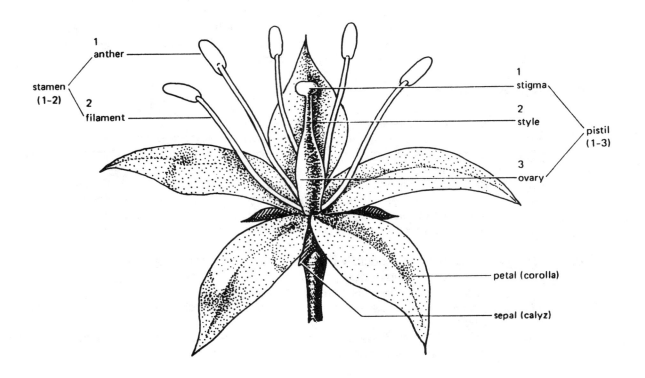

4. Coevolution of flowers and pollinators.

A. <u>General facts</u>.
    1) Wind pollinated flowers are inconspicuous and unscented (oaks and maples).
    2) Animal pollinated flowers have elaborate modifications used to:
        a. Attract useful animal pollinators.
        b. Frustrate undesirable visitors.
        c. Ensure cross-fertilization.

B.　Flowers may provide <u>food</u> for animal pollinators.
　　1)　<u>Beetle-</u> and <u>fly-</u>pollinated flowers smell like rotting flesh or dung, since these insects need protein. They are simple and open in structure since these insects are clumsy.
　　2)　<u>Bee-</u>pollinated flowers are sweet smelling and brightly colored (white, yellow, or blue) with ultraviolet patterns as well. Nectar is produced at the bottom of a short tube to force the bee to stick her head into the flower and pick up pollen on their bodies.
　　3)　<u>Butterfly-</u> and <u>moth-</u>pollinated flowers have deep nectar tubes to allow only insects with long tongues to reach the nectar and pick up pollen. These flowers are brightly colored and fragrant.
　　4)　<u>Hummingbird-</u>pollinated flowers are not fragrant but produce much nectar in deep tubular flowers of red or orange colors, which birds see but bees don't.

C.　Some flowers attract insects <u>sexually</u>, by superficially resembling females of certain species. Some orchids mimic female wasps in scent and shape, attracting males that attempt to copulate with the flower, picking up pollen in the process.

D.　Some insects use the fruits of flowers as <u>nurseries</u> and <u>food</u> for their young. Example: yuccas and yucca moths:
　　1)　A female moth rolls yucca pollen into a compact ball.
　　2)　She flies with the ball to another yucca flower, drills a hole into its ovary and lays eggs there.
　　3)　She then smears pollen from the ball all over the flower stigma.
　　4)　Thus, the young moths have lots of seeds to eat, with plenty left over for successful yucca reproduction as well.

5.　**Gametophyte development** in flowering plants:

A.　<u>Pollen</u> <u>grains</u> (male) and <u>embryo</u> <u>sacs</u> (female) are haploid gametophytes that develop within sporophyte flowers. They are quite small and cannot live on their own.

B.　<u>Pollen</u> develops within the chambers (pollen sacs) of flower anthers.
　　1)　Diploid <u>microspore</u> <u>mother</u> <u>cells</u> develop and divide by meiosis to make
　　2)　Haploid <u>microspores</u>, which divide by mitosis to make
　　3)　Haploid male gametophytes called <u>pollen</u> <u>grains</u> consisting of:
　　　　a.　A smaller <u>generative</u> <u>cell</u> in the cytoplasm of
　　　　b.　A larger <u>tube</u> <u>cell</u>.
　　4)　Pollen grains are released when mature.

C.　<u>Embryo</u> <u>sacs</u> develop within the ovaries of carpels.
　　1)　Diploid <u>ovules</u> develop. Each has outer layers called <u>integuments</u>, surrounding
　　2)　A single diploid <u>megaspore</u> <u>mother</u> <u>cell</u>, which divides by meiosis to produce
　　3)　Haploid <u>megaspores</u>. One enlarges to fill the space within the integuments. Three nuclear divisions occur by mitosis to produce
　　4)　A haploid <u>embryo</u> <u>sac</u> (female gametophyte) with 7 cells:
　　　　a.　Three small cells at each end, each having one haploid nucleus. One is the <u>egg</u> <u>cell</u>.
　　　　b.　One large central cell with two haploid nuclei. This is the <u>primary</u>

endosperm cell.

6.   **Pollination and fertilization.**

   A.   Pollination: a pollen grain lands on a carpel's stigma. Then,
      1)   The tube cell grows down the style towards the ovary, and
      2)   The generative cell divides to form 2 sperm cells.
      3)   The male gametophyte is now mature (has 3 cells).

   B.   Fertilization. After a pollen tube penetrates the embryo sac, releasing 2 sperm cells, double fertilization occurs:
      1)   One sperm cell fertilizes the egg cell, forming the diploid zygote (start of the new sporophyte).
      2)   One sperm cell enters the primary endosperm cell and fuses with its 2 nuclei, forming a triploid cell that develops into endosperm food in the seed.

7.   **Development of seeds and fruits.**

   A.   Seed development.
      1)   The ovule integuments develop into the diploid seed coat that protects the young embryo.
      2)   The triploid endosperm food develops by mitosis.
      3)   The diploid zygote develops into an embryo with a shoot, root, and seed-leaves called cotyledons.
         a.   Dicots have 2 cotyledons which absorb the endosperm food and become fleshy.
         b.   Monocots have one cotyledon which absorbs very little of the endosperm food.
      4)   The embryo's shoot has 2 regions.
         a.   Hypocotyl region below the cotyledons.
         b.   Epicotyl region above the cotyledons, containing the apical meristem of the shoot at its tip.

   B.   Fruit development.
      1)   The ovary wall develops into a diploid fruit.
      2)   By eating fruits, animals dispense seeds (in their feces) to points distant from the parent plant.

   C.   Seed dormancy.
      1)   Seeds have a period of dormancy during which they will not sprout.
      2)   This solves two problems.
         a.   Seeds can't sprout within fruits, otherwise the seeds wouldn't reach the ground.
         b.   Seeds produced in the late summer will not sprout until the next spring, avoiding the harsh winter.
      3)   Mechanisms of seed dormancy. Before sprouting, seeds may need:
         a.   Initial desiccation (will not sprout within juicy fruits).
         b.   Exposure to cold (will not sprout until after winter).
         c.   Disruption of the seed coat: by water in desert plants or digestive

enzymes in animals that eat fruits.

8.   **Germination and growth of seedlings**.

> A.   When a seed <u>germinates</u>:
> 1)   It absorbs water, swells, and bursts the seed coat.
> 2)   The root grows rapidly down into the soil, protected by a root cap.
> 3)   The stem lengthens, pushing up through the soil into the air. The stem tip is protected by:
>> a.   A tough sheath, the <u>coleoptile</u>, in monocots.
>> b.   The shoot being bent into a hook in dicots.
> B.   Endosperm food stored in the seed provides the initial energy for seed sprouting.

****

REVIEW QUESTIONS:

1.   Why is it more advantageous for plants to reproduce sexually than to reproduce only asexually?

2.-8.   **MATCHING TESTS**: alternation of generations

Choices:	A.	Sporophyte	C.	Both of these
	B.	Gametophyte	D.	Neither of these

2.   Begins as a diploid zygote.
3.   Not seen in mosses and ferns.
4.   Multicellular.
5.   Produce haploid spores by meiosis.
6.   Pollen grains.
7.   Occurs in seed plants.
8.   Produce haploid gametes by mitosis.
----

9.   Describe how the evolution of flowers occurred.

10.-17.   **MATCHING TESTS**: flower structure

Choices:	A.	Sepals	C.	Stamens
	B.	Petals	D.	Carpels

10.   Develop into fruits.
11.   Have filaments and anthers.
12.   Surround and protect the flower bud.
13.   Have stigma, style, and ovary.
14.   Usually brightly colored and fragrant.
15.   Produce pollen grains.

16.   Attract insects by their color.
17.   Produce female gametophytes with eggs.

----

18.   Give several examples of coevolution between flowers and their pollinating animals allowing food
        for the correct pollinator while discouraging other animals.

19.-25.   **MATCHING TEST**: gametophytes in flowering plants

            Choices: A.   Male gametophytes   C.   Both of these
                                B.   Female gametophytes   D.   Neither of these

19.   Contains an egg cell and a primary endosperm cell.
20.   Develop within flowers.
21.   Develop from megaspores in the ovules of ovaries.
22.   Pollen grains.
23.   Contains a generative cell and a tube cell.
24.   Develop from microspores in the anthers of stamens.
25.   Embryo sacs.

----

26.   Describe the process of "double fertilization" in flowering plants.

27.-35.   **MATCHING TEST**: seeds and fruits

            Choices: A.   Seed coats   D.   Fruit
                                B.   Endosperm   E.   None of these
                                C.   Embryo

27.   The young sporophyte.
28.   Parts of a seed.
29.   Diploid cells.
30.   Triploid cells.
31.   Protects seeds and helps to disperse them.
32.   Protects the young sporophyte.
33.   Haploid cells.
34.   Food for the young sporophyte.
35.   Develop from the walls of the ovary.

----

36.   What are the advantages for seeds needing a period of dormancy?
37.   What mechanisms exist to ensure seed dormancy?
38.   Describe what happens when a seed germinates.

****

****

ANSWERS TO REVIEW QUESTIONS:

1.   see 1.A.1)-2)

2.-8. [see 1.B.1)-2); 3)a.]

2.	A	6.	B
3.	D	7.	C
4.	C	8.	B
5.	A		

9.   see 2.A.-B.

10.-17. [see 3.A.1)-4)

10.	D	14.	B
11.	C	15.	C
12.	A	16.	B
13.	D	17.	D

18.   see 4.B.1)-4)

19.-25. [see 5.A.-C.]

19.	B	23.	A
20.	C	24.	A
21.	B	25.	B
22.	A		

26.   see 6.A.-B.

27.-35. [see 7.A.-B.]

27.	C	32.	AD
28.	ABC	33.	E
29.	ACD	34.	B
30.	B	35.	D
31.	D		

36.   see 7.C.2)a.-b.
37.   see 7.C.3)a.-c.
38.   see 8.A.-B.

****

CHAPTER 29: CONTROL OF THE PLANT LIFE CYCLE

CHAPTER OUTLINE.

This chapter concentrates on the roles of various plant hormones in the control and integration of plant growth and development, flowering, seed set, fruit development, senescence, and dormancy.

1. **General Facts**:

    A. Plants perceive features of the environment crucial to them such as:
        1) Direction of gravity.
        2) Direction, intensity, and duration of sunlight.
        3) Strength of the wind.

    B. Plants respond to these stimuli by regulating their growth and development in appropriate ways through the action of simple chemicals called <u>plant hormones</u> or growth regulators.

2. Discovery of plant hormones affecting **phototropism** (the bending of plants towards sunlight).

    A. Charles and Francis Darwin's studies showed:
        1) The region a few mm below the tip of a grass coleoptile bends towards light until the tip faces the light source.
            a. Covering the tip stops the bending.
            b. Covering the bending region does not stop the bending.
        2) Conclusions:
            a. The tip detects the direction of light.
            b. Bending occurs further down the coleoptile.
            c. The tip transmits information down to the bending region.

B.   Growth in the bending region occurs entirely due to elongation of pre-existing cells.
1)   Coleoptiles bend due to <u>differential</u> <u>elongation</u> of cells: cells away from the light elongate more than cells towards the light.
2)   Information from the tip causes greater elongation of cells on the coleoptile side away from light.

C.   P. Boysen-Jensen then showed that the information from the tip is chemical in nature:
1)   Cutting off the tip stops elongation and bending of the coleoptile.
2)   Replacing the tip, the coleoptiles:
      a.   Elongated straight up in the dark.
      b.   Showed normal phototropism in the light.
3)   Placing a thin layer of porous gelatin between the severed tip and shaft still allowed normal phototropism, but a thin layer of non-porous mica eliminated all responses.
4)   Conclusion: a chemical is produced in the tip and moves down the shaft, causing cell elongation.

D.   Isolation and identification of the chemical:
1)   F. Went (1920s) placed the severed tips of oat coleoptiles on agar cubes to absorb the chemical.
2)   Placing the agar cubes on coleoptile shafts caused normal phototrophic responses.
3)   When he replaced the severed tips off center and kept the coleoptiles in the dark, the coleoptiles bent away from the side with the tip.
4)   He called the elongation-promoting chemical <u>auxin</u> ("to increase").
5)   K. Thimann later purified auxin and determined its molecular structure.

3.   **Plant hormones** and their actions (see Table 29-1).

A.   <u>Hormones</u> are chemicals produced in one location and transported to other regions where they exert specific effects.

B.   The <u>5 major types</u> of plant hormones are:
1)   <u>Auxins</u>: regulate plant responses to light and gravity.
      a.   Promote cell elongation in shoots and prevent sprouting of lateral buds.
      b.   Stimulate differentiation of vascular tissues and fruits.
2)   <u>Gibberellins</u>:
      a.   Promote cell elongation in stems.
      b.   Stimulate flowering, fruit development, seed germination, and bud sprouting.
3)   <u>Cytokinins</u>:
      a.   Promote cell division.
      b.   Stimulate overall metabolism, delaying the aging of leaves.
4)   <u>Ethylene</u>, the only known gaseous hormone:
      a.   Causes fruit to ripen.
      b.   Causes breakdown of cell walls in abscission layers, allowing leaves, fruit, and flowers to drop off at appropriate times.
5)   <u>Abscisic</u> <u>acid</u> is an inhibitory hormone. It:
      a.   Causes stomata to close when water is scarce.

        b.   Maintains dormancy in buds and seeds in bad weather by inhibiting the action of gibberellin.

4.   **Plant life cycle**: reception, response, and regulation.

    A.   Seed dormancy and germination.
        1)   Abscisic acid enforces seed dormancy by slowing embryo metabolism. The acid is eliminated either by a soaking rain (in desert areas) or a prolonged period of cold (in northern plants).
        2)   Germination is stimulated by gibberellin.
            a.   Synthesis is triggered by heavy rain or bitter cold.
            b.   Gibberellin causes genes to make enzymes that break down endosperm food into subunits needed for embryo growth and germination.

    B.   Seedling growth. Auxins control the responses of both shoots and roots to light and gravity.
        1)   Shoot growth.
            a.   Gravity causes auxins to accumulate on the lower side of the stem where cell elongation then occurs.
            b.   Light causes the same auxin distribution as gravity. So, gravitropism and phototropism augment each other.
        2)   Root growth.
            a.   Auxin is transported from the shoot down to the root.
            b.   The root cap causes the auxin to accumulate on the lower side of a non-vertical root.
            c.   Since shoot-concentrations of auxin inhibit cell elongation in roots, cell elongation on the lower side of the root is inhibited while it is not affected on the upper side.
            d.   Result: the root bends downward.

    C.   Development of the mature plant body.
        1)   The amount of growth in shoot and root systems must always be kept in balance.
            a.   Shoots must be large enough to supply roots with sugars.
            b.   Roots must be large enough to provide shoots with water and minerals.
        2)   Interactions between auxin and cytokinin regulate root and shoot branching.
        3)   In stems, there is apical dominance (growing tip suppresses the sprouting of lateral buds). Explanation:
            a.   Shoot tips make auxin (suppresses lateral buds) while root tips make cytokinins (stimulate sprouting of lateral buds).
            b.   The relative amounts of these hormones control the activity of the buds. Those closest to the tip have auxin > cytokinin, preventing sprouting, while those further down the stem have auxin < cytokinin, stimulating sprouting.
        4)   In roots, even low concentrations of auxin stimulate branching of lateral roots from the pericycle.
        5)   So, more roots make more cytokinin which stimulates more lateral buds to branch. Fewer roots have the opposite effect. And, more stems make more auxin, stimulating more root branching.

D.   Differentiation of <u>vascular tissue</u> is caused by high concentrations of auxins and perhaps gibberellin released by root and shoot meristem cells.

E.   <u>Control of flowering</u>.
    1)   <u>Daylength</u> is the environmental cue plants use to determine when flowering occurs.
        a.   <u>Day-neutral</u> <u>plants</u> flower as soon as they mature, independent of daylength (Examples: tomatoes, corn, snapdragons).
        b.   <u>Long-day</u> <u>plants</u> require a day length longer than some critical value in order to flower (Example: spinach requires > 13 hours of daylight).
        c.   <u>Short-day</u> <u>plants</u> require a day length shorter than some critical value in order to flower (Example: cocklebur needs < 15.5 hours of daylight).
    2)   <u>Mechanisms for measuring daylength</u>:
        a.   To measure daylength, a plant needs:
            i.   A <u>clock</u> to measure time (how long it has been light or dark).
            ii.   A <u>light-detecting system</u> to set the clock.
        b.   All organisms have internal <u>biological</u> <u>clocks</u> that measure time without environmental cues.
        c.   Activities that recur about every 24 hours are called <u>circadian</u> <u>rhythms</u>.
        d.   Plants detect light by using <u>phytochrome</u> (plant color) <u>pigments</u> of 2 interchangeable forms.
            i. P_r form: strongly absorbs red light.
            ii. P_{fr} form: strongly absorbs far-red (almost infrared) light. P_{fr} stimulates or inhibits physiological processes.
            iii.   When P_r absorbs red, it becomes P_{fr}; when P_{fr} absorbs far-red, it becomes P_r; P_{fr} reverts to P_r in total darkness.
            iv.   During the day, plants have roughly equal amounts of P_r and P_{fr}. After dark, P_r gradually replaces P_{fr}.
    3)   <u>Daylength and flowering</u>.
        a.   The length of <u>continuous darkness</u> is more important than the length of daylight in determining flowering. So,
            i.   Spinach is really a <u>short-night</u> plant.
            ii.   Cocklebur is really a <u>long-night</u> plant.
        b.   In short-day (long-night) plants, P_{fr} may <u>inhibit</u> flowering:
            i.   During short nights, P_{fr} never completely disappears, so flowering is inhibited.
            ii.   During long nights, P_{fr} levels reach zero and flowering occurs.
        c.   In long-day (short-night) plants, P_{fr} may <u>promote</u> flowering:
            i.   During short nights, P_{fr} never completely disappears, so flowering occurs.
            ii.   During long nights, P_{fr} levels reach zero, inhibiting flowering.
    4)   Other <u>phytochrome-mediated</u> processes.
        a. P_{fr} inhibits elongation of seedlings, so shaded seedlings grow rapidly (P_{fr} converts to P_r in shade).
        b. P_{fr} stimulates the straightening of the stem hook in dicot seedlings, leaf growth, and chlorophyll synthesis.

F.   <u>Development of seeds and fruits</u>.
    1)   Auxin and/or gibberellin, released from pollen or developing seeds, stimulate ovaries to develop into fruits.

        2)   When seeds are mature, fruits ripen, becoming brightly colored, softer, sweeter, and more attractive to animals:

          a.   Ethylene stimulates fruit ripening.

          b.   Ethylene is produced by fruit cells in response to a surge of auxin released by mature seeds.

G.   Senescence and dormancy.

      1)   In autumn, fruits and leaves undergo rapid aging called senescence. At this time, the abscission layer at the base of the petioles develops, allowing leaves and fruit to fall off.

      2)   Cytokinins produced by roots prevent senescence.

          a.   As winter approaches, roots make less cytokinin and leaves and fruits make less auxin.

          b.   Much organic material in leaves is broken down into simple molecules, absorbed by the plant, and stored.

          c.   Ethylene is released by aging leaves and ripening fruit, weakening the abscission layer, allowing leaves and fruit to fall.

      3)   Bud dormancy is enforced by abscissic acid and metabolism slows down during the winter months.

****

REVIEW QUESTIONS:

1.   What evidence led the Darwins to conclude that the stem tip transmits information down to the bending region of the stem?

2.   How did Boysen-Jensen demonstrate that the information from the stem tip that influences bending is chemical in nature?

3.-16.   **MATCHING TEST**: plant hormones

        Choices: A.   Auxins

                          B.   Gibberellins

                          C.   Cytokinins

                          D.   Ethylene

                          E.   Abscisic Acid

                          F.   None of these

3.   Promotes cell division.

4.   Causes fruit to ripen.

5.   Promotes cell elongation in shoots and stems.

6.   Inhibits the action of gibberellin.

7.   Delays the aging of leaves.

8.   Allows fruits and leaves to fall off at appropriate times.

9.   Prevents sprouting of lateral buds.

10.   Inhibits bud sprouting.

11.   Stimulates flowering.

12.   Enforces seed dormancy.

13. Inhibits cell elongation in roots.
14. The only gaseous plant hormone known.
15. Stimulates root branching
16. Stimulates bud sprouting.

----

17. Describe how seed dormancy and germination are controlled by abscisic acid and gibberellin.
18. Describe how auxins influence shoot growth and root growth.
19. Describe how the interplay of auxins and cytokinins regulate branching in root and stem systems.
20. Define day-neutral, long-day, and short-day plants in relation to flowering.
21. Describe the interplay of P_r and P_{fr} phytochrome pigments.
22. Why is the length of continuous darkness more important than the length of daylight in determining when plants will flower?
23. **True or False?** (Explain your choice in each case.)
    A.   In short-day plants, P_{fr} inhibits flowering.
    B.   In long-day plants, P_{fr} stimulates flowering.
24. Explain why fruits ripen after their seeds mature.
25. Describe the physiology of senescence and dormancy.

****

ANSWERS TO REVIEW QUESTIONS:

1.	see 2.A.1)-2)				19.	see 4.C.2)a.-c.
2.	see 2.C.1)-4)				20.	see 4.E.1)a.-c.
3.-16.	[see 3.B.1)-5)]				21.	see 4.E.2)c.i.-iv.

3.	C	10.	E		
4.	D	11.	B	22.	see 4.E.3)a.-c.
5.	AB	12.	E		
6.	E	13.	A	23.	see 4.E.3)b.-c.
7.	C	14.	D		A and B are true
8.	D	15.	A		
9.	A	16.	B	24.	see 4.F.1)-2)

17.   see 4.A.1)-2)   25.   see 4.G.1)-3)

18.   see 4.B.1)-2)

****

UNIT IV EXAM:

Chapters 25 - 29. All questions are dichotomous. Circle the correct choice in each case.

Chapter 25.

1. Plants generally have (faster / slower) metabolism than animals.
2. Higher (plants / animals) have completely eliminated the need of liquid for any kind for reproduction.
3. The skeletons of (plants / animals) are composed of cellulose.
4. Skeletons of insects are made of (cellulose / chitin).
5. Energy enters natural life systems through (plants / animals).
6. (Mosses / Reptiles) require water for reproduction.
7. (Plants / Animals) make their own amino acids from scratch.
8. (Plants and animals / Only animals) produce hormones to coordinate growth and function.
9. (Amphibians / Insects) have waterproof eggs and internal fertilization.
10. (Conifers / Flowering plants) protect their embryos with seeds and fruits.

Chapter 26.

1. Food is conducted by the (xylem / phloem).
2. Wood is made up of (phloem / xylem).
3. Xylem conducts materials (up / down) the stem.
4. Meristem cells are described as (embryonic / differentiated).
5. The (apical / lateral) meristem is located at the tips of stems and roots.
6. Increase in width is due to (primary / secondary) growth.
7. Monocots are characterized by (taproots / fibrous roots).
8. The light portion of an annual ring is formed during the (spring / summer).
9. The roots of (carrots / strawberries) store much food.
10. The stems of (sweet potatoes / white potatoes) store much food.

Chapter 27.

1. The cohesion-tension theory explains the movement of (food / water) in a plant.
2. At maturity, some (phloem / xylem) cells are still alive.
3. An increase in the turgor of guard cells causes a stoma to (close / open).
4. Water will (enter / leave) guard cells if their cytoplasm has a higher concentration of solutes than the cytoplasm of surrounding cells.
5. (Lowering / Raising) carbon dioxide levels in a leaf triggers active transport to pump in potassium ions to the guard cells.
6. (Decreasing / Increasing) potassium concentrations within guard cells is a response to high carbon dioxide concentrations in a leaf.
7. The hormone involved in stomatal opening is (abscisic acid / gibberellin).
8. The source of energy for making ATP in plants and animals is (quite similar / very different).
9. Most minerals are moved into a root by (active transport / simple diffusion within water).
10. Plant nutrients from the soil are called (minerals / food).

Chapter 28.

1. (Asexual / Sexual) reproduction is highly efficient but results in lack of genetic variation among the offspring.
2. Primitive plants have a predominant (gametophyte / sporophyte) generation.
3. The zygote is the start of a new (gametophyte / sporophyte) generation.
4. Pollination is more efficient in (gymnosperms / angiosperms).
5. Stamens have (filaments and anthers / stigmas, styles, and ovaries).
6. Exclusively male flowers are (complete / incomplete).
7. Pollen grains and embryo sacs are (gametophyte / sporophyte) structures.
8. Megaspores produce (pollen grains / embryo sacs).
9. Double fertilization produces (zygotes and endosperm food cells / zygotes and embryo sacs).
10. The wall of an ovary develops into a (seed / fruit).

Chapter 29.

1. The stimulus that causes the coleoptile to bend is produced in the (region where it bends / tips).
2. Auxin accumulates on the (shady / sunny) side of the coleoptile.
3. Phototropism is due to cell (division / elongation).
4. Dormancy is promoted by (high / low) concentrations of abscisic acid.
5. The root is (negatively / positively) phototropic and is (negatively / positively) geotropic.
6. Destruction of (lateral / terminal) bud(s) causes the plant to become bushier.
7. Auxin (inhibits / stimulates) lateral buds.
8. Auxins stimulate branching in the (roots / stems).
9. Branches of the root arise from division of the (endodermis / pericycle).
10. Most biological clocks are responsive to (photoperiod / temperature).
11. (Far red / Red) phytochrome is more stable in the dark.
12. The duration of (light / dark) controls flowering.
13. In (long / short) day plants, the far red phytochrome promotes flowering.

14. Immature fruits are (conspicuously / inconspicuously) colored.
15. Ethylene (inhibits / stimulates) fruit ripening.

****

ANSWER KEY:

Chapter 25:

1. slower
2. plants
3. plants
4. chitin
5. plants
6. mosses
7. plants
8. plants and animals
9. insects
10. flowering plants

Chapter 26:

1. phloem
2. xylem
3. up
4. embryonic
5. apical
6. secondary
7. fibrous roots
8. spring
9. carrots
10. white potatoes

Chapter 27:

1. water
2. phloem
3. open
4. enter
5. lowering
6. decreasing
7. abscisic acid
8. very dissimilar
9. active transport
10. minerals

Chapter 28:

1. asexual
2. gametophyte
3. sporophyte
4. angiosperms
5. filaments and anthers
6. incomplete
7. gametophyte
8. embryo sacs
9. zygotes and endosperm cells
10. fruit

Chapter 29:

1. tips
2. shady
3. elongation
4. high
5. negatively, positively
6. terminal
7. inhibits
8. roots
9. pericycle
10. photoperiod
11. red
12. dark
13. long
14. inconspicuously
15. stimulates

****

UNIT V: ANIMAL ANATOMY AND PHYSIOLOGY

CHAPTER 30:   CIRCULATION

CHAPTER OUTLINE.

This chapter covers the vertebrate circulatory system, which is necessary to provide cells with food molecules and oxygen, and allow these cells to eliminate organic wastes and carbon dioxide. Concentrating on humans, discussion of the circulatory system includes the heart, blood vessels, blood cells, and the lymphatic system.

1.   **General facts**.

    A.   The first living cells were nurtured in the sea which provided nutrients that diffused into the cells and washed away wastes that diffused out.
        1)   Diffusion is a slow process (random movement of molecules along a concentration gradient) and diffusion distances must be kept short.
        2)   Diffusion alone is adequate only for simple organisms living in a moist environment (bacteria, protists, sponges, nematode worms).

    B.   Circulatory systems evolved in larger more complex animals:
        1)   Individual cells can be far from the skin surface and would starve or drown in wastes without a system bringing food and removing wastes.
        2)   Circulatory systems are "internal seas", bringing each cell close to a source of food and oxygen, and being a sink for wastes and CO_2.

    C.   All circulatory systems have three major parts:
        1)   Fluid or blood for transport.
        2)   Channels or vessels to conduct the fluid.
        3)   A pump or heart to keep the blood moving.

D.   Major types of circulatory systems in animals:
1)   <u>Open</u> <u>circulatory</u> <u>systems</u>: have an open space (<u>hemocoel</u>) within the body where the vessels empty into and pick up blood.
     a.   Within the hemocoel, tissues are bathed in blood.
     b.   Found in arthropods and most molluscs (clams, snails).
2)   <u>Closed</u> <u>circulatory</u> <u>systems</u>: blood is confined to the heart and a continuous series of vessels.
     a.   This allows for more rapid and efficient flow of blood.
     b.   Found in earthworms, the advanced molluscs (squids, octopuses), and all vertebrates.

2.   Vertebrate (human) circulatory systems:

A.   <u>Important functions</u>:
1)   Transport of oxygen from the lungs to tissues and CO_2 from tissues to the lungs.
2)   Distribution of nutrients from the digestive system to body cells.
3)   Transport of wastes and toxins to the liver (for detoxification) and kidneys (for excretion).
4)   Distribution of hormones from producing organs to sites of action.
5)   Partial regulation of body temperature (e.g., more blood flows to the skin if hot).
6)   Defense against blood loss and attack by microbes.

B.   <u>The heart</u>.
1)   <u>Structure</u>:
     a.   Consists of muscular chambers capable of strong contractions that circulate blood.
     b.   Birds and mammals (warm blooded vertebrates that maintain a constant body temperature) have highly efficient 4-chambered hearts arranged as two separate pumps each with 2 chambers. For each pump:
          i.   An <u>atrium</u> receives and briefly stores blood, then passes it to
          ii.   A <u>ventricle</u> that pumps the blood to the body.
     c.   <u>Functions</u> of the two heart pumps:
          i.   Right pump is for <u>pulmonary</u> <u>circulation</u>: oxygen depleted blood is collected from the body and pumped to the lungs.
          ii.   Left pump is for <u>systemic</u> <u>circulation</u>: newly oxygenated blood is collected from the lungs and pumped to the body.
2)   The <u>cardiac cycle</u>: alternating contraction and relaxation of the heart chambers.
     a.   The two atria contract in synchrony, emptying their contents into the ventricles.
     b.   A fraction of a second later, the two ventricles contract simultaneously, forcing blood into arteries leading from the heart.
     c.   Then, both chambers relax briefly before the cycle is repeated.
     d.   <u>Systole</u>: period of ventricular contraction (b. above).
     e.   <u>Diastole</u>: the rest of the cycle (a. and c. above).
3)   <u>Coordination</u> of heart activity.
     a.   Blood must be pumped through the heart in a single direction: vein → atrium → ventricle → artery.
     b.   The heart has four simple one way valves to ensure unidirectional blood flow.

   i. Pressure in one direction opens them, while reverse pressure closes them tightly.

   ii. <u>Atrioventricular valves</u> separate the atria from the ventricles:

    a) A <u>tricuspid valve</u> ("three pointed") separates the right ventricle and atrium.

    b) A <u>bicuspid valve</u> ("two pointed") lies between the left ventricle and atrium.

   iii. <u>Semilunar valves</u> ("half moon") allow blood to enter the pulmonary artery and the aorta when the ventricles contract, but prevent blood from returning when the ventricles contract.

  c. The heart muscles must beat in smooth coordinated contractions in order to avoid random fibrillation which does not pump blood.

   i. Coordinated contraction requires a <u>pacemaker</u>, an area of muscle that sets the pace for other muscle cells.

   ii. The heart's primary pacemaker is the <u>sinoatrial</u> node (<u>SA</u> node), a small mass of muscle cells in the wall of the right atrium.

   iii. The SA node is spontaneously active, contracting faster than surrounding cells. The SA signal spreads to both atria, causing them to contract in smooth synchrony.

  d. Coordinated contraction of the 4 chambers:

   i. The atria must contract first to push blood into the ventricles and then refill as the ventricles contract.

   ii. There is a delay between contraction of the atria and the ventricles since a barrier of inexcitable tissue between them blocks the SA signal from reaching the ventricles.

   iii. Instead, the SA signal is channeled through a second mass of cells, the <u>atrioventricular</u> node (<u>AV</u> node) which delays the ventricular contraction for about 1/10 second after contraction of the atria.

   iv. From the AV node, the signal to contract travels to the base of both ventricles along tracts of excitable fibers. Then, the signal spreads to ventricular muscles causing the ventricles to contract in unison.

 4) <u>Outside influences</u> on heart rate:

  a. The SA node pacemaker maintains a steady beat of 100/minute.

  b. Influences of nervous impulses and hormones on heart rate:

   i. At rest, the <u>parasympathetic nervous system</u> slows heart rate to about 70 beats/minute.

   ii. During exercise or stress, the <u>sympathetic nervous system</u> accelerates the heart rate.

   iii. The hormone <u>epinephrine</u> increases heart rate.

C. The <u>blood vessels</u>.

 1) Sequence of blood flow: heart → arteries → arterioles → capillaries → venules → veins → heart.

 2) <u>Arteries</u> and <u>arterioles</u>.

  a. Arteries carry blood away from the heart.

  b. Arteries have thick, muscular, elastic walls that expand very little.

  c. Arterioles are smaller diameter branches from the arteries.

3) <u>Capillaries</u>:
   a. Exchange of wastes, nutrients, gases, and hormones between blood and body cells occurs in the capillaries.
   b. Are thin tubes, with walls only one cell thick so that dissolved substances readily diffuse through.
   c. Are microscopically narrow: red blood cells must pass through them in single file.
   d. No body cell is more than 10 micrometers away from a capillary, allowing efficient exchange of materials.
   e. Are quite numerous: total length in humans is over 50,000 miles.
   f. Rate of capillary blood flow is relatively slow.

4) <u>Venules</u> and <u>Veins</u>:
   a. Function: provide low pressure pathways for blood back to the heart.
   b. Have walls much thinner and more distensible than arteries.
   c. Return of blood to the heart is aided by muscle contractions during exercise and breathing which squeeze the veins. One-way valves in the veins prevent back flow of blood.
   d. If blood pressure falls, the sympathetic nervous system stimulates contraction of smooth muscles in the vein walls, decreasing their diameter (blood pressure rises).

5) Distribution of blood flow.
   a. The muscular walls of arterioles are influenced by nerves, hormones, and local chemical changes so that they contract or relax in response to changing needs. Examples:
      i. When frightened, blood leaves the face and is redirected to heart and muscles.
      ii. On a hot day, blood rushes to the skin ("flushing"), cooling the blood.
      iii. On a cold day, blood leaves the extremities (possibly leading to frostbite) to conserve body warmth.
   b. Capillary walls are only a single cell thick: lacking muscles, they act as passive tubes.
      i. The flow of capillary blood is regulated by tint rings of smooth muscle (<u>precapillary sphincters</u>) that surround the junctions between arterioles and capillaries.
      ii. These open and close in response to local changes. Example: CO_2 buildup causes the sphincters to relax, increasing blood flow through capillaries.

D. The <u>blood</u>.
   1) General facts:
      a. Blood has two <u>components</u>:
         i. Specialized <u>cells</u> (red cells, white cells, platelets).
         ii. Fluid <u>plasma</u> (55-60% of blood volume).
      b. A typical human has 5-6 liters of blood, about 8% of total body weight.
   2) <u>Plasma</u>: straw colored fluid.
      a. 90% water with many dissolved substances.
      b. Dissolved substances include:
         i. Proteins and hormones.
         ii. Nutrients (glucose, vitamins, amino acids, lipids).
         iii. Gases (CO_2, O_2).

    iv. Ions (Na^+, Cl^-, Ca^+, K^+, Mg^{++}).

    v. Wastes (urea).

  c. Plasma proteins are the most abundant dissolved molecules, including:

    i. Albumins: help maintain osmotic pressure.

    ii. Globulins: help transport nutrients and function in immunity.

    iii. Fibrinogen: major factor in blood clotting.

3) <u>Red Blood Cells</u> (<u>rbcs</u> or <u>erythrocytes</u>): 99% of all blood cells and 40-45% of total blood volume.

  a. 1 ml of blood has over 5 billion rbcs.

  b. Contain the red pigment <u>hemoglobin</u> (Hb) containing iron; 97% of the blood's O_2 is bound to Hb.

    i. Hb picks up O_2 where the concentration is high (lungs) and releases it where the concentration is low (body cells).

    ii. After releasing O_2, some Hb picks up CO_2 from body cells for transport back to the lungs.

    iii. Iron from dead rbcs is recycled from the liver and spleen to the bone marrow.

  c. Formed in bone marrow of chest, upper arms, legs, and hips.

  d. Lose nuclei as they mature and cannot divide; life span is 120 days with 2 million replaced each second.

  e. Dead or damaged rbcs are removed from circulation (by the liver and spleen) and broken down to release their iron.

    i. Blood carries the salvaged iron to the bone marrow where it is used to make more Hb and packaged into new rbcs.

    ii. Small amounts of iron are excreted daily and must be replenished by the diet, as must be rbcs lost due to injury or menstruation.

  f. The number of rbcs is maintained through a negative feedback system involving the hormone <u>erythropoietin</u>, produced by the kidneys in response to oxygen deficiency.

    i. The hormone stimulates rapid production of new cells by the bone marrow.

    ii. When tissue oxygen levels become adequate, erythropoietin production ceases and the rate of rbc production returns to normal.

4) <u>White Blood Cells</u> (<u>wbcs</u> or <u>leukocytes</u>):

  a. Five types exist (comprising less than 1% of blood cells), distinguished by size, shape of nucleus, and staining characteristics.

  b. Most function to fight infections. Examples:

    i. <u>Monocytes</u> travel to wounds with bacterial infections, leave the capillaries by amoeboid movement, becoming <u>macrophages</u> which engulf bacterial invaders.

    ii. <u>Neutrophils</u> also eat bacteria.

    iii. Pus: accumulated dead macrophages and neutrophils at sites of infection.

    iv. <u>Lymphocytes</u>: help provide for immunity against disease.

  c. <u>Basophils</u> and <u>eosinophils</u> are less abundant wbcs:

    i. Basophils prevent blood clotting and participate in inflammatory and allergic reactions.

    ii. Eosinophils are stimulated by parasitic infections, releasing substances that kill the invaders.

5)   <u>Platelets</u> and blood clotting.
    a.   Large cells (<u>megakaryocytes</u>) that remain in the bone marrow pinch off fragments called <u>platelets</u> into the blood stream.
    b.   Platelets are important in blood clotting:
        i.   Platelets stick to irregular surfaces (damaged blood vessels), forming a plug over the rupture if small enough.
        ii.   Then, blood clotting (coagulation), the body's main defense against bleeding, occurs.
    c.   <u>Blood clotting</u> is a complex series of chemical events, culminating in the production of the enzyme <u>thrombin</u>:
        i.   Thrombin converts <u>fibrinogen</u> protein into stringy <u>fibrin</u>.
        ii.   Fibrin fibers stick together, forming a web that causes blood plasma to solidify like gelatin and traps rbcs in the clot.
        iii.   Platelets then adhere to the fibrous mass.
        iv.   Within 30 minutes, the platelets contract, forcing fluid out and making the clot denser. This also constricts the wound and promotes healing.

3.   **The Lymphatic System**.

A.   <u>General facts</u>.
    1)   The lymphatic system consists of:
        a.   Network of <u>lymphatic capillaries</u> that drain into larger <u>lymphatic vessels</u> that drain into large blood veins just before they enter the heart.
        b.   Numerous small <u>lymph nodes</u>.
        c.   The <u>thymus</u> and the <u>spleen</u>.
    2)   Important <u>functions</u> of the lymphatic system:
        a.   Removal of excess fluid and dissolved substances that leak from the blood capillaries.
        b.   Transport of fats from the intestine to the bloodstream.
        c.   Defense from bacteria and viruses by exposing them to wbcs.

B.   <u>Structure</u> of lymphatic capillaries and vessels.
    1)   Like blood capillaries, lymph capillaries form a complex network of thin-walled vessels into which substances move readily.
    2)   Unlike blood capillaries:
        a.   Lymph capillary walls have cells with openings between them that act as one-way valves: large particles can enter but not leave.
        b.   Lymph capillaries "dead end" in tiny spaces between cells.
    3)   Large lymph vessels have somewhat muscular walls, but most lymph flow occurs due to contraction of nearby muscles.
    4)   Direction of flow is regulated by one-way valves.

C.   Return of fluids to the blood.
    1)   Exchange of materials between capillary blood and nearby cells occurs through <u>interstitial fluid</u> that:
        a.   Bathes nearly all body cells.
        b.   Is derived from blood plasma.

      c.   Leaks through permeable walls of capillaries.

      d.   Is primarily water with dissolved materials from the blood.

  2)   In an average human, about 3 liters more fluid leaves the blood capillaries than is reabsorbed each day.

      a.   As interstitial fluid accumulates, its pressure forces the fluid through openings in the lymph capillaries.

      b.   The lymphatic system transports this fluid, now called <u>lymph</u>, back to the circulatory system.

D.   Transport of fats.

  1)   The small intestine is richly supplied with blindly ending lymph capillaries.

  2)   Intestinal cells release fat globules into the interstitial fluid.

  3)   The fat globules move into lymph capillaries.

  4)   Once in the lymph, they are transported to the veins leading to the heart.

E.   Defense of the body.

  1)   The lymphatic system helps defend against bacteria and viruses.

  2)   The large lymph vessels are interrupted periodically by 1 inch long kidney bean shaped structures called <u>lymph nodes</u>.

      a.   Lymph is forced through channels in the nodes that are lined with macrophages.

      b.   Nodes also produce lymphocytes.

      c.   These wbcs recognize and destroy microbes but are killed in the process.

      d.   Painful swelling of lymph nodes in certain diseases (mumps) results from accumulation of dead lymphocytes, macrophages, and microbes.

  3)   Though not directly connected to lymph vessels, thymus and spleen are functional parts of the lymphatic system.

      a.   <u>Thymus</u>:

         i.   Is located beneath the sternum near the heart.

         ii.   Is quite active in infants and young children.

      b.   <u>Spleen</u>:

         i.   Is located in the left side of the abdomen between the stomach and diaphragm.

         ii.   Filters blood, exposing it to macrophages and lymphocytes that destroy foreign particles and aged rbcs.

****

REVIEW QUESTIONS:

1.   Why is the circulatory system necessary in large complex animals?

2.   What are the three major parts of a circulatory system?

3.   Compare open and closed circulatory systems and list which type is possessed by arthropods, molluscs, earthworms, fish, and humans.

4.   List six functions of the human circulatory system.

5.   Explain what is meant by "pulmonary circulation" and "systemic circulation."

6.   Describe blood flow through the human heart.

7.   Describe the cardiac cycle and include the terms "diastole" and "systole."

8. Name all the heart valves and where they are located within the heart.
9. Explain how coordinated contractions of heart muscle are controlled in humans, mentioning the SA and AV nodes.
10. Explain the effects of the sympathetic and parasympathetic nervous systems on heart rate.

11.-22. **MATCHING TEST:** blood vessels

Choices:
A.	Arteries	D.	Venules and veins
B.	Arterioles	E.	All of these
C.	Capillaries	F.	None of these

11. Thin tubes with walls only one cell thick.
12. Have thick, muscular walls that expand very little.
13. Have thin, highly distensible walls.
14. Have muscular walls that contract in response to nervous or chemical signals.
15. Where exchange of wastes, nutrients, and gases occurs between blood and body cells.
16. In these vessels, blood rushes to the skin on hot days.
17. Low pressure pathways for carrying blood to the heart.
18. No farther than 10 micrometers from any body cell.
19. Blood leaves the extremities on a cold day through these vessels.
20. Contains valves to prevent back-flow of blood.
21. Red blood cells must pass through them in single file.
22. Contract when blood pressure falls.

---

23. Name three groups of blood plasma proteins and briefly state their functions.

24.-33. **MATCHING TEST:** red and white blood cells

Choices:
A.	Red blood cells	C.	Both of these
B.	White blood cells	D.	Neither of these

24. Production stimulated by the hormone erythropoietin.
25. Contains hemoglobin for oxygen transport.
26. Leukocytes.
27. 99% of all blood cells.
28. Play a central role in blood clot formation.
29. Contain nuclei when mature.
30. Lose nuclei as they mature.
31. Function to fight infection.
32. Contain iron.
33. Become pus.

34.-40. **MATCHING TEST:** white blood cells

Choices:
A.	Monocytes	D.	Basophils & eosinophils
B.	Neutrophils	E.	None of these
C.	Lymphocytes		

34. Help provide immunity against disease organisms.
35. The least abundant white blood cells.
36. Become macrophages.
37. Help prevent clots from forming in blood vessels.
38. Leave capillaries by amoeboid movement.
39. Play a crucial role in blood clotting.
40. Eat bacteria.

----

41. Describe the process of blood clotting in humans.
42. What are the three functions of the lymphatic system?
43. What is lymph and how is it returned to the bloodstream?
44. Describe the functions of lymph nodes.
45. What system does the spleen belong to and what does the spleen do?

****

ANSWERS TO REVIEW QUESTIONS:

1. see 1.B.1)-2)

2. see 1.C.1)-3)

3. see 1.D.1)-2)

4. see 2.A.1)-6)

5. see 2.B.1)c.i.-ii.

6. see 2.B.

(pulmonary circulation)
right atrium → right
ventricle → lungs →

(systemic circulation)
left atrium → left
ventricle → body

7. see 2.B.2)a.-e.

8. see 2.B.3)b.i.-iii.

9. see 2.B.3)c.-d.

10. see 2.B.4)b.i.-ii.

11.-22. [see 2.C.1)-4)]

11.	C	17.	D
12.	A	18.	C
13.	D	19.	B
14.	B	20.	D
15.	C	21.	C
16.	B	22.	D

23. see 2.D.2)c.i.-iii.

24.-33. [see 2.D.3)-4)]

24.	A	29.	B
25.	A	30.	A
26.	B	31.	B
27.	A	32.	A
28.	D	33.	B

34.-40. [see 2.D.4)b.-c.]

34.	C	38.	A
35.	D	39.	E
36.	A	40.	AB
37.	D		

41. see 2.D.5)b.-c.
42. see 3.A.2)a.-c.
43. see 3.C.1)-2)
44. see 3.E.2)
45. see 3.E.3)b.

****

CHAPTER 31: RESPIRATION

CHAPTER OUTLINE.

In this chapter, the evolution, structure, and functioning of respiratory systems are covered. Respiratory systems using gills and lungs are compared, and the human respiratory system is thoroughly described.

1. **General facts**.

   A. In the animal body, the production of energy by cellular respiration requires a steady supply of O_2 and generates CO_2 as a waste product.
      1) In most animals, these gases are transported in the blood and exchanged with cells by diffusion.
      2) Gas exchange between blood and external environment (air or water) occurs by diffusion.

   B. Respiratory systems support cellular respiration by bringing a large, moist surface into contact with the blood and the external environment so that O_2 and CO_2 may be exchanged.

2. **Evolution of respiratory systems**.

   A. Two features shared by all respiratory systems to facilitate diffusion:
      1) A sufficiently large surface area in contact with the environment for gas exchange by diffusion to meet the needs of the body.
      2) A moist exchange surface since gases must be dissolved in fluid when they enter or leave cells.

   B. Respiration without specialized respiratory structures.
      1) Respiration by simple diffusion is sufficient if:
         a. The animal lives in a moist environment.
         b. The animal is extremely small (e.g., nematode worms).

      c.   The animal has a flattened body with a large surface area (e.g., flatworms).

      d.   Body cells have low energy demands (e.g., jellyfish).

      e.   The animal can bring the environment close to all cells by circulating seawater throughout a perforated body (e.g., sponges).

2)   Diffusion is supplemented by a <u>circulatory system</u> in the earthworm.

      a.   Earthworms have well-developed closed circulatory systems to carry gases throughout their bodies.

      b.   Gases diffuse in and out of the circulatory system through the moist skin surface area which is large in this elongated animal.

      c.   The earthworm's sluggish metabolism reduces its need for O_2.

3.   Respiratory systems and gas exchange.

A.   Transfer of gases from environment → blood → cells and back occurs in stages that alternate between bulk flow and diffusion.

     1)   During <u>bulk flow</u>, fluid or gas molecules move in unison (in "bulk") through relatively large spaces, from areas of higher to areas of lower pressure.

     2)   With <u>diffusion</u>, individual molecules move from areas of higher to areas of lower concentration.

B.   <u>Stages</u> in gas exchange in respiratory systems:

     1)   Muscular breathing movements move air or water containing O_2 across a respiratory surface by bulk flow.

     2) O_2 and CO_2 are exchanged through the respiratory surface by diffusion.

     3)   Bulk flow of blood, pumped by the heart, transports gases between the respiratory system and tissues.

     4)   Gases are exchanged between tissues and the circulatory system by diffusion.

C.   <u>Respiration using gills</u> in large, active animals.

     1)   For life in water, a wide variety of animals have evolved <u>gills</u> for gas exchange.

     2)   Just beneath a gill's outer membrane, a dense profusion of capillaries brings blood close to the surface, enhancing diffusion.

     3)   Gills may be a sac opening into the water (some molluscs and amphibians), or an elaborately branched structure, and may be protected inside a rigid exoskeleton (crabs).

     4)   Fish gills are covered by a flap (<u>operculum</u>).

        a.   Opercula protect the delicate gills and streamline the body for faster swimming.

        b.   Fish create a continuous current past the gills by gulping water into the mouth and sending it out the opercular openings.

     5)   Land animals can't use gills since they collapse and rapidly desiccate in dry air. Land animals have evolved 3 solutions: book lungs, tracheae, and lungs.

D.   <u>Book lungs</u> in arachnids and <u>tracheae</u> in insects.

     1)   Some arachnids (spiders and scorpions) evolved a series of "pagelike" membranes within a chamber of the exoskeleton, forming <u>book lungs</u>.

     2)   Insects convey air to body cells through a system of elaborately branched tubes called <u>tracheae</u>.

a.   Tracheae subdivide into tiny channels so that each body cell is close to a tube, minimizing diffusion distances.

b.   Tracheae open to the outside through pores (spiracles) along the sides of the abdomen.

c.   Muscular pumping movements of the abdomen help air move through the tracheae in large insects.

E.   Respiration using lungs occurs in land snails and vertebrates.

1)   Lungs are delicate, moist, respiratory chambers deep within the body, using the body wall for support.

2)   In the vertebrates:

a.   Lungs evolved in fresh water fish as outpockets of the digestive tract.

b.   Some amphibians use gills and lungs at different stages of their life cycles (frog tadpoles vs. terrestrial adults with moist skin).

c.   Reptiles rely more exclusively on lungs since they have dry, scaly skin.

d.   Birds and mammals are exclusively lung-breathers. The lungs of birds are more efficient, having air tubes instead of blind sacs for air flow and having inflatable air sacs to store fresh air that passes through the lungs during expiration.

4.   The human (mammalian) respiratory system.

A.   Mammalian respiratory systems have 2 parts.

1)   Conducting portion: passageways that carry air to and from the lungs.

2)   Gas exchange portion: sacs called alveoli.

B.   The conducting portion.

1)   Air flow:

a.   Mouth and nose →

b.   Pharynx →

c.   Larynx with vocal cords (sound is created when exhaled air causes them to vibrate) →

d.   Trachea (rigid tubes with cartilage rings for support and flexibility) →

e.   2 large branches called bronchi →

f.   Lungs with repeated smaller branchings called bronchioles →

g.   Microscopic alveoli (tiny air pockets where gas exchange occurs).

2)   As air moves towards the lungs:

a.   It is warmed and moistened.

b.   Most dust and microbes are trapped in mucous from lining cells.

i.   The mucus is swept towards the pharynx by ciliated cells of the bronchioles, bronchi, and trachea (smoking interferes with this process).

ii.   In the pharynx, mucus is coughed up or swallowed into the digestive tract.

C.   Gas exchange in the alveoli.

1)   Each lung has 1.5 million microscopic (0.2 mm) alveolar chambers, providing about 160 m^2 (500 ft^2) surface area for diffusion in an adult.

2) The alveoli cluster about the end of each bronchiole like bunches of grapes and are enmeshed in capillaries.
    a. O_2 diffuses from the moist air through the one-cell thick alveoli and capillary walls into the bloodstream (which is low in O_2) where it is picked up by red blood cells.
    b. CO_2 does the reverse.
3) Blood is pumped to the lungs by the heart after circulating through body tissues.
4) Blood from the lungs, oxygenated and purged of CO_2, returns to the heart, which pumps it to the body tissues.
    a. In the tissues, O_2 concentration is lower than in blood, and O_2 diffuses into the cells.
    b. CO_2, which has built up in cells, diffuses into the blood.

D. <u>Transport of gases in the blood</u>.
1) Inside capillaries, O_2 binds loosely and to hemoglobin in the red blood cells.
2) Each hemoglobin molecule can bind 4 O_2 molecules.
3) As hemoglobin binds O_2, it changes shape slightly resulting in a color change.
    a. Deoxygenated blood is dark maroon-red (appears blue through the skin).
    b. Oxygenated blood is bright cherry-red.
4) Carbon monoxide (CO) can fool hemoglobin and bind in place of and 200 times as tightly as O_2.
    a. The resulting hemoglobin is bright red but incapable of transporting O_2.
    b. The lips and nails of victims of CO poisoning are brighter red than usual.
5) Nearly all O_2 in blood is bound to hemoglobin.
    a. Due to hemoglobin, blood can carry 70 times as much O_2 as compared to O_2 dissolved in blood plasma.
    b. Hemoglobin maintains a high concentration gradient of O_2 from air to blood by removing O_2 from solution in the plasma, enhancing diffusion of O_2 from alveoli into the blood.
6) About 20% of the CO_2 is returned to the lungs bound to hemoglobin.
    a. Most of the CO_2 (70%) is carried as bicarbonate ion in the plasma.
    b. A small amount is dissolved in the plasma.

E. <u>Mechanics of breathing</u>.
1) Stages of breathing.
    a. <u>Inspiration</u>: air is actively inhaled.
    b. <u>Expiration</u>: air is passively exhaled.
2) <u>Inspiration</u> is accomplished by making the chest cavity larger.
    a. The <u>diaphragm</u> muscle contracts, drawing it downward.
    b. Rib muscles (<u>intercostals</u>) contract, lifting the ribs upward and outward.
    c. The lungs expand with the chest cavity due to a vacuum between lungs and chest cavity walls.
3) <u>Expiration</u> occurs when diaphragm and intercostals relax. Lungs hold some air after expiration.

F. <u>Control of breathing</u>.
1) Impulses to move the diaphragm and intercostals originate in the <u>breathing</u> (respiratory) <u>center</u> of the brainstem.
2) Receptor neurons in the brainstem monitor CO_2 concentration in the blood.
    a. If CO_2 levels rise, breathing rate and depth increase.

b. The respiratory center is regulated to maintain a constant level of CO_2 in the blood.

3) When blood O_2 levels fall, receptors in the aorta and carotid arteries stimulate the breathing center to increase breathing rate.

4) When the brain activates muscles during heavy exercise, it also stimulates the breathing center to increase breathing rate.

****

REVIEW QUESTIONS:

1. What is the function of the respiratory system?
2. What are the two features shared by all respiratory systems?
3. Under what conditions is respiration by simple diffusion sufficient?
4. Why are circulatory and respiratory systems necessary in larger active animals like earthworms?
5. In what ways are gills and lungs different?
6. Trace the flow of air through the human respiratory system.
7. Describe how inspiration and expiration occur in humans.
8. In what ways does the respiration (breathing) center in the brain affect the rate and depth of breathing?

****

ANSWERS TO REVIEW QUESTIONS:

1. see 1.A.-B.

2. see 2.A.1)-2)

3. see 2.B.1)a.-e.

4. see 2.B.2)a.-c.

5. see 3.C.,E.

6. see 4.B.1)a.-g.

7. see 4.E.1)-3)

8. see 4.F.1)-4)

****

CHAPTER 32: NUTRITION AND DIGESTION

CHAPTER OUTLINE.

This chapter deals with animal nutrition and digestion. Sources of energy for cells are discussed and the uses of amino acids, minerals, and vitamins in animal nutrition are covered. Types of digestive systems are enumerated, but most of the chapter concentrates on the steps of digestion in humans carried on in each part of the digestive system.

1. **Nutrition.**

    A. The <u>five major categories</u> of animal nutrients are:
        1) Lipids.
        2) Carbohydrates.
        3) Proteins.
        4) Minerals.
        5) Vitamins.

    B. Sources of energy.
        1) Animal cells require a continuous supply of energy to stay alive.
        2) Three nutrients provide dietary energy for animals: lipids, carbohydrates, and proteins (if the others are lacking).
            a. They are broken down during cellular respiration, providing chemical energy stored in ATP.
            b. Nutrient energy is measured in <u>calories</u> (amount of energy needed to raise the temperature of 1 gram of water by 1° C).
                i. Food energy content measured in <u>kilocalories</u> (1000 calories = <u>C</u>alorie).
                ii. The human body at complete rest burns about 1500 Calories per day.

iii.   Exercise greatly boosts caloric requirements.

3)   Lipids: fats, phospholipids, and cholesterol.

    a.   Are converted to certain hormones, and are components of cell membranes and nerve cell coverings.

    b.   The chief form of long-term energy storage in animals.

    c.   Advantages of fat storage: more calories with less weight (3600 Calories/pound):

        i.   Contains over twice the energy per unit weight (9 Calories/gram) of either carbohydrates or proteins (4 Calories/gram).

        ii.   Are hydrophobic: do not cause extra accumulation of water in the body.

    d.   Provide about 45% of the energy intake for a typical North American.

    e.   Mammals use fat deposits as stored energy and as insulation: fat conducts heat at only 1/3 the rate of other body tissues.

4)   Carbohydrates: sugars, starches, cellulose, and glycogen.

    a.   Principal energy storage molecules in plants.

    b.   Typically provides 45% of our dietary energy, though Eskimos survive well with little carbohydrates.

    c.   Liver and muscles store some glycogen (animal starch) which is used as a quick energy source; runners "hit the wall" as they exhaust their glycogen reserves.

5)   Proteins.

    a.   Inefficient energy source, but provides amino acids.

    b.   Typically supplies about 10% of our dietary energy.

    c.   Protein breakdown produces <u>urea</u> which must be filtered out of the body by the kidneys.

C.   Amino acids:

1)   Are used to make proteins, certain hormones and neurotransmitters.

2)   In the digestive tract, proteins are broken down into amino acids.

3)   In body cells, the amino acids are linked to form new proteins unique to the individual.

4)   Of the 20 amino acids commonly found in proteins, 11 (the <u>essential amino acids</u>) must be supplied to humans in the diet (see text Table 32-2).

D.   Minerals: Small inorganic molecules (see text Table 32-3).

1)   All must be obtained from food or dissolved in drinking water.

2)   Some essential minerals:

    a.   Calcium, magnesium, and phosphorus for bones and teeth.

    b.   Sodium and potassium for nerve impulse conduction and muscle contraction.

    c.   Iron for hemoglobin.

    d.   Iodine to make thyroid gland hormones.

    e.   Small amounts of <u>trace elements</u> (become parts of enzymes) like zinc and selenium.

E.   Vitamins (see Table 32-4).

1)   Diverse organic compounds needed in small amounts.

2)   Humans cannot make any vitamins except vitamin D in the skin by exposure to sunlight.

3) <u>Water-soluble</u> <u>vitamins</u>:
    a.   Vitamin C and the 11 B vitamins.
    b.   Dissolve in blood and excreted by the kidneys; cannot be stored in the body.
    c.   Usually work in union with enzymes.

4) <u>Fat-soluble</u> <u>vitamins</u>:
    a.   Include vitamins A (forms pigments for vision), D, E, and K (needed for normal blood clotting).
    b.   Stored in body fat and accumulate in the body over time.
    c.   Vitamins A and D are toxic if dietary intake is excessive.

2. The Challenge of Digestion.

A. The complex fats, carbohydrates, and proteins in the bodies of organisms eaten by animals must first be broken down into smaller subunits before being absorbed into the consumer's bloodstream and used by its cells.

B. <u>Functions</u> of digestive systems:
1) <u>Ingestion</u>: usually through a mouth opening.
2) <u>Mechanical breakdown</u>: food is broken into smaller pieces to increase the surface area for attack by digestive enzymes. Accomplished by:
    a.   Gizzards.
    b.   Teeth.
    c.   Churning action of digestive cavity.
3) <u>Chemical breakdown</u>: by digestive fluids and enzymes.
4) <u>Absorption</u>: transport of small nutrient molecules out of digestive system and into cells.
5) <u>Elimination</u>: expelling indigestible materials from the body.

C. Types of digestive systems in animals.
1) Digestion <u>within</u> <u>single</u> <u>cells</u> (<u>intracellular</u> <u>digestion</u>).
    a.   Is seen in sponges and single celled protists that consume microscopic food particles.
    b.   Once engulfed by a cell (endocytosis), food is enclosed in membrane-bound <u>food</u> <u>vacuoles</u> that fuse with <u>lysosomes</u> containing digestive enzymes.
    c.   The food breaks down into smaller molecules and is absorbed into the cytoplasm.
    d.   Undigested remnants are eliminated from the cell (exocytosis).
2) Digestion <u>in</u> <u>a</u> <u>simple</u> <u>sac</u> (<u>extracellular</u> <u>digestion</u>).
    a.   Larger organisms evolved a special chamber where food is broken down by enzymes acting outside the cells.
    b.   Cnidarians.
        i.   Have a <u>gastrovascular</u> <u>(GV)</u> <u>cavity</u> with a single opening (mouth/anus).
        ii.   Food is captured by stinging tentacles, passed into the GV cavity, and reduced to small particles by digestive enzymes.
        iii.   Cells lining the GV cavity engulf the small particles by phagocytosis and digestion is completed intracellularly.

    c. This type of digestive system is inefficient and cannot support active animals.

  3) Digestion <u>in a tube</u> open at both ends and running through the body.
    a. Seen in animals from nematode worms (with simple, unspecialized tubes) through vertebrates (with tubes having specialized regions).
    b. Allows active animals to eat continuously.
    c. These animals utilize extracellular digestion to dismantle the food.
    d. In earthworms:
      i. A muscular <u>pharynx</u> draws in bits of food;
      ii. Food passes through the <u>esophagus</u> to a storage area, the <u>crop</u>;
      iii. Food then enters the muscular <u>gizzard</u> where it is pulverized by sand grains and muscular contractions;
      iv. In the <u>intestine</u>, enzymes break down food into simple molecules that are absorbed by cells in the intestinal wall.

3. **Human digestion**.

  A. <u>Mouth</u>: mechanical and chemical breakdown of food begins.
    1) 32 adult <u>teeth</u> cut and grind the food into smaller pieces.
    2) 3 pairs of <u>salivary glands</u> pour out <u>saliva</u> containing <u>amylase</u> enzymes to chemically break down starches into sugars.
    3) Saliva also:
      a. Contains antibiotics to kill bacteria.
      b. Lubricates food for easier swallowing.
      c. Dissolves some food molecules so the taste buds in the tongue can recognize them.
    4) The muscular <u>tongue</u> manipulates the food into a mass and passes it back to the pharynx cavity.
    5) The <u>pharynx cavity</u>:
      a. Connects mouth and esophagus.
      b. Connects nose and mouth with tracheal tubes of the lungs.
      c. The swallowing reflex causes a flap of tissue (the <u>epiglottis</u>) to block the opening to the trachea while directing the food to the esophagus.

  B. <u>Esophagus and Stomach</u>.
    1) <u>Esophagus</u>: a muscular tube connecting mouth and stomach.
      a. It actively moves the food along due to <u>peristalsis</u> (sequential contractions of circular muscles pushing the food along).
      b. Esophagus mucus further lubricates the food.
    2) <u>Stomach</u>: an expandable muscular sac.
      a. Stomach <u>sphincters</u> (rings of muscles):
        i. The <u>gastroesophageal</u> <u>sphincter</u> opens to allow food to enter the stomach.
        ii The <u>pyloric sphincter</u> opens to allow food to leave the stomach and enter the small intestine.
      b. Three major functions:
        i. Storage chamber which regulates passage of <u>chyme</u> (partially digested food and digestive secretions) into the small intestine by peristalsis and regulation of the pyloric sphincter.

        ii.   Mechanical breakdown of food by churning its muscular walls.

        iii.   Chemical digestion.

              a)   <u>Gastrin</u> hormone stimulates secretion of hydrochloric acid by specialized stomach cells.

              b)   <u>Pepsin</u> enzyme breaks proteins into smaller peptides (pepsin is produced as inactive <u>pepsinogen</u> which is converted by hydrochloric acid).

              c)   The stomach doesn't self-digest because gland cells produce mucus which coats the stomach lining.

    c.   Peristaltic waves, about 3 per minute, propel chyme towards small intestine.

        i.   Pyloric sphincter allows only a teaspoon to pass through per contraction.

        ii.   It takes about 2-6 hours to completely empty the stomach.

    d.   Only a few substances can enter the bloodstream through the stomach wall, including water, alcohol, and some drugs. A full stomach slows alcohol absorption.

C.   <u>Small Intestine</u>.

    1)   Coiled, narrow (1-2 inch diameter) and 9 feet long in adults.

    2)   <u>Two major functions</u>:

        a.   Digestion of food into smaller molecules.

        b.   Absorption of small molecules into the bloodstream.

    3)   Chyme is digested by secretions from the liver, pancreas, and small intestine cells.

    4)   <u>Liver</u>: the largest organ in the human body.

        a.   Functions:

             i.   Storage of fats and carbohydrates.

             ii.   Regulation of glucose levels in blood.

             iii.   Synthesis of blood proteins.

             iv.   Storage of iron and some vitamins.

             v.   Detoxification of harmful substances (nicotine, alcohol, ammonia into urea).

        b.   Role in digestion:

             i.   Produces <u>bile</u>, stores it in the <u>gall bladder</u>, and releases it into small intestine through the <u>bile duct</u>.

             ii.   <u>Bile</u>: complex mixture of bile and other salts, water, and cholesterol.

             iii.   Bile salts are not enzymes. They act as detergents or emulsifying agents to disperse fats, exposing a large surface area for attack by lipases (fat digesting enzymes from the pancreas).

    5)   <u>Pancreas</u>: has 2 cell types.

        a.   One cell type produces the hormones <u>insulin</u> and <u>glucagon</u> for blood sugar regulation.

        b.   The other cell type produces digestive pancreatic juices which empty into the small intestine, including:

             i.   <u>Sodium bicarbonate</u> that neutralizes the acidic chyme.

             ii.   Three types of digestive enzymes:

                 a)   <u>Amylases</u>: break down carbohydrates.

b)   <u>Lipases</u>: break down fats.

c)   <u>Proteases</u>: break down proteins and peptides.

6)   <u>Intestinal</u> <u>wall</u> <u>cells</u>:

    a.   Produce digestive enzymes:

        i.   <u>Proteases</u>: convert peptides into animo acids.

        ii.   <u>Sucrase</u>, <u>lactase</u>, and <u>maltase</u>: convert disaccharides into monosaccharides.

        iii.   Small amounts of <u>lipase</u>.

    b.   Enzymes occur on the external cell membranes that line the intestine; these enzymes work as nutrients are being absorbed into the cells.

7)   <u>Absorption</u> in the small intestine: where most nutrients are absorbed into the body.

    a.   To facilitate absorption, the internal surface area of small intestine is increased over 600-fold (to 250 m^2, the size of a tennis court) by various foldings and projections of its wall due to:

        i.   Folding of the wall itself.

        ii.   Minute fingerlike projections called <u>villi</u> which sway back and forth. Each villus has:

            a)   A rich supply of blood capillaries to absorb nutrients.

            b)   A single lymph capillary (<u>lacteal</u>) to carry off fat droplets.

        iii.   Microscopic <u>microvilli</u> projections on the villi.

    b.   <u>Segmentation</u> <u>movements</u> (rhythmic contractions, not synchronized like peristalsis) slosh the fluid back and forth to enhance absorption.

    c.   When absorption is complete, peristalsis moves the remnants into the large intestine.

D.   <u>Large Intestine</u>.

1)   Large in diameter (2.5 inches), but not in length (5 feet in adults).

2)   Receives leftovers from digestion which are used by bacteria living there. In return, the bacteria produce vitamins needed by the body (B$_{12}$, thiamine, riboflavin, K).

3)   Final absorption of water and salts occurs here.

4)   The end result, semisolid <u>feces</u> consisting of indigestible wastes and dead bacteria, are transported by peristalsis to the <u>rectum</u> where defecation occurs.

E.   <u>Control of digestion</u>: coordinated by both nerves and hormones.

1)   First, signals from the head (smell, taste, chewing) stimulate:

    a.   Saliva production.

    b.   Nervous stimulation of the stomach wall to secrete acid.

    c.   Production of the hormone <u>gastrin</u> to stimulate further acid secretion (a negative feedback system).

2)   Arrival of food in the stomach triggers:

    a.   Mucus in response to stomach wall irritation.

    b.   Pepsinogen → pepsin → protein digestion.

    c.   More acid production.

3)   As chyme enters the small intestine:

    a.   Cells there secrete <u>secretin</u> hormone, causing the pancreas and liver to release bicarbonate to neutralize the acidic chyme.

    b.   Cells also secrete <u>cholecystokinin</u> hormone, causing release of pancreatic digestive enzymes and contraction of the gall bladder, releasing bile.

c.   Also, <u>gastric</u> <u>inhibitory</u> <u>peptide</u> hormone is produced which inhibits acid production and peristalsis in the stomach, slowing down the rate of chyme movement into the small intestine.

****

REVIEW QUESTIONS:

1.-11.   **MATCHING TEST**: sources of dietary energy

Choices:

A.   Lipids
B.   Carbohydrates
C.   Proteins

D.   All of these
E.   None of these

1.   Provides amino acids.
2.   Principal energy storage molecule in plants.
3.   Fats.
4.   Glycogen.
5.   Genetic material.
6.   Most concentrated energy source.
7.   Inefficient energy source.
8.   Provides about 45% of our dietary energy.
9.   Produces urea when broken down.
10.   Components of cell membranes and nerve cell coverings.
11.   Principal energy storage molecules in animals.

12.-18.   **MATCHING TEST**: essential minerals

Choices:

A.   Calcium and phosphorus
B.   Sodium and potassium
C.   Iron

D.   Iodine
E.   Zinc and selenium

12.   For hemoglobin.
13.   For bones and teeth.
14.   Trace elements.
15.   For nerve conduction.
16.   For thyroid gland hormones.
17.   For muscle contraction.
18.   Works in conjunction with enzymes.

19.-28.   **MATCHING TEST**: types of digestive systems

Choices:

A.   Digestion within single cells
B.   Digestion within a simple sac

C.   Digestion within a tube

19. Separate mouth and anal openings.
20. Gastrovascular cavity.
21. Sponges.
22. Most efficient type of digestive system.
23. Exclusively intracellular digestion.
24. Cnidarians.
25. Humans.
26. Wastes are eliminated by exocytosis.
27. Common mouth/anal opening.
28. Digestion is exclusively extracellular.

----

29. Name three ways that water-soluble vitamins differ from fat-soluble vitamins.
30. List, in sequence, the names and functions of all regions of the digestive system in an earthworm.
31. List four functions of saliva.
32. List three major functions of the stomach.
33. Name the two stomach sphincters and describe their functions.
34. Name five functions of the liver.
35. Describe the functions of the pancreas.
36. Describe how the structure of the small intestine facilitates absorption of nutrients.
37. When food is smelled, tasted, and chewed, what digestive functions are stimulated?
38. As chyme enters the small intestine, what digestive functions are stimulated?

****

ANSWERS TO REVIEW QUESTIONS:

1.-11. [see 1.B.1)-5)]

1.	C	7.	C	
2.	B	8.	AB	
3.	A	9.	C	
4.	B	10.	A	
5.	E	11.	A	
6.	A			

12.-18. [see 1.D.2)a.-e.]

12.	C	16.	D
13.	A	17.	B
14.	E	18.	E
15.	B		

19.-28. [see 2.C.1)-3)]

19.	C	24.	B
20.	B	25.	C
21.	A	26.	A
22.	C	27.	B
23.	A	28.	C

29. see 1.E.3)-4)

30. see 2.C.3)d.

31. see 3.A.2)-3)

32. see 3.B.2)b.i.-iii.

33. see 3.B.2)a.i.-ii.

34. see 3.C.5)a.-b.

35. see 3.C.4)a.-b.

36. see 3.C.7)a.-b.

37. see 3.E.1)a.-c.

38. see 3.E.3)a.-c.

****

CHAPTER 33: EXCRETION

CHAPTER OUTLINE.

This chapter deals with the role of excretory systems in maintaining internal homeostasis within animals, particularly humans. The structure and function of the human kidney are explained, including filtration by the glomerulus, urine formation in the nephrons, and water balance.

1. **Homeostasis**: maintenance of precise internal regulation despite differences in diet and temperature.

    A. Homeostasis results from the coordinated activity of the nervous, endocrine and circulatory systems, and of the liver, and is aided by the action of lungs, skin, digestive system, and excretory organs.

    B. The digestive system is unselective in what crosses the intestinal wall and enters the cells (excessive water, nutrients, salts, minerals, drugs). The excretory system in particular restores and maintains the proper balance of these materials in the body.

    C. <u>Excretory systems</u>.
        1) <u>Major functions</u>:
            a. Excretion of cellular waste products (urea).
            b. Regulation and maintenance of body fluid composition.
        2) Both functions occur simultaneously as the blood is selectively filtered by the kidneys of vertebrates and simpler excretory organs of invertebrates.
            a. Excretory systems collect blood fluids.
            b. Water and nutrients return to the blood.
            c. Wastes, toxins, excessive water and nutrients are collected and eliminated from the body as urine.

2. **Simple excretory systems**.

A. <u>Flame</u> <u>Cells</u> in flatworms:
   1) Flatworms have a simple excretory system consisting of:
      a. A network of tubes branching throughout the body.
      b. The tubes end blindly in hollow bulbs called <u>flame</u> <u>cells</u> (beating cilia look like flames).
   2) Water and dissolved wastes are filtered into the bulbs. There, beating cilia direct the fluid through tubes where water and nutrients are absorbed, and wastes are added from body cells.
   3) Eventually, wastes reach one of numerous pores in the skin and are released to the outside.

B. <u>Nephridia</u> (simple kidneys) in earthworms and molluscs.
   In the earthworm:
   1) Coelomic fluid fills the body cavity, collecting wastes and nutrients from blood and tissues.
   2) The fluid enters nephridia, through funnel shaped openings called <u>nephrostomes,</u> and passes through a narrow twisted tube where useful substances are absorbed back into the blood, leaving <u>urine</u> (water and wastes) behind.
   3) Urine is stored in bladder-like portions of nephridia and excreted through excretory pores in the body wall.
   4) Nearly every segment in earthworms contains a pair of nephridia.

3. **Vertebrate (Human) Excretion.**

   A. <u>Kidneys</u>: complex organs resembling dense collections of nephridia.
      1) Are part of the excretory system which produces, transports, stores, and eliminates urine.
      2) Are paired, bean-shaped organs, each measuring about 5 x 3 x 1 inch.
      3) Unfiltered blood enters through <u>renal</u> <u>arteries</u> and filtered blood leaves through <u>renal</u> <u>veins</u>.
      4) Urine leaves through tubes called <u>ureters</u> (each 12 inches long and 1/2 inch in diameter).
      5) Ureters pass urine (by peristalsis) to the hollow muscular <u>bladder</u> where it is collected and stored.
      6) Urine is retained in the bladder by two sphincter muscles located at the base, just above the junction with the <u>urethra</u> (narrow tube to the outside).
         a. Receptors in the walls of a distended bladder trigger reflexive contractions and the sphincter nearest the bladder opens involuntarily.
         b. However, the lower, or external, sphincter is under voluntary control.
         c. The average adult bladder holds about a pint of urine, but the desire to urinate is triggered by much smaller accumulations.
      7) Upon bladder distention (stretching due to fullness), <u>urination</u> occurs when:
         a. Both sphincter muscles at the bladder's base relax, and
         b. The bladder contracts, forcing urine down the urethra which is about 1.5 inches long in human females and 8 inches long in males.

   B. <u>Human</u> <u>kidney</u> structure.
      1) In cross-section, we see:
         a. A solid outer layer, where urine forms.

       b.   A hollow inner chamber (<u>renal</u> <u>pelvis</u>) that funnels urine into the ureter.

2)   The solid outer layer has 2 parts:
       a.   An inner <u>medulla</u>.
       b.   An outer <u>cortex</u>. Here, over one million individual filters (<u>nephrons</u>) are packed side by side with many extending into the medulla.

3)   Each nephron has 3 major parts:
       a.   <u>Glomerulus</u>: a pressure filter for the blood.
       b.   <u>Bowman's capsule</u>: collects the filtrate (fluid filtered from the blood).
       c.   A long, twisted <u>tubule</u> that is subdivided into:
          i.   First, the <u>proximal tubule</u>.
          ii.   Then, the <u>loop of Henle</u>.
          iii.   Finally, the <u>distal tubule</u> that leads to the collecting duct.

4)   In the tubule, nutrients are selectively reabsorbed from the filtrate back into the blood, while wastes and some water are left behind to form urine.

C.   <u>Filtration</u> by the glomerulus.

1)   Blood is conducted to each nephron by an arteriole that branches from the renal artery.
       a.   Within Bowman's capsule (cup shaped region of the nephron), this vessel subdivides into numerous microscopic capillaries that form an intertwined mass, the <u>glomerulus</u>.
       b.   The glomerulus capillary walls are extremely permeable to water and dissolved substances.

2)   Past the glomerulus, the capillaries reunite into an arteriole whose diameter is smaller than the incoming artery. This causes high pressure in the glomerulus, driving water and most dissolved substances through the capillary walls and out of the blood, a process called <u>filtration</u>.

3)   The watery filtrate, resembling blood plasma minus its proteins, is collected in the Bowman's capsule for transport through the nephron.

4)   Blood leaving the glomerulus in the arteriole is concentrated, containing blood cells, proteins and fat droplets too big to be filtered out, and little fluid.

5)   The arteriole then branches into smaller highly porous capillaries that surround the tubule. Here water and nutrient reabsorption occurs during urine formation.

D.   <u>Urine formation</u> in the nephron.

1)   The blood filtrate in Bowman's capsule contains wastes, nutrients, and water.

2)   Tubular <u>reabsorption</u> occurs: cells of the tubule remove water and nutrients from the filtrate to be reabsorbed by the blood.
       a.   Reabsorption of salts and other nutrients is by active transport via tubule cells that expend energy to move molecules out of the tubule and into the surrounding fluid where they enter the blood by diffusion.
       b.   Water is reabsorbed by passive osmosis.
       c.   Wastes (urea) remain in the tubule and become concentrated as water leaves.

3)   Tubular <u>secretion</u>: waste materials (hydrogen and potassium ions, foreign substances) remaining in the blood are actively secreted <u>into</u> the tubule by tubule cells.

E.   <u>Production and concentration of urine</u>.

1)   Urine becomes highly concentrated by the action of nephrons and the <u>collecting</u>

                    <u>duct</u> into which several nephrons empty.

    2)   Each nephron has 5 major parts: fluid flows from the (1) glomerulus into (2) Bowman's capsule, then into the (3) <u>proximal</u> tubule, the (4) <u>Loop</u> of <u>Henle</u>, and the (5) <u>distal</u> tubule and finally into the collecting duct.

    3)   By active transport and diffusion, Henle's loop creates an osmotic concentration gradient in the fluid surrounding it, with the highest concentration at the loop's bottom.

        a.   While the filtrate moves through the nephron, water and nutrients are reabsorbed into the blood while wastes remain.

        b.   During its passage through the collecting duct, additional water leaves the filtrate by osmosis until the urine reaches an equilibrium with the highly concentrated surrounding fluid.

        c.   <u>Antidiuretic</u> <u>hormone</u> (<u>ADH</u>) regulates how concentrated the urine becomes.

4.   Kidneys as organs of **homeostasis**.

    A.   Kidneys filter a human's blood 350 times daily, fine-tuning blood composition. Kidney failure rapidly leads to death.

    B.   <u>Water balance</u> is regulated by the kidneys:

    1)   Over 45 gallons of water daily enter Bowman's capsule. Most is reabsorbed.

    2)   Reabsorption is passive: water follows salt and nutrient reabsorption by osmosis, the exact amount depending on <u>ADH</u> (also called <u>vasopressin</u>) levels in the blood.

        a.   ADH increases permeability of the distal tubule and collecting duct to water.

        b.   ADH is made by the hypothalamus in response to receptor cells there (that monitor osmotic levels of the blood) and in the heart (monitor blood volume).

        c.   When the osmotic concentration of blood rises or blood volume falls, more ADH is released and more water is reabsorbed.

        d.   Drinking beer causes dilution of blood and increase in blood volume: less ADH is released, less water is reabsorbed from the distal tubules and collecting ducts, the urine retains more water, and much weak urine results.

    C.   Regulation of <u>dissolved substances</u>.

    1)   The kidneys regulate water balance and the concentrations of glucose, amino acids, vitamins, urea, and several ions.

    2)   The kidneys regulate a constant blood pH by adjusting the content of hydrogen and bicarbonate ions.

    3)   The kidneys also eliminate harmful ingested substances (some drugs, food additives, pesticides, and cigarette toxins).

****

REVIEW QUESTIONS:

1. Trace the flow of blood from the artery leading into the kidney to the vein leading out of the kidney.
2. How do the functions of the digestive system and the excretory system play roles in the process of homeostasis?
3. Name and briefly describe the excretory organs possessed by flatworms and earthworms.
4. Describe the functions of the following structures associated with kidneys:
    A. Glomerulus.
    B. Bowman's capsule.
    C. Loop of Henle.
5. Explain how water balance is regulated by the effect of anti-diuretic hormone (ADH) and the brain hypothalamus on kidney function.

****

ANSWERS TO REVIEW QUESTIONS:

1. [see 3.A.-B.]

    Renal artery → capillaries of the glomerulus → arterioles → capillaries surrounding Bowman's capsule → capillaries surrounding the proximal tubules, loops of Henle, and distal tubules → renal vein

2. see 1.A.-C.

3. see 2.A.-B.

4. A. see 3.C.
    B. see 3.C.-D.
    C. see 3.E.

5. see 4.B.

****

CHAPTER 34:   DEFENSE AGAINST DISEASE: THE IMMUNE SYSTEM

CHAPTER OUTLINE.

This chapter focuses on the human body's defenses against microbial attack, including physical barriers and chemical barriers (non-specific inflammatory responses and specific immune responses). Particularly emphasized is the immune system, including mechanisms for recognizing invading microbes, attacking the microbes, and remembering microbes in case of future attack. The medical implications of the immune system are discussed, as are AIDS and cancer.

1.   **Defenses against microbial attack.**

    A.   The body's <u>three lines of defense</u>:
        1)   <u>External barriers</u> to keep microbes out of the body.
        2)   <u>Non-specific internal defenses</u> to combat all invading microbes.
        3)   <u>Immune response</u> directed against specific microbes.

    B.   <u>External barriers</u> (the first line of defense).
        1)   The intact <u>skin</u> is both:
            a.   A physical barrier to microbial entry.
            b.   An inhospitable environment for microbial growth with:
                i.   Dry, dead cells at the surface.
                ii.   Sweat and sebaceous glands secreting acids and natural antibiotics.
        2)   The <u>mucous membranes</u> of the respiratory and digestive tracts are well-defended:
            a.   Their mucous secretions have antibacterial enzymes.
            b.   Mucus physically traps microbes entering through nose or mouth.

        i.   Membrane cilia sweep up mucus and microbes and they are coughed or sneezed out of the body.

        ii.   If microbes are swallowed, stomach acid and protein-digesting enzymes destroy them.

        iii.   The intestine contain bacteria that secrete substances to destroy invading foreign microbes.

        iv.   Antibody IgA, residing in the respiratory and digestive tracts, binds to the surfaces of microbes:

            a)   Preventing them from invading mucous membrane cells.

            b)   Allowing them to be swept out by ciliary action.

C.   <u>Nonspecific internal defenses</u> (second line of defense).

    1)   Attack a wide variety of microbes.

    2)   <u>The three nonspecific internal defenses</u> are:

        a.   Eaters and killers: <u>Phagocytic cells</u> (destroy microbes) and <u>natural killer cells</u> (destroy body cells infected by viruses).

        b.   <u>Inflammatory response</u>, caused by large-scale microbial invasion through an injury causing tissue damage. This response:

           i.   Recruits killer cells and natural killer cells.

           ii.   Walls off the injured area, isolating infected tissue from the rest of the body.

        c.   <u>Fever</u>, response to microbes that succeed in establishing a major infection. Fever:

           i.   Slows down microbial reproduction.

           ii.   Enhances the body's own fighting abilities.

    3)   <u>Eaters and killers</u>: amoeboid cells.

        a.   <u>Macrophages</u> ("big eaters"): White blood cells (wbc) in the extracellular fluid that:

           i.   Ingest microbes by phagocytosis.

           ii.   Help in immune response by "presenting" parts of the microbe to cells of the immune system

        b.   <u>Natural killer cells</u>: wbcs that attack body cells that have been invaded by viruses.

           i.   Also recognize and kill cancer cells.

           ii.   Secrete, into the cell membranes of infected or cancerous cells, proteins that open up large pores, causing the target cell to die.

    4)   <u>The inflammatory response</u>:

        a.   Damaged cells release <u>histamine</u> into the wounded area.

        b.   Histamine increases blood flow to the wound by relaxing arterioles and making capillary walls leaky.

        c.   The wound becomes red, swollen, and warm.

        d.   Blood clotting occurs, "walling off" the wounded area.

        e.   Other chemicals attract phagocytic wbcs which engulf bacteria, dirt, and tissue debris.

        f.   As the wbcs die, they collect as pus.

        g.   Those microbes that escape into the bloodstream are eaten by wbcs in the blood vessels and lymph nodes.

        h.   The inflammatory response is insufficient:

           i.   If the wound is large or the invading microbes reproduce too quickly.

ii.   If microbes enter through the respiratory tract.

iii.   In such cases, the highly specific immune system takes over.

5)   <u>Fever</u>.

    a.   Severe fevers are dangerous, but low-grade fevers (100-102°F) are beneficial.

    b.   The brain hypothalamus has temperature-sensing nerve cells that act as a body thermostat.

        i.   When disease organisms invade, certain wbcs release hormones called <u>endogenous pyrogens</u> ("self-produced fire makers") that travel in the bloodstream to the hypothalamus and raise the thermostat's set point, triggering shivering, increased fat metabolism, and the feeling of coldness.

        ii.   Pyrogens also cause other cells to reduce iron and zinc concentrations in the blood.

    c.   Fever has good effects on the body and bad effects on invading microbes:

        i.   Microbes need more iron to reproduce at higher temperatures, so fever and less iron in the blood slow bacterial reproduction.

        ii.   Fever increases activity of phagocytic wbcs that attack bacteria.

    d.   Fever helps fight viral infections.

        i.   When invaded by viruses, certain body cells make and release a protein called <u>interferon</u> that travels to other cells and increases resistance to viral attack.

        ii.   Fevers increase interferon production.

2.   **The Immune System** (the third line of defense).

A.   The immune response attacks one specific type of microbe at a time, overcomes it, and provides future protection against this microbe type but no others. How?

    1)   Two types of wbcs (B and T cells made from precursors in bone marrow) are involved.

        a.   <u>T cells</u>, early in embryonic development, migrate to the thymus and mature.

        b.   <u>B cells</u> differentiate in the bone marrow itself.

    2)   <u>Steps in the immune response</u>:

        a.   Recognizing the invader.

        b.   Launching a successful attack.

        c.   Remembering the invader to ward off future attacks.

B.   <u>Recognition</u>.

    1)   Recognizing foreign substances:

        a.   B and T cells produce <u>antibodies</u>, complex Y-shaped proteins made of 2 pairs of peptides, one large (heavy) and one small (light) peptide on each side of the Y.

        b.   Each peptide chain consists of both:

            i.   A <u>constant region</u> which is very similar in all antibodies.

            ii   A <u>variable region</u>, which differs widely among antibodies, is at the tips of each arm of the Y. These regions form a specific binding site for other molecules.

2)   The antibody arms: Specific binding to antigens.

      a.   Variable regions of the Y-arms form highly specific binding sites for large molecules.

      b.   Each binding site has its own peculiar shape and electrical charge, so only one or a few types of molecules can fit in and bind.

      c.   Only large complex molecules (called antigens) can bind to antibodies.

      d.   The binding of antigen to antibody triggers the immune response.

3)   The antibody stem: Determining antibody function.

      a.   The Y-stem consists of part of the constant regions of the two heavy chains.

      b.   The stem determines the activity of the antibody, allowing it to attach to the membranes of certain cells or to certain molecules.

4)   Sources and functions of antibodies.

      a.   B cells make antibodies and T cells make antibody-like molecules called T-cell receptors.

      b.   Antibodies and T-cell receptors are receptors.

            i.   The Y-stem attaches to the cell plasma membrane with the Y-arms protruding outward.

            ii.   Antigens bind to the Y-arm receptor regions, triggering responses in the lymphocyte cells bearing the antibodies.

      c.   Antibodies and T-cell receptors are effectors.

            i.   Certain B-cell descendants secrete antibodies into the bloodstream where they:

                  a)   Neutralize poisonous antigens.

                  b)   Promote destruction of microbes with antigens.

            ii.   Stem differences define 5 types of true antibodies:

                  a)   IgG: Y-stems form a coating on bound microbes, making them easier to ingest by macrophages.

                  b)   IgA: Y-stems promote secretion into respiratory and digestive tracts.

                  c)   IgE: Y-stems bind to "mast cells" lining respiratory and digestive tracts, releasing histamine during allergic responses.

5)   Recognizing a multitude of foreign molecules.

      a.   Paradox: The immune system recognizes and responds to many types of antigens because it produces millions of different antibodies, each capable of binding a different antigenic molecule. But humans don't have that many kinds of antibody genes.

      b.   Mechanism of producing antibody diversity:

            i.   First, genes encoding parts of antibodies recombine during immune cell development, allowing each immune cell to have a unique combination of antibody genes that it and its descendent cells use to make antibodies:

                  a)   Each cell originally has a few constant region (C) genes and several hundred variable region (V) genes.

                  b)   During cell maturation, these genes move around, the chromosome ending up with one V gene adjacent to each C gene.

                      i)   Thus, each cell comes to possess one "C+V" light chain gene and one "C+V" heavy chain gene.

      ii)   Which <u>V</u> gene winds up next to which <u>C</u> gene is random.

  c)   Only adjacent <u>C+V</u> genes are transcribed. Thus, each immune cell produces an antibody, specified by the chance recombination of <u>C+V</u> genes, that is different from the antibody produced by other immune cells.

  ii.   <u>Second</u>, the genes for certain antibody parts are incredibly prone to mutate, constantly generating new antibody genes.

  c.   The immune system does not custom-design antibodies to fit invading antigens.

    i.   Instead, it randomly makes billions of different antibodies.

    ii.   By chance, there will be one or more properly shaped antibodies for virtually any invading antigen.

6)   Distinguishing "<u>self</u>" from "<u>non-self</u>".

  a.   As an embryo develops, some immune cells produce, by chance, antibodies against the body's own molecules.

  b.   However, if <u>immature</u> immune cells contact molecules that bind to their surface antibodies, the cells die. So, "self" antigens eliminate "self" immune cells.

  c.   The immune system distinguishes "self" from "non-self" by retaining only those immune cells that do not respond to the body's own molecules.

C.   <u>Attack</u>.

  1)   The immune system can mount 2 types of attack:

    a.   <u>Humoral immunity</u> by B cells, mediated by free antibodies in the bloodstream.

    b.   <u>Cell-mediated immunity</u> by T cells killing the microbes directly.

  2)   <u>Humoral immunity</u>.

    a.   Microbes in the bloodstream encounter B cells, each with a specific surface antibody. Those B cells with the appropriate antibody bind to a microbe antigen.

    b.   Antigen-antibody binding triggers massive changes in these specific B cells, initiating rapid cell division.

    c.   The resulting population of cells differentiates into 2 types:

      i.   <u>Plasma cells</u>: release large amounts of antibodies into the bloodstream.

      ii.   <u>Memory cells</u>: don't release antibodies but allow for future immunity to the microbe.

    d.   Four effects of circulating antibodies on antigen molecules and cells bearing antigens:

      i.   <u>Neutralization</u> of toxins by binding with them.

      ii.   <u>Promotion of phagocytosis</u> by coating the microbe and making it easier for wbcs to engulf them.

      iii.   <u>Agglutination</u>: clumping of microbes due to each antibody being able to attach to several antigen molecules. This seems to enhance phagocytosis by wbcs.

      iv.   <u>Complement reactions</u>: The antigen-antibody complex causes other blood proteins to attract phagocytic wbcs to the site to engulf the foreign cells.

3) <u>Cell-mediated immunity</u>.
   a. Destroys the body's own cells when they become cancerous or infected with viruses.
   b. When T cell-surface antibodies (called "T-cell receptors") bind antigens, T cells divide rapidly, producing 2 types of cells:
      i. <u>Effector cells</u>: attack cells but do not release antibodies into the bloodstream. There are 3 types of effector cells:
         a) <u>Helper T cells</u>: release hormone-like chemicals that stimulate cell division and differentiation in both killer T cells and B cells to fight the microbial invasion. <u>AIDS</u> disrupts this by destroying helper T cells.
         b) <u>Killer T cells</u>: bind to antigens on the surface of infected cells and release proteins that disrupt the infected cells' plasma membranes.
         c) <u>Suppressor T cells</u>: appear after the infection has been conquered to shut off the immune response in both B and killer T cells.
      ii. <u>Memory cells</u>: protect the body against future infections.

D. <u>Memory</u>.
   1) Retaining immunity to future infection is the function of <u>memory cells</u>:
      a. Plasma B cells and killer T cells immediately fight infection but live only a few days.
      b. B and T memory cells survive for years. If foreign cells with the same antigen reenter the bloodstream, memory cells will recognize them, multiply rapidly to make huge populations of plasma cells and killer cells, to evoke a second immune response.
   2) Memory cells respond more rapidly than the B and T cells that originally made them. Memory cells react so rapidly that often there are no noticeable symptoms of the reinfection.

3. **Medical implications** of the immune system.

A. <u>Antibiotics</u> and immune responses.
   1) Some microbes can kill because they produce particularly toxic products or divide so rapidly that full activation of the immune system comes too late.
   2) Antibiotics slow down the growth and multiplications of many microbes, allowing the immune system enough time to finish the job.

B. <u>Vaccinations</u>.
   1) Jenner and Pasteur discovered that injections of weakened or dead microbes produced "<u>immunization</u>" (immunity to reinfection by deadly strains).
   2) <u>Vaccination</u>: injection of weakened or killed microbes to cause immunity.
   3) Through genetic engineering, it is possible to produce tailor-made vaccines.

C. <u>Allergies</u>.
   1) <u>Allergies</u> are reactions, in some people, to substances that are not harmful (e.g., pollen, dust, mold spores, bee stings, etc.).
   2) Allergies are a form of immune response. Example:

          a.   A pollen grain enters the bloodstream.

          b.   It is recognized as a foreign antigen by a B cell.

          c.   The B cell proliferates, making plasma cells that pour out IgE into the bloodstream.

          d.   The antibodies bind to histamine-containing "mast cells" in connective tissue.

          e.   When they make contact with pollen grains, mast cells release <u>histamine</u>, causing increased mucus secretion, leaky capillaries, and other symptoms of inflammation ("<u>hay</u> <u>fever</u>").

          f.   <u>Antihistamine</u> drugs block some of the histamine effects.

     3)   People without allergies either lack the genes for the allergy-causing antibodies, or produce less of the antibody.

     4)   Many parasites invade the body through mouth, nose, or anus.

          a.   These parasites attach to the lining of the nasal passages, mouth, or intestine.

          b.   Probably, typical allergy symptoms are properly directed to dislodge and expel these parasites, not to attack pollen.

D.   <u>Autoimmune Diseases</u>.

     1)   An abnormal immune system may produce antibodies against antigens on the body's own cell surfaces, causing an <u>autoimmune</u> <u>disease</u> which destroys the body cells. Examples:

          a.   Some forms of anemia destroy self-rbcs.

          b.   Some forms of juvenile-onset diabetes are due to self-destruction of insulin-secreting cells of the pancreas.

     2)   There is no cure for autoimmune disease:

          a.   Replacement therapy can alleviate symptoms.

          b.   Some autoimmune responses may be suppressed with drugs.

E.   <u>Immune Deficiency Diseases</u>.

     1)   Rarely, a child is born with the ability to make few or no immune cells, and common bacterial infections may be fatal.

     2)   These children must live in isolated germ-proof "bubbles".

     3)   In some of these children, bone marrow transplants have resulted in antibody production.

     4)   In the last decade, "acquired immune deficiency syndrome" (AIDS), caused by a virus, has become a public health threat. Read the text for details of AIDS.

4.   **Cancer**.

A.   About 1/3 of all Americans will develop a cancer.

     1)   <u>Cancer</u> is a malfunctioning of the growth controls of the body's own cells, causing self-destruction.

     2)   A cancer is a population of cells that has escaped from normal regulatory processes and grows without control. Questions:

          a.   What changes occur that allow cells to escape normal control?

          b.   What agents (genetic, viral, environmental) initiate these cellular changes?

B.   <u>Causes of cancer</u>.
1)   "<u>Oncogenes</u>" are genes that cause cancer.
2)   Two ways that oncogenes may cause cancer:
    a.   A potentially dangerous oncogene may be present in all cells but causes cancer only when activated by some external trigger.
        i.   Some oncogenes may normally act only during embryonic development but have erroneously been turned on to full speed in adults.
        ii.   In some cancers, chromosomes may become rearranged so that an embryonic growth gene is transferred to a region that is normally transcribed rapidly in adults, initiating abnormal growth and cell division.
    b.   A harmless gene for normal cell replacement may mutate into an oncogene that accelerates the division rate.
3)   Some types of cancer are "caused" by infections by viruses containing RNA as genes:
    a.   The viral RNA forces the cell to "reverse transcribe" it into DNA.
    b.   The new viral DNA integrates into a host chromosome where it and adjacent host genes may be rapidly transcribed to make RNA.
    c.   If the nearby host DNA is a previously inactive oncogene, the cell may become cancerous.
4)   Probably, most cancers are caused by environmental "insults" such as chemicals and radiation. Exposure to such agents early in life may cause a gene mutation that predisposes cells to react to viral infection later in life by becoming cancerous.
5)   Most cancers require two or more distinct steps:
    a.   Mutation of a pre-oncogene to a true oncogene early in life, perhaps by exposure to radiation.
    b.   Many years later, a viral infection or exposure to chemicals may move the oncogene to an active region of DNA, and cancer begins.
6)   <u>Cancer-suppressor genes</u>.
    a.   These genes suppress cell growth and division.
        i.   Their protein products may turn off pre-oncogene transcription.
        ii.   They may block hormonal signals to stimulate cell division.
    b.   Many cancers begin when tumor-suppressor genes are damaged or lost.

C.   <u>Defenses against cancer</u>.
1)   Probably, cancer cells form in our bodies every day but are detected and destroyed by killer T cells which detect "non-self" changes in surface proteins of the cancer cells. When this defense fails, cancer develops.
2)   Medical approaches to treat cancer:
    a.   Burn the cancer out with <u>radiation</u>.
    b.   Cut the cancer out with <u>surgery</u>.
    c.   Poison the cancer off with <u>drugs</u> ("chemotherapy").
        i.   The most common chemotherapies include drugs that are <u>nucleotide mimics</u> that enter DNA during chromosome replication but don't allow for further division. Problems arise since normal cells (hair follicles and intestinal cells) also divide and these will also be affected by the drugs.
        ii.   <u>Monoclonal antibodies</u>: a future therapy involving fusion of anti-

cancer B cells with myeloma cells (cancerous wbcs).

a) The fused cells will have antibody production against the patient's cancer and rapid cell division.

b) Lethal drugs may be attached to the monoclonal antibody cells which are then injected into the patient.

c) The antibody cells bind only to cancer cells and the drug destroys them without harming normal body cells.

****

REVIEW QUESTIONS:

1. Name the body's three lines of defense against microbial attack.
2. Why are the skin and membranes of the respiratory and digestive tracts good barriers to the entry of microbes?
3. List the events that occur during an inflammatory response.
4. What conditions render the inflammatory response insufficient?
5. What are the three steps involved in an immune response?

6.-19.   **MATCHING TEST**: B cells and T cells

Choices: A.   B cells
                B.   T cells
                C.   Both of these
                D.   Neither of these

6. Suppressor cells.
7. Involved in humoral immunity.
8. Involved in allergic reactions.
9. Are white blood cells.
10. Mature in the thymus.
11. Involved in cell-mediated immunity.
12. Produce antibodies.
13. Differentiate in bone marrow.
14. Helper cells.
15. Antibodies are attached to their cell membranes.
16. Produce histamines.
17. Release antibodies into the bloodstream.
18. Produce memory cells.
19. Killer cells.

-----

20. Describe the role of antibodies in recognition of foreign substances.
21. Describe how antibody genes are produced in white blood cells.
22. Describe how a developing human's immune system learns to distinguish "self molecules" from "non-self molecules".
23. Describe the mechanism of humoral immunity.

24. Describe four effects that circulating antibodies have on cells containing foreign antigens.
25. Describe the mechanism of cell-mediated immunity.
26. Define vaccination and how is it helpful in combatting disease organisms?
27. Describe an allergic reaction and explain how it is really a type of immune response.
28. Define autoimmune disease and tell why it is so hard to "cure".
29. What is "cancer"?
30. What is an "oncogene" and what can trigger oncogene action?
31. List the three approaches taken by medical science to "cure" cancer.
32. What are monoclonal antibodies and how might they be used to fight cancer?

****

ANSWERS TO REVIEW QUESTIONS:

1.	see 1.A.1)-3)
2.	see 1.B.1)-2)
3.	see 1.C.4)a.-g.
4.	see 1.C.4)h.
5.	see 2.A.2)a.-c.

6.-19.   [see 2.A.-D.]

6.	B	13.	A	
7.	A	14.	B	
8.	A	15.	C	
9.	C	16.	A	
10.	B	17.	A	
11.	B	18.	C	
12.	C	19.	B	

20.   see 2.B.1)-4)

21.   see 2.B.5)a.-c.

22.	see 2.B.6)
23.	see 2.C.2)a.-c.
24.	see 2.C.2)d.
25.	see 2.C.3).a)-c)
26.	see 3.B.
27.	see 3.C.1)-4)
28.	see 3.D.
29.	see 4.A.1)-2)
30.	see 4.B.
31.	see 4.C.2)a.-c.
32.	see 4.C.2)c.ii.

****

CHAPTER 35: CHEMICAL CONTROL OF THE ANIMAL BODY: THE ENDOCRINE SYSTEM

CHAPTER OUTLINE.

This chapter considers how the body's internal chemistry is controlled by hormones. The four classes of animal hormones are discussed as are the endocrine glands that produce them. Hormone regulation by negative feedback is outlined and the major mammalian endocrine systems are presented in detail.

1.   **Communicating by cell contact** in multicellular organisms.

    A.   <u>General facts</u>.
        1)   Occurs by means of molecules protruding from cell surfaces.
        2)   Important in the development of embryos as cells migrate to their proper destinations.
        3)   Plays a key role in defense against disease organisms since immune cells recognize invaders by their surface molecules.

    B.   <u>Limitations</u>:
        1)   Restricted to the number of cells that can touch one another simultaneously.
        2)   Communication over even small distances is difficult or impossible unless cells release chemicals for communication into the environment.

2.   **Chemical communication** within the animal body.

    A.   <u>Similarities</u> between hormone control cells and nervous control cells: both cell types make "messenger" chemicals that they release into extracellular spaces.

    B.   <u>Differences</u> between hormone control and nervous control.

1) Distance.
    a.   Nerve cells release the chemical ("neurotransmitter") very close to the cells they influence.
    b.   Hormone producing cells release their chemicals into the bloodstream which carries them great distances to the target cells.

2) Number of cells contacted.
    a.   Nerve cells quite precisely affect small numbers of cells.
    b.   Hormone cells indiscriminately bathe millions of cells.

3) Speed.
    a.   Nerve cells speed information quickly via electrical signals.
    b.   Hormones travel more slowly in the bloodstream.
    c.   However, some nerve cells secrete hormones (neurohormones) into the bloodstream.

3.   **Hormone functions in animals**.

A.   Hormone: a chemical secreted by cells in one part of the body and transported in the bloodstream to other body parts where it affects particular target cells.

B.   Hormone action on target cells.
1) Target cells for a hormone have appropriate receptors on their surfaces to interact with that hormone.
    a.   Cells without the proper receptors will not be affected by the hormone.
    b.   Receptors for hormones are found in three general locations on target cells:
        i.   In the cell membrane.
        ii.   In the cytoplasm.
        iii.   In the nucleus.

2) Hormones may bind to surface receptors.
    a.   Most modified amino acid and peptide hormones are water-soluble but not lipid soluble and cannot cross the cell membrane.
    b.   These hormones must react with protein receptors on the cell surface which can trigger rapid short-term internal cellular changes.
    c.   Many of these hormones stimulate synthesis of cyclic AMP (c-AMP) that activates cellular enzymes.
    d.   c-AMP is called the second messenger since it transfers information from the first messenger (the hormone) to target molecules in the cell.
    e.   Calmodulin is another second messenger.
        i.   Some hormones bind to surface receptors to open calcium channels allowing calcium to flow into the cell.
        ii.   Calmodulin protein in the cytoplasm binds to the calcium, and the calmodulin-calcium complex activates cellular enzymes.

3) Hormones may bind to intracellular receptors.
    a.   Steroid hormones are lipid-soluble and easily pass through the cell membranes.
    b.   In the cytoplasm, they combine with soluble protein receptors and enter the nucleus, some exerting their effects over long periods of time.
    c.   The steroid-receptor complexes bind to specific places on chromosomes, activating genes to make RNA which then produces specific proteins by

translation.

C.   Hormones maintain homeostasis through <u>negative feedback</u>.
 1)   Hormone secretion must be regulated so that just the right amounts are released at the right times.
 2)   Animals regulate most hormone release through <u>negative feedback</u>: hormone secretion causes effects in target cells that inhibit further secretion of the hormone.
   a.   This helps maintain homeostasis (keeping body conditions relatively constant over time).
   b.   Negative feedback works like a thermostat does to keep your house a constant temperature in winter:

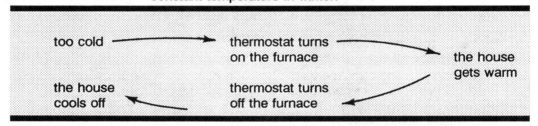

   c.   Negative feedback in regulating water levels in the blood via antidiuretic hormone (ADH):

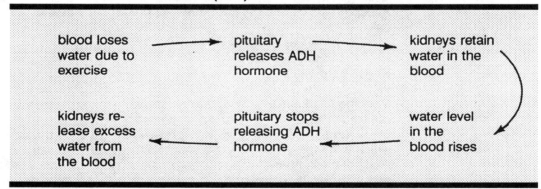

D.   <u>Four classes of animal hormones</u> (see Table 35-1).
 1)   <u>Modified amino acids</u>: adrenalin, noradrenalin, thyroxine.
 2)   <u>Proteins</u> or peptides: most hormones.
 3)   <u>Steroids</u> made from cholesterol: sex hormones.
 4)   <u>Prostaglandins</u>: made from modified fatty acids in most cells.

4.   **Mammalian endocrine systems**.

A.   <u>Gland structure and function</u>. Mammals have 2 types of glands:
 1)   <u>Exocrine glands</u> (sweat, mammary and parts of the liver and pancreas): release secretions into ducts leading outside the body or into the digestive tract.
 2)   <u>Endocrine glands</u>: called "ductless glands"
   a.   Consist of clusters of hormone-producing cells embedded within a

                network of capillaries.

b.   Secrete hormones into extracellular fluid surrounding the capillaries.

c.   The hormones enter the capillaries by diffusion and are carried to other parts of the body.

B.   Seven major endocrine systems exist (see Figure 35-3 and Table 35-2): Hypothalamus/pituitary complex, thyroid, parathyroid, pancreas, adrenal cortex, adrenal medulla, and gonads.

C.   <u>Hypothalamus/Pituitary complex</u>.

    1)   <u>Pituitary</u>: pea-sized gland, with posterior and anterior regions, that hangs by a stalk from the brain <u>hypothalamus</u> which controls hormone release for the entire pituitary.

    2)   <u>Posterior pituitary</u>: really part of the brain, containing nerve endings originating in the hypothalamus.

        a.   Contains the nerve endings of 2 types of <u>neurosecretory</u> cells that make, store, and release peptide hormones into the bloodstream when stimulated.

        b.   Types of hormones produced:

            i.   <u>ADH</u> (antidiuretic hormone, "hormone that prevents urination"): increases water permeability of the kidney nephrons so that less water is lost by the blood through excretion, thus reducing dehydration.

            ii.   <u>Oxytocin</u>. Causes:

                a)   Contraction of the uterus during childbirth.

                b)   Contraction of breast muscle cells during lactation, allowing milk to flow.

                c)   Maternal effects in virgin female rats.

                d)   Muscle contraction of tubes conducting sperm to the penis in males, causing ejaculation.

    3)   <u>Anterior pituitary</u>: a true endocrine gland with the hormone secreting cells enmeshed in capillaries.

        a.   Six different hormones are produced, 4 of which regulate release of hormones in other glands.

            i.   <u>FSH</u> (<u>follicle stimulating hormone</u>) and <u>LH</u> (<u>luteinizing hormone</u>): affect sex cells and sex hormone production and release in both sexes (Details in Chapter 39).

            ii.   <u>TSH</u> (<u>thyroid stimulating hormone</u>): stimulates the thyroid to release its hormones.

            iii.   <u>ACTH</u> (<u>adrenocorticotropic hormone</u>): causes release of hormones from the adrenal cortex.

            iv.   <u>Prolactin</u>: stimulates development of the mammary glands during pregnancy and milk production after delivery of the baby.

            v.   <u>Growth hormone</u> (<u>somatotropin</u>): regulates growth of the body, especially bone growth. In adults, it helps regulate protein, fat, and sugar metabolism.

        b.   The hypothalamus regulates the release of the 6 anterior pituitary hormones.

            i.   The hypothalamus makes at least 9 types of peptides (in neurosecretory cells) that affect the anterior pituitary.

        ii.   Some are <u>releasing</u> <u>hormones</u> and some are <u>inhibitory</u> <u>hormones</u> depending on their action in the anterior pituitary.

        iii.   They are produced in minute amounts.

D.   <u>Thyroid and Parathyroid.</u>

    1)   The thyroid gland lies around the pharynx in the front of the neck and the 4 parathyroid glands are embedded in the back of the thyroid.

    2)   The <u>thyroid</u> produces 2 hormones:

        a.   <u>Thyroxine</u>: an iodine-containing modified amino acid that raises the metabolic rate of most body cells.

           i.   In juvenile mammals, thyroxine and growth hormones regulate growth.

           ii.   In adults, elevated metabolic rate helps regulate body temperature and stress reactions.

           iii.   Thyroxine release is stimulated by TSH (from the anterior pituitary) which is stimulated by a releasing hormone (from the hypothalamus) which is regulated by thyroxine levels in the blood. This is a negative feedback system:

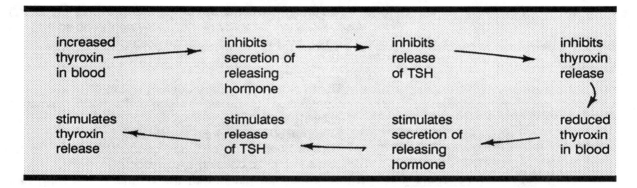

        b.   <u>Calcitonin</u> (from thyroid) and <u>parathormone</u> (from parathyroids): control concentration of calcium in the blood and other body fluids. These hormones regulate calcium absorption and release by the bones:

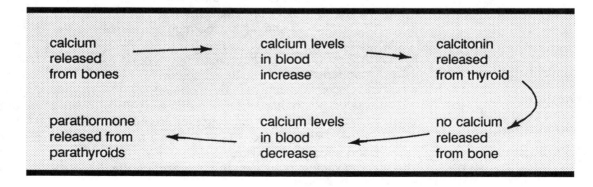

E. <u>Pancreas</u>.
    1) The pancreas is a double gland.
        a. The exocrine part makes digestive enzymes that flow, via the pancreatic duct, to the small intestine.
        b. The endocrine part contains clusters of cells called <u>islets</u>.
    2) Each <u>islet</u> contains 2 types of cells.
        a. One cell type makes <u>insulin</u> (reduces blood sugar levels).
        b. The other type makes <u>glucagon</u> (increases blood sugar levels).
    3) These hormones work in opposite ways to regulate carbohydrate and fat metabolism:

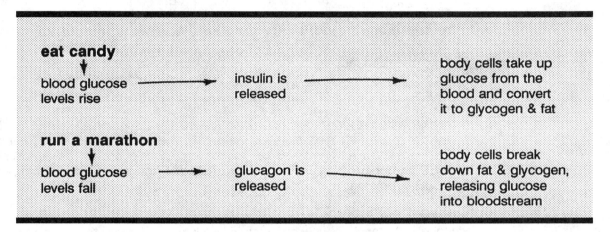

    4) Defects in insulin production, release, or reception by target cells results in <u>diabetes</u> <u>mellitus</u>.
        a. Blood glucose levels are high and fluctuate wildly with sugar in the diet.
        b. Lipid metabolism also is impaired, causing fat deposition in blood vessels and the heart. Diabetes is a major cause of heart attacks in the US.

F. <u>Sex organs</u>: male testis and female ovary.
    1) The <u>testis</u> secretes <u>androgen</u> hormones, the most important of which is testosterone.
    2) The <u>ovary</u> secretes two types of hormones, <u>estrogen</u> and <u>progesterone</u>, as discussed in Chapters 39 and 40.

G. <u>Adrenal</u> ("on the kidney") <u>glands</u>.
    1) Are located atop the kidneys and act as double glands.
    2) The center of the gland is the <u>adrenal</u> <u>medulla</u>, a group of secretory cells derived from nerve cells (the nervous system controls secretion of the adrenal hormones).
        a. Makes two hormones in response to stress: <u>adrenalin</u> (epinephrine) and <u>noradrenalin</u> (norepinephrine).
        b. These hormones prepare the body for action (increased heartbeat and breathing, increased blood glucose levels, direct blood flow towards the brain and muscles).
    3) The outer layer of the gland, the <u>adrenal</u> <u>cortex</u>, secretes several types of lipid steroid hormones:
        a. Three types of <u>glucocorticoids</u>: help to control glucose metabolism.

        i.   Release is stimulated by ACTH (from anterior pituitary) which in turn is stimulated by hypothalamus releasing hormones that are produced in response to stress.

        ii.   They act like glucagon, stimulating increased glucose levels in the blood.

    b.   <u>Aldosterone</u>: secretion is regulated by sodium levels in the blood.

        i.   If blood sodium falls, adrenal cortex releases aldosterone which causes kidneys and sweat glands to retain sodium.

        ii.   As blood sodium levels rise, further aldosterone secretion ceases.

    c.   <u>Testosterone</u> (male sex hormone): produced in very small amounts. Tumors of the adrenal cortex can lead to excessive testosterone release, causing masculinization of women (the "bearded ladies" of circus fame).

H.   <u>Prostaglandins</u>: modified fatty acids.

    1)   Produced by nearly all body cells in a diversity of forms and for a variety of functions.

    2)   Some cause inflammation of joints and stimulate pain receptors. Aspirin (acetylsalicylic acid) inhibits production of these prostaglandins.

    3)   Some cause smooth muscle to contract and, like oxytocin, stimulate uterine contractions during childbirth. When these contractions occur excessively during menstruation, "menstrual cramps" occur. Aspirin or ibuprofen reduce cramping by blocking prostaglandin production.

I.   Other endocrine organs.

    1)   <u>Pineal gland</u>: located between cerebral hemispheres behind the brainstem.

        a.   Is smaller than a pea.

        b.   It's functions are poorly understood.

        c.   It produces the hormone <u>melatonin</u>, a modified amino acid, secreted in a daily rhythm regulated by the mammalian eye.

        d.   By responding to daylength characteristics of different seasons, the pineal may regulate seasonal reproductive cycles.

        e.   In humans, it may cause a depression during the short days of winter.

    2)   <u>Kidney</u>.

        a.   If blood O_2 levels drop, the kidneys produce <u>erythropoietin</u> which increases red blood cell production (see Chapter 33).

        b.   In response to low blood pressure, the kidneys produce <u>renin</u> enzyme that catalyzes production of the blood protein hormone <u>angiotensin</u> to raise blood pressure by constricting arterioles.

    3)   The <u>heart</u> makes a protein hormone called <u>atrial natriuretic peptide</u> (ANP) that increases the output of salt and water by the kidneys.

    4)   The <u>stomach</u> and <u>small intestine</u> produce a variety of peptide hormones that help regulate digestion (see Chapter 32).

****

****

REVIEW QUESTIONS:

1.   Name two limitations of communication by cell contact.

2.-8.   **MATCHING TEST**: hormone vs. neuron control.

Choices: A.   Nerve cells   C.   Both cell types
             B.   Hormone producing cells   D.   Neither cell type

2.   Indiscriminately affect millions of cells.
3.   Speed transmissions quickly using electrical signals.
4.   Release neurotransmitters very close to the cells they influence.
5.   Their chemicals travel slowly to the sites of action.
6.   Release chemicals into the bloodstream that carries them great distances.
7.   Precisely affect small numbers of cells.
8.   Some release neurohormones.

----

9.   Name four classes of animal hormones.

10.-14. **MATCHING TEST**: types of glands.

Choices: A.   Endocrine glands
             B.   Exocrine glands

10.   Clusters of hormone-producing cells embedded in capillaries.
11.   Some release hormones into ducts leading to the outside of the body.
12.   Sweat glands.
13.   Release hormones into extracellular spaces surrounded by capillaries.
14.   Mammary glands.

----

15.   Why is cyclic-AMP (c-AMP) called the "second messenger"?
16.   Describe how a negative feedback system works to heat a home in the winter.

17.-26. **MATCHING TEST**: pituitary hormones.

Choices: A.   ACTH   E.   Growth hormone
             B.   ADH   F.   Oxytocin
             C.   FSH   G.   Prolactin
             D.   LH   H.   TSH

17.   Affect sex cell release in both sexes.
18.   Produced by the posterior pituitary.

19. Regulates growth of bones.
20. Increases water permeability by the kidney nephrons, reducing dehydration.
21. Stimulates development of the mammary glands during pregnancy.
22. Causes uterine contraction during childbirth.
23. Stimulates release of hormone by the thyroid gland.
24. Allows milk to flow from the breasts during lactation.
25. Causes release of hormones from the cortex of the adrenal glands.
26. Allows ejaculation to occurs in males.

----

27. Describe the relationship between the hypothalamus, TSH, and thyroxine hormone release from the thyroid gland.
28. Describe how the action of the hormones calcitonin and parathormone regulate calcium absorption and release from bones.
29. Name the two types of hormones produced by the islet cells of the pancreas and explain how these hormones interact to regulate carbohydrate and fat metabolism.

30.-36. **MATCHING TEST**: adrenal gland hormones.

         Choices: A.   Adrenalin
                      B.   Aldosterone
                      C.   Glucocorticoids
                      D.   Noradrenalin
                      E.   Testosterone

30. Produced by the adrenal medulla.
31. Release is stimulated by ACTH.
32. Help control glucose metabolism.
33. Male sex hormone.
34. Acts like glucagon.
35. Secretion regulated by sodium levels in the blood.
36. Over production, caused by certain tumors, can result in "bearded ladies".

----

37. Name two functions of prostaglandins and how a person can deal with lessening their effects.
38. Describe how the pineal gland, the kidney, and the heart are endocrine organs.

****

****

ANSWERS TO REVIEW QUESTIONS:

1.   see 1.B.1)-2)

2.-8.   [see 2.A.-B.1)-3)]

2.	B	6.	B
3.	A	7.	A
4.	A	8.	A
5.	B		

9.   see 3.D.1)-4)

10.-14. [see 4.A.1)-2)]

10.	A	13.	A
11.	B	14.	B
12.	B		

15.   see 3.B.2)a.-d.

16.   see 3.C.2)a.-b.

17.-26. [see 4.C.1)-3)]

17.	C,D	22.	F
18.	B,F	23.	H
19.	E	24.	F
20.	B	25.	A
21.	G	26.	F

27.   see 4.D.2)a.iii.

28.   see 4.D.2)b.

29.   see 4.E.2)-3)

30.-36 [see 4.G.2)-3)]

30.	A,D	34.	C
31.	C	35.	B
32.	A,C,D	36.	E
33.	E		

37.   Pain and uterine contractions;
        treated with aspirin
        or ibuprofen for menstrual
        cramps; [see 4.H.2)-3)]

38.   see 4.I.1)-3)

****

CHAPTER 36: INFORMATION PROCESSING: THE NERVOUS SYSTEM

CHAPTER OUTLINE.

This chapter covers the structure and function of nerve cells, the nature of resting potentials in nerves, and how action potentials are generated and conducted. Also, the human nervous system is discussed in great detail, including the central nervous system (brain and spinal cord), the peripheral nervous system, neurotransmitters, and neuromodulators. Finally, learning, memory, and retrieval are covered briefly.

1.   **Nerve cell structure.**

     A.   <u>Functions of neurons</u> (individual nerve cells):
           1)   Receive information.
           2)   Integrate the information received and produce an appropriate output signal.
           3)   Conduct the signal to its output terminal.
           4)   Transmit the signal to other cells.
           5)   Coordinate their own metabolic activities.

     B.   A typical neuron has <u>four distinct structural regions</u>.
           1)   <u>Dendrite</u>: a tangle of fibers that branch from the cell body. Dendrites:
                 a.   Receive information from the environment.
                 b.   Convert the information to electrical signals.
           2)   <u>Cell body</u>: the cell's integration center. Cell bodies:
                 a.   Receive various signals from dendrites.
                 b.   Produce <u>action</u> <u>potential</u> output signals.
                 c.   Coordinate metabolic activities of the cells.
           3)   <u>Axons</u>: long thin fibers extending from the cell body. Axons:
                 a.   Carry action potentials to output terminals.

> b.   Are bundled together into <u>nerves</u>, conducting action potentials undiminished in strength.

4)   <u>Synaptic</u> <u>terminals</u>: sites at which signals are transmitted to other cells (glands, muscles, or dendrites of a second neuron).

5)   <u>Diagram</u> of a nerve cell: the arrows indicate that a nerve impulse travels from dendrties → cell body → axons.

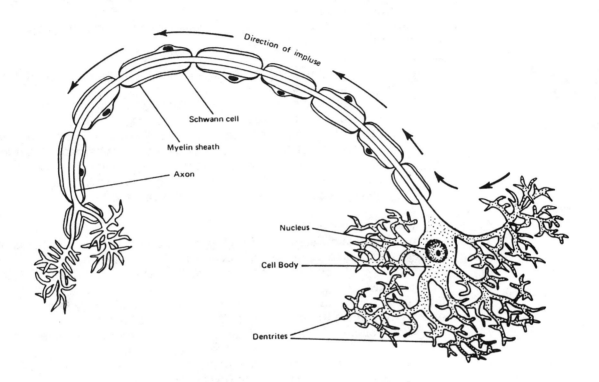

2.   **Mechanisms of neural activity**.

A.   Like a battery, unstimulated, resting neurons maintain a constant electrical potential across their cell membranes.

1)   This <u>resting</u> <u>potential</u> is always negative within a cell and measures from -40 to -90 millivolts (.001 volt).

2)   This potential is altered when the nerve cell is stimulated.

a.   It can become either more or less negative.

b.   If it becomes sufficiently less negative, it reaches a <u>threshold</u>.

c.   At the threshold, the potential suddenly reverses, becoming +20 to +50 millivolts positive inside. This is the <u>action</u> <u>potential</u> which lasts a few milliseconds (.001 sec.) before reverting back to the negative resting potential.

B.   <u>Origin of resting potentials</u>.

1)   Maintaining a resting potential involves the properties of diffusion, electrical attraction, and differential permeability.

2)   Ions inside and outside the cells:

    a.   The ions of the cytoplasm are mainly K^+ (positive potassium ions) and large negatively charged organic molecules.

    b.   The extracellular fluid contains mainly Na^+ (positive sodium ions) and Cl^- (negative chlorine ions).

3)   Ions cross cell membranes through tunnel-shaped proteins called <u>channels</u> that extend through the membranes.

    a.   In unstimulated neurons:

        i.   Only K^+ can cross the membrane, travelling through specific K channels.

        ii. K^+ diffuse out of the cell, leaving the large negatively charged organic molecules behind.

    b.   As more K^+ leaves, the inside of the cell becomes more negative and this tends to pull the K^+ back inside (opposite charges will attract).

    c.   When a balance exists between the forces of diffusion pulling K^+ out of the cell and electrical attraction pulling them back in, a stable resting potential occurs (about -60 millivolts) within the cells.

C.   <u>Action potentials</u>: long distance messages.

1)   Nervous information is encoded in <u>changes</u> in potential. Two examples of such changes are action potentials and postsyanptic potentials.

2)   <u>Action potential</u>: a wave of positive charge that travels, undiminished in strength, along the axon.

3)   If the potential inside the neuron cell body is made sufficiently less negative, the neuron may reach threshold, triggering an action potential.

4)   An action potential is usually initiated at the point where the axon leaves the cell body and travels along the axon to the synaptic terminal.

5)   As soon as the action potential passes any point on the axon, the negative resting potential is restored to that point.

6)   A resting neuron is like a loaded musket: charged and ready to fire if triggered.

    a.   The "charge" is the concentration gradient of sodium ions, higher outside the cell than inside.

    b.   The "trigger" is a set of sodium channel membrane proteins.

    c.   The sodium channels are closed in a resting neuron, but open suddenly to admit sodium when threshold is reached.

    d.   At threshold, sodium channels open and Na^+ ions flood into the cell, making its interior momentarily positive.

    e.   Then the sodium channels spontaneously close and a different set of K^+ channels open.

    f. K^+ now flow out of the cell through both types of K^+ channels, being driven out by both their diffusion gradient and by electrical repulsion from the Na^+ recently entered.

    g.   So many K^+ leave that the inside again becomes negative, re-establishing the resting potential.

7)   So, the action potential is brief: the neuron first becomes positive as Na^+ enter, and then negative as K^+ flow out.

8)   Action potentials are <u>all-or-none</u>: they do not vary in strength.

a. If the neuron does not reach threshold, there will be no action potential.

b. If threshold is reached, a full-sized action potential occurs and travels the entire length of the neuron.

9) <u>Role of the sodium-potassium pump</u>.

a. To re-establish the resting potential, the Na-K pump uses energy from ATP to pump Na^+ out and K^+ in, maintaining the concentration gradients of these ions across the cell membrane.

b. Thus, the "charge" is created by the pump using ATP energy.

10) <u>Conduction of the action potential</u>.

a. To be effective, the action potential must be transmitted along the axon to cells specialized to receive the message, like other neurons, muscles, or glands.

b. The action potential does not diminish in strength or die out along the way because it is renewed at various points along the axon:

i. The action potential begins when threshold is reached, sodium channels open and Na^+ enter the cell, making it positive inside.

ii. Some of the positive charge spreads rapidly and passively along the inside of the axon, making the adjacent region less negative.

iii. When the adjacent region reaches threshold, its sodium channels open, allowing Na^+ to enter and creating an action potential there.

iv. The positive charge spreads still further, generating another action potential in the adjacent membrane.

v. This continues along the entire axon length.

vi. Meanwhile, the Na channels at the site of the original action potential close, and the resting potential is restored there.

11) <u>Saltatory</u> ("jumping") <u>conduction</u>: the transmission of an action potential along a myelinated axon.

a. Action potentials must travel rapidly.

b. The opening and closing of ion channels is slow, so axons must open and close as few as possible as action potentials travel.

c. In vertebrates, rapid conducting axons are wrapped with insulating layers of membrane called <u>myelin</u> (formed from specialized Schwann cells), interrupted at intervals with naked areas called <u>nodes of Ravier</u>.

d. Rapid conduction occurs because ion channels are concentrated only at the nodes and the action potential moves from node to node.

D. <u>Communication between neurons</u>: synapses and postsynaptic potentials.

1) Once the action potential has been conducted to the synaptic terminal of the neuron, it is transmitted to the next cell at specialized regions called <u>synapses</u>.

2) The signals transmitted at synapses are called <u>postsynaptic potentials</u>.

3) <u>The synapse</u>.

a. Electrical signals reach a synaptic terminal at a synapse region where two neurons are close together but do not touch one another.

b. A small gap (<u>synaptic cleft</u>) separates the synaptic terminal of the first (<u>presynaptic</u>) neuron from the dendrite of the second (<u>postsynaptic</u>) neuron.

c. When action potential reaches a synaptic terminal:

i. Inside of the synaptic terminal becomes positive, triggering it to release a chemical neurotransmitter into the synaptic cleft.

327

ii.   The neurotransmitters rapidly diffuse across the gap and bind to receptor proteins in the dendrite membrane of the postsynaptic cell.

4)   <u>Postsynaptic potentials</u>.
   a.   Receptors in postsynaptic membranes have two roles:
   - i.   Each type of receptor binds to a specific neurotransmitter.
   - ii.   The bound receptor causes specific types of ion channels in the postsynaptic neuron membrane to open, allowing ions to flow, along their concentration gradients, across the postsynaptic neuron membrane.

   b.   The ion flow in the postsynaptic neuron causes a <u>postsynaptic potential</u> in the dendrites or cell body.
   c.   Synaptic potentials can be either:
   - i.   <u>Excitatory</u> (EPSP), caused by an excitatory synapse, making the neuron less negative inside and more likely to fire an action potential.
   - ii.   <u>Inhibitory</u> (IPSP), caused by an inhibitory synapse, making the neuron more negative inside and less likely to fire an action potential.

   d.   Postsynaptic potentials don't travel far, but do go far enough to reach the cell body where they determine if an action potential will occur.

5)   <u>Integration of synaptic potentials</u>.
   a.   The dendrite and cell body of a single neuron can receive EPSPs and IPSPs from synaptic terminals of thousands of presynaptic neurons.
   b.   All the postsynaptic potentials are then added up or <u>integrated</u> in the postsynaptic neuron cell body.
   c.   Only if the total raises the electrical potential inside the neuron above threshold level will an action potential occur.

6)   <u>Fate of neurotransmitters</u>.
   a.   All act briefly on the postsynaptic cell.
   b.   Some are destroyed by enzymes in the synaptic cleft.
   c.   Some are removed from the synaptic cleft by active transport and sent back into the presynaptic neuron.
   d.   Some diffuse away into the extracellular fluid.

3.   Building and operating a nervous system.

A.   <u>Information processing</u> in the nervous system. Four operations are necessary:
   1)   <u>Signal the intensity of a stimulus</u> (how loud, how hot, etc.) in 2 ways:
   - a.   The <u>rate</u> of action potentials in a single neuron. The more intense the stimulus, the faster the neuron fires.
   - b.   The <u>number</u> of similar neurons firing at the same time. The more intense the stimulus, the greater the number of neurons firing.

   2)   <u>Determine the type of stimulus</u> (light, touch, sound, etc.): the nervous system monitors <u>which</u> neurons are firing (e.g., optic nerve stimulation indicates light, olfactory nerve stimulation indicates smell, etc.).
   3)   <u>Integrate</u> (in the brain) <u>information from many sources</u>.
   - a.   Information is integrated by <u>convergence</u> (many neurons funnel their signals to fewer neurons).

      b.   The brain then sums the postsynaptic potentials and produces the appropriate response.

    4)   <u>Initiate and direct the response</u> through the output of the integrating cells.

      a.   The action directed by the brain may involve many body parts.

      b.   This requires <u>divergence</u> (flow of electrical signals from relatively fewer brain cells to many different neurons controlling various body parts).

B.   <u>Neural networks</u>.

    1)   Most behaviors are controlled by nervous-muscular pathways made of 4 elements:

      a.   <u>Sensory neurons</u>: respond to stimulus.

      b.   <u>Association neurons</u>: "decide" what to do.

      c.   <u>Motor neurons</u>: receive instructions from association neurons and activate muscles.

      d.   <u>Effectors</u> (muscles or glands): perform the behavior.

    2)   The <u>reflex</u> (simplest type of behavior).

      a.   Involuntary movement not involving the brain.

      b.   Examples: the knee jerk or the withdrawal (from heat) reflex.

    3)   <u>Interconnected nervous pathways</u>: have all 4 elements and control complex behaviors.

C.   <u>Nervous system design</u>. Two types exist:

    1)   <u>Diffuse nervous systems</u> (in Cnidarians).

      a.   No centralized head or brain (radial symmetry and slow moving life style).

      b.   Composed of a network of neurons (<u>nerve net</u>) connecting the tissues of the animal.

      c.   Occasional <u>ganglia</u> (clusters of neurons) occur, but no real brain.

    2)   <u>Centralized nervous systems</u> (in bilaterally symmetrical, mobile animals).

      a.   Sense organs concentrated in the head region with the ganglia often organized into a brain to integrate information and initiate appropriate responses.

      b.   Seen in all animals from molluscs to vertebrates.

4.   The Human nervous system.

A.   <u>Parts</u>.

    1)   <u>Central nervous system</u> (CNS) consisting of:

      a.   <u>Brain</u>, protected by a bony skull.

      b.   <u>Spinal cord</u>, protected by vertebrae.

    2)   <u>Peripheral nervous system</u>: nerves leading to all body parts from the brain and spinal cord.

B.   <u>Peripheral nervous system</u>.

    1)   <u>Divisions</u>:

      a.   <u>Somatic nervous system</u>:

         i.   Composed of nerves having bundles of axons of sensory and motor neurons.

        ii.   These conduct messages between the environment and the CNS.

      iii. Neurons synapse on skeletal muscles and control voluntary movements.

      iv. Cell bodies are located in the gray matter of the spinal cord.

    b. <u>Autonomic nervous system</u>:

      i. Regulates various internal organs (heart, intestines, kidneys, etc.).

      ii. Is regulated by the medulla and hypothalamus of the brain.

      iii. Motor neurons control involuntary responses.

  2) <u>Autonomic nervous system</u> has two parts, both innervating the same organs but producing opposite actions:

    a. <u>Sympathetic nervous system</u>.

      i. Acts on internal organs to prepare the body for stressful or highly energetic activity ("flight or fight" situations).

      ii. Examples: it curtails activity of the digestive system, speeds up heart rate, opens the eye pupils to let in more light, and expands air passages in the lungs for deeper breathing.

      iii. Axons are found in nerves that originate from the middle and lower portions of the spinal cord.

      iv. Synapses occur in larger ganglia near the spinal cord.

    b. <u>Parasympathetic nervous system</u>.

      i. Governs maintenance activities associated with "rest and rumination".

      ii. Examples: it activates the digestive tract, slows down heart rate, and increases urine production.

      iii. Axons are found in nerves originating from the brain and from the base of the spinal cord.

      iv. Synapses occur in smaller ganglia located near each target organ.

 C. <u>Central nervous system</u> (CNS): between 10-100 billion association neurons in the brain and spinal cord.

  1) <u>General facts</u>.

    a. CNS receives and processes sensory information, generates thoughts, and directs responses.

    b. CNS consists primarily of between 10 and 100 billion association neurons.

    c. CNS is protected in three ways:

      i. Bony armor: skull protects the brain, and vertebral column surrounds the spinal cord.

      ii. Beneath the bones is a triple layer of connective tissue called <u>meninges</u>.

      iii. Between the layers of the meninges is a clear lymphlike fluid cushion, the <u>cerebrospinal fluid</u>, produced in the <u>ventricle</u> spaces of the brain.

  2) <u>Spinal cord</u>.

    a. The neural cable (as thick as the little finger) running from brain to hips and protected by the vertebral column.

    b. <u>Contents</u>:

      i. Between vertebrae, <u>dorsal root</u> and <u>ventral root</u> nerves arise from dorsal and ventral portions of the spinal cord, respectively, and merge to form the spinal nerves.

    ii.   Neuron cell bodies ("gray matter") in the center of the cord, surrounded by

    iii.   Bundles of axons ("white matter") with white insulating myelin sheaths.

    c.   Events in a single spinal reflex:

        i.   A nerve signal from the skin travels to the cell body of a skin sensory cell found just outside the spinal cord (in a dorsal root ganglion).

        ii.   Both association and motor neuron cell bodies are found in the gray matter:

            a)   Axons of the white matter communicate with the brain.

            b)   Association neurons for the pain reflex synapse with motor neurons and also have axons extending to the brain.

        iii.   An association neuron for the pain reflex sends signals to the brain through its axons while signalling the motor neuron to withdraw from the pain.

        iv.   The brain then sends impulses down axons of the white matter to cells in the gray matter which can modify the spinal reflexes if desired.

    d.   All the neurons and interconnections needed to walk and run are found within the spinal cord. The brain's role is to integrate and guide the spinal activity. This increases speed and coordination of the activity.

    e.   Motor neurons of the spinal cord control the muscles involved in conscious, voluntary activities (e.g., eating, writing, playing tennis, etc.).

3)   The brain: all vertebrates have the same general structure.

    a.   The brain begins as a simple tube in the embryo, then develops into 3 parts:

        i.   Hindbrain: controls autonomic behavior such as breathing and heart rate in primitive vertebrates.

        ii.   Midbrain: controls vision in primitive vertebrates.

        iii.   Forebrain: controls smell in primitive vertebrates.

    b.   The human hindbrain has three parts:

        i.   Medulla: like an extension of the spinal cord, controlling several autonomic functions (breathing, heart rate, blood pressure, swallowing).

        ii.   Pons: above the medulla; influences transitions between sleep and wakefulness, and the rate and patterns of breathing.

        iii.   Cerebellum: coordinates movements of the body and body position (best developed in animals requiring fine coordination like birds in flight).

    c.   The human midbrain is reduced in size but contains the reticular activating formation center that:

        i.   Receives input from every sense organ and body part and many areas of the brain.

        ii.   "Decides" which stimuli require attention, suppressing unimportant stimuli (e.g., a sleeping mother hears her baby whimper but ignores traffic loud noise outside).

    d.   The human forebrain has 3 functional parts:

    i.   <u>Thalamus</u>: a relay center that channels sensory information to other forebrain parts.

    ii.   <u>Limbic system</u> (instinctive/emotional center) with several regions, including:

        a)   <u>Hypothalamus</u>: activates the autonomic nervous system and controls unconscious body functions like body temperature, pituitary hormone secretion, salt and water balance, and menstrual cycle, and emotions like fear.

        b)   <u>Amygdala</u>: area where feelings such as rage, anger, and sexual arousal originate.

        c)   <u>Hippocampus</u>: involved in long-term memory and learning.

    iii.   <u>Cerebrum</u> (information processing/control center): largest part of the brain, containing half of the brain's nerve cells.

        a)   Is split into 2 halves, the <u>cerebral hemispheres</u>, that communicate via a large band of axons called the <u>corpus callosum</u>.

        b)   The cerebral hemispheres have 50-100 billion neurons packed into the thin convoluted <u>cortex</u> surface layer that acts as the brain's most sophisticated information processing center.

        c)   The functions of the cerebral hemispheres are localized in discrete regions of the cortex. Brain damage to one area causes specific deficits such as loss of speech, reading, or partial paralysis.

5.   **Neurotransmitters and neuromodulators.**

    A.   Over 50 have been identified.

    B.   <u>Acetylcholine</u>: a neurotransmitter.
        1)   Found in many areas of the brain.
        2)   The only transmitter at the synapses between motor neurons and skeletal muscles; always excitatory.
        3)   Effects of drugs:
            a.   Curare blocks acetylcholine receptors on postsynaptic membranes, preventing muscle contraction and causing paralysis.
            b.   Many insecticides poison insects by inhibiting an enzyme that breaks down acetylcholine, causing muscles to contract uncontrollably, producing seizures and death.
        4)   In human CNS, degeneration of some groups of acetylcholine-producing neurons is found in patients with Alzheimer's disease.

    C.   <u>Dopamine</u>: a neurotransmitter.
        1)   Found in the brain, with inhibitory effects.
        2)   Degeneration of dopamine-producing neurons causes Parkinson's disease.
        3)   The drug L-dopa can relieve symptoms of Parkinson's disease since it is used by unharmed neurons to make dopamine.
        4)   Schizophrenia may be due to overabundance of receptors to dopamine.

D.   Serotonin: a neurotransmitter.
1)   Acts in the brain and spinal cord.
2)   Inhibits pain sensory neurons in the spinal cord.
3)   Affects sleep and mood adversely.
4)   Too little may cause depression.
5)   LSD may block serotonin receptors.

E.   Noradrenalin (norepinephrine): similar to adrenalin.
1)   Is released by neurons of the parasympathetic nervous system into many organs.
2)   Effects may be either inhibitory or excitatory, but generally prepare the body to respond to stressful situations.
3)   Cocaine and amphetamines act by stimulating release of noradrenalin and dopamine.

F.   Neuromodulators: peptides made and released by neurons.
1)   Modify the properties of synapses, making them more or less effective.
2)   Can cause long-term changes in excitability of neurons.
3)   Function over a longer time period than neurotransmitters, and may influence many neurons at once.
4)   Examples are the opioids such as endorphin and enkephalin.
a.   Have pain-relieving effects.
b.   Bind to specific receptors on CNS neurons.
c.   Some act with serotonin in spinal cord to block pain perception, especially in times of extreme stress (battlefield, sports field).
d.   Released during strenuous exercise; may account for "runner's high."
e.   Participate in maintenance of body temperature and blood pressure.

6.   Brain and mind.

A.   It is hard to equate brain cells with human thoughts, insights, emotions, and memories ("brain vs. mind"). How does the brain create the mind?

B.   Left brain - right brain: The 2 cerebral hemispheres, although similar in structure, control quite different functions.
1)   The left hemisphere controls speech, writing, language comprehension, mathematical ability, and logical problem solving.
2)   The right hemisphere controls musical and artistic skills, and abilities to recognize faces and express emotions.
3)   Axons from each optic nerve cross over so that the left eye's images travel to the right hemisphere and vice versa.
4)   The right hemisphere has some latent language capabilities, especially in females.

C.   Learning and Memory: cellular mechanisms not well understood.
1)   The time course of learning. Two phases occur in learning:
a.   Short-term memory (e.g., looking up a phone number and remembering it only long enough to dial).
i.   Is electrical in nature, activating a particular neural pathway in the

brain.

    ii.   When the circuit becomes inactive (by distraction or concussion), the memory is lost.

  b.   <u>Long-term memory</u>.

    i.   Is <u>structural</u>, probably involving formation of new permanent synaptic connections between specific neurons or strengthening existing but weak synaptic connections.

    ii.   These synapses last indefinitely unless brain structures are destroyed.

  c.   How short-term memory is converted to long-term memory remains a mystery.

2)   <u>Learning, memory, and retrieval</u>. Each is mediated by a separate area of the brain.

  a.   The <u>hippocampus</u> (part of the limbic system) is involved in learning. Hippocampal damage results in an inability to learn or remember new things.

  b.   <u>Temporal lobes</u> of the cerebral hemispheres are involved with retrieval (recall) of long-term memories.

  c.   The site of storage of complex long-term memories is unclear. Perhaps a memory is stored in numerous distinct places simultaneously.

****

REVIEW QUESTIONS:

1.-7. **MATCHING TEST:**   structural regions of neurons.

    Choices: A.   Cell bodies
    B.   Synaptic terminals
    C.   Dendrites
    D.   Axons

1.   Carry action potentials to output terminals.
2.   The cell's integration center.
3.   Receive information from the environment.
4.   Sites where signals are transmitted to other cells.
5.   Convert environmental information into electrical signals.
6.   Bundled together into "nerves".
7.   Initiate action potentials.

8.-10. **MATCHING TEST:** nerve cell function.

    Choices: A.   Threshold
    B.   Action potential
    C.   Resting potential

8.   Always negative (-40 to -90 millivolts) within a nerve cell.
9.   A sudden positive charge within a nerve cell.

10. Occurs when a nerve cell becomes sufficiently less negative inside.

----

11. Describe how a nerve cell achieves a negative resting potential involving the movement of potassium ions (K^+).
12. In what way are excitatory and inhibitory potentials different?
13. Describe how an action potential develops.
14. Describe the conduction of action potentials, including the concept of "saltatory conduction".
15. Describe the steps involved in the transfer of impulses between neurons at a synapse.
16. Name two ways that the intensity of a stimulus is signalled to the brain.
17. Name the four elements of a typical nervous-muscular pathway.
18. Describe the difference between a diffuse and a centralized nervous system.
19. Name the two parts of the central nervous system.
20. Describe the difference between the somatic and autonomic portions of the peripheral nervous system.

21.-27.   **MATCHING TEST**: autonomic nervous system.

    Choices: A.   Sympathetic nervous system
                B.   Parasympathetic nervous system
                C.   Both of these
                D.   Neither of these

21. Prepares the body for "flight or fight" situations.
22. Conducts messages between the environment and the central nervous system.
23. Regulates various internal organs and is regulated by the hypothalamus.
24. Speeds up heart rate.
25. Associated with "rest and rumination" activities.
26. Opens the eye pupils.
27. Increases urine production.

----

28. Compare the contents of the white matter and the gray matter of the spinal cord.
29. Describe the events in a single spinal reflex.
30. In primitive vertebrates, list the functions of the hindbrain, midbrain, and forebrain.

31.-42.   **MATCHING TEST**: the human brain.

    Choices: A.   Cerebellum
                B.   Cerebrum
                C.   Limbic system
                D.   Medulla
                E.   Reticular activating formation center
                F.   Thalamus

31. Midbrain.
32. Like an extension of the spinal cord.
33. Channels sensory information to other forebrain parts.

34. Hindbrain.
35. Controls several autonomic functions.
36. Hypothalamus.
37. Forebrain.
38. Receives input from all sense organs and "decides" which require attention.
39. Controls learning, emotions, and the autonomic nervous system.
40. Largest part of the brain.
41. Coordinates body movements and body positions.
42. Controls speech, reading, math ability, and musical skills.

----

43. Which of the neurotransmitters is thought to be most closely related in some way to each of the following? Alzheimer's disease, Parkinson's disease, schizophrenia, LSD, cocaine.
44. List the major functions controlled by the left and right cerebral hemispheres.
45. Define and compare short-term and long-term memory.

****

ANSWERS TO REVIEW QUESTIONS:

1.-7.	[see 1.B.1)-4)]				20.	see 4.B.1)a.-b.		
	1.	D	5.	C				
	2.	A	6.	D	21.-27.	[see 4.B.2)a.-b.]		
	3.	C	7.	A		21.	A	25. B
	4.	B				22.	D	26. A
						23.	C	27. B
8.-10.	[see 2.A.1)-2)]					24.	A	
	8.	C	10.	A				
	9.	B			28.	see 4.C.2)b.i.-iii.		

11. see 2.B.1)-3)

12. see 2.D.4)c.

13. see 2.C.6)a.-g.

14. see 2.C.10)-11)

15. see 2.D.

16. see 3.A.1)a.-b..

17. see 3.B.1)a.-d.

18. see 3.C.1)-2)

19. see 4.A.1)a.-b.

29. see 4.C.2)c.i.-iv.

30. see 4.C.3)a.i.-iii.

31.-42. [see 4.C.3)b.-d.]

31.	E	37.	B,C,F
32.	D	38.	E
33.	F	39.	C
34.	A,D	40.	B
35.	D	41.	A
36.	C	42.	B

43. see 5.B.-E.

44. see 6.B.1)-4)

45. see 6.C.1)a.-b.

****

CHAPTER 37:   PERCEPTION: THE SENSES

CHAPTER OUTLINE.

In this chapter, the ways animals perceive and respond to nervous stimulation are discussed, and the structures and functions of the major sense organs are covered.

1.   **Receptor mechanisms.**

   A.   <u>Receptors</u>: structures that change when acted upon by stimuli from the surroundings, causing signals to be produced.
      1)   All receptors are <u>transducers</u>: structures that convert signals from one form to another.
      2)   Each type of receptor produces electrical signals, but is specialized to produce its signal only in response to a particular type of stimulus.

   B.   <u>Receptor potentials</u>:
      1)   Produced by stimulation of a sensory receptor in a size proportional to the strength of the stimulus.
      2)   Are translated into action potentials by one of two mechanisms:
         a.   In neurons with axons that carry the signal for relatively long distances:
            i.   If the stimulus is sufficiently strong, the receptor potential will exceed threshold, initiating action potentials.
            ii.   The frequency of the action potentials is related to the size of the receptor potential.
         b.   In receptor cells that do not have axons or produce action potentials (e.g., those that detect taste, sounds, and light):
            i.   Receptor potentials cause release of transmitter molecules onto a postsynaptic neuron, producing a postsynaptic potential.
            ii.   The postsynaptic neuron signals the brain using action potentials.

337

C.   Ways stimuli produce receptor potentials.
1)   <u>Mechanical deformation</u> of the cell membrane of <u>mechanoreceptor cells</u>. Mechanisms of initiating receptor potentials:
      a.   Stretching the membrane of the receptor cell (e.g., touch receptors in the skin).
      b.   Bending "hairs" that project from the receptor membrane (e.g., reception of sound, motion, and gravity).
2)   <u>Interaction with membrane receptor molecules</u> in receptors for light, chemicals, and pain. When a stimulus hits a receptor molecule, it changes shape and this alters the permeability of the cell membrane, generating a receptor potential.

D.   Other unknown receptor mechanisms exist for perceiving temperature, electrical fields, and magnetic fields.

2.   The major senses.

A.   <u>Thermoreception</u>.
1)   Thermoreceptors sense temperature by responding to infrared radiation.
2)   Changes in the metabolic rate of the receptor, which is directly affected by temperature, cause the receptor potential.
3)   The most sensitive thermoreceptors are in the pit organs of viper snakes; these are so sensitive that they respond to the body heat of a mouse 16 inches away.

B.   <u>Mechanoreception</u> in the skin: sensory endings of mechanoreceptors produce a receptor potential when their membranes are stretched.
1)   <u>Types</u>:
      a.   Free nerve endings (respond to touch and pressure).
      b.   Endings wound around hairs (detect hair movement).
      c.   Endings enclosed in capsules (e.g., pressure sensitive Pacinian corpuscles).
2)   Receptor density varies in different skin areas (more in the fingertips than on the back).
3)   Mechanoreceptor endings of <u>proprioreceptor cells</u> in the joints and muscles sense orientation and direction of movement of body parts.
4)   Mechanoreceptors in the walls of the stomach, rectum, and bladder signal fullness when stretched.

3.   The auditory system: a specialized sensitivity to vibration.

A.   All animals have receptors sensitive to vibration, and sound is merely high-frequency vibration of air or water.

B.   All sound receptors have a flexible membrane with receptor cells connected to the membrane. Sound waves cause the membrane to vibrate, stimulating the receptor cells.

C.   <u>Lateral line organs</u>.
1)   The vertebrate organs that detect sound, gravity, and movement evolved from the <u>lateral line organ</u> found in all fish and aquatic amphibians.
2)   It consists of a series of clusters of hair cells located in pits or tubes that form a

strip beginning in the head and extending along either side of the body.

    a.   The "hairs" are embedded in a gelatinous cap (<u>cupula</u>) which is deflected by water currents, causing the hairs to bend.

    b.   The receptor potentials caused by the bending produce action potentials that travel to the brain.

  3)   Lateral line organs detect

    a.   Water currents.

    b.   Water movements caused by prey or predators.

    c.   Possibly low-frequency sounds.

D.   <u>Human ear structure</u>: the human ear has 3 parts:

  1)   <u>Outer ear</u>: all structures outside the eardrum (<u>tympanic</u> <u>membrane</u>). The <u>external</u> <u>ear</u> funnels the sound into the <u>auditory</u> <u>canal</u>.

  2)   <u>Middle ear</u> (air-filled) transmits vibrations through membranes and bones.

    a.   First, sound vibrates the tympanic membrane which vibrates 3 bones (the <u>hammer</u>, <u>anvil</u>, and <u>stirrup</u>).

    b.   These bones vibrate the smaller <u>oval</u> <u>window</u> membrane that covers the opening to the inner ear.

    c.   The <u>Eustachian</u> <u>tube</u> connects middle ear to the pharynx, allowing air pressure in the ear and the atmosphere to equalize.

  3)   <u>Inner ear</u> (fluid-filled) contains the <u>cochlea</u> and the <u>semicircular</u> <u>canals</u>.

E.   <u>Cochlea</u> of the inner ear: spiral shaped chamber containing receptor cells for hearing. Parts:

  1)   An outer U-shaped canal.

  2)   An inner straight <u>central</u> <u>canal</u> containing two parts:

    a.   <u>Basilar</u> <u>membrane</u> with sound receptors called <u>hair</u> <u>cells</u>.

    b.   Gelatinous <u>tectorial</u> <u>membrane</u> in which hairs of the hair cells are embedded.

F.   <u>How sound is perceived</u>.

  1)   Sound waves enter the ear and vibrate → tympanic membrane → hammer, anvil, and stirrup bones → oval window membrane → fluid of the cochlea.

  2)   Then the waves vibrate the basilar membrane but not the tectorial membrane, bending the hairs which cause receptor potentials in the hair cells.

  3)   The hair cells release transmitter molecules into neurons of the <u>auditory</u> <u>nerve</u>, triggering action potentials in the auditory nerve axons that travel to the brain.

  4)   <u>Degree of loudness</u> is perceived by the amount of bending the hairs undergo.

  5)   <u>Pitch</u> is perceived by different portions of the basilar membrane resonating to different pitches due to a gradual widening of the membrane as it increases in distance from the oval window.

    a.   High notes cause greater vibration near the oval window end.

    b.   Low notes cause greater vibration away from the oval window end.

    c.   The brain interprets signals from the receptors near the oval window as high-pitched sounds, and signals from receptors away from the oval window as lower in pitch.

G.   <u>Vestibular apparatus</u> of the inner ear: detecting motion and gravity.

  1)   The vestibular apparatus is housed in a set of bony canals adjacent to the

cochlea.

2) Three curved, fluid-filled <u>semicircular canals</u> detect motion by the bending of hairs on cells in the canals.

3) Two chambers (the <u>utricle</u> and the <u>saccule</u>) at the base of the semicircular canals detect gravity.

    a. Each chamber contains hair cells atop whose hairs rests a gelatinous mass (the <u>otolith membrane</u>) containing tiny grains of limestone.

    b. The grains are pulled downward by gravity, bending the hairs according to how the head is bent.

4. Vision.

A. All forms of vision utilize <u>photoreceptor</u> <u>cells</u> containing <u>photopigments</u> that change chemically as they absorb light, causing receptor potentials to occur.

B. <u>Types of eyes</u>.

1) <u>Eyespots</u> in flatworms.

    a. Lack lenses.

    b. Can distinguish light from dark.

    c. Can perceive direction and intensity of light.

2) <u>Compound eyes</u> in insects.

    a. Contain many <u>ommatidia</u> (light-sensitive subunits) each of which acts as an on/off, bright/dim detector.

    b. The image is formed as a mosaic of signals from all the ommatidia.

3) <u>Camera eye</u> of molluscs and, independently evolved, the vertebrates. Parts:

    a. A light-sensitive layer (<u>retina</u>).

    b. A <u>lens</u> for focusing light.

    c. Muscles for adjusting focus by changing the position or shape of the lens.

C. <u>Human eye structure</u>.

1) Light first hits the transparent <u>cornea</u> covering the front of the eyeball.

2) Next, a chamber contains the watery <u>aqueous humor</u> which nourishes the lens.

3) Next, the muscular <u>iris</u> adjusts how much light enters by expanding or constricting the iris opening, the <u>pupil</u>.

4) Light then strikes the proteinaceous <u>lens</u>, attached by muscles and ligaments that regulate its shape.

5) Behind the lens is a large chamber filled with clear gelatinous <u>vitreous humor</u> which maintains the shape of the eye.

6) Light finally hits the <u>retina</u>, a multilayered nervous tissue where light is converted to electrical nerve impulses sent to the brain.

7) The <u>choroid</u> is behind the retina which it nourishes with its rich blood supply.

8) Surrounding the entire eyeball is the <u>sclera</u>, a tough connective tissue layer (the "white" of the eye).

D. <u>Image focusing</u> in the human eye.

1) The visual image is focused most sharply on a small area of the retina called the <u>fovea</u>.

2) The <u>cornea</u> does most of the bending of incoming light but the cornea's shape

cannot be adjusted.

3) The <u>lens</u> is responsible for final, sharp focusing of the image onto the retina. A rounded lens allows the eye to focus on nearby objects.

4) Nearsighted people can't focus on distant objects due to abnormally long eyeballs.

5) Farsighted people can't focus on nearby objects due to abnormally short eyeballs.

E. <u>Neural processing in the retina</u>.

1) In vertebrates, the retina is built "backwards," with its photoreceptor cells facing away from the light.

2) The outermost retinal neurons are <u>ganglion cells</u> whose axons are part of the <u>optic nerve</u> connected to the brain. The eye has a blind spot called the <u>optic disc</u> where these axons meet and pass through the retina.

3) Photoreceptors, called <u>rods</u> and <u>cones</u>, lie behind the ganglion cells.

    a. Photoreception in both types of cells begins with absorption of light by a photopigment molecule.

    b. Light changes the shape of the photopigment molecule and initiates a series of biochemical reactions inside the photoreceptor.

    c. These result in a change in the permeability of the receptor membrane to ions, producing a receptor potential in the photoreceptor cell.

4) The pigmented regions of rods and cones are shaped like their cellular names.

    a. 125 million <u>rods</u> are scattered uniformly in the retina, allowing for peripheral vision.

        i. These are very sensitive to light, even dim light.

        ii. They do not distinguish color, so in the moonlight, the world looks black and white.

    b. Many of the 5 million <u>cones</u> are aggregated in one small retinal area, the <u>fovea</u> area of sharpest focus.

        i. They are less sensitive to light than rods and work only in the daytime.

        ii. Cones distinguish color because they exist in 3 forms (each with a different type of photopigment: red, green, and blue sensitive).

        iii. Different colors are sensed by the relative stimulation of different cone types (i.e., yellow = equal stimulation of both red and green cones).

        iv. "Colorblind" males make abnormal red photopigment and have difficulty distinguishing red from green.

    c. Day-active/night-inactive animals have all-cone retinas, while night-active animals or those in dimly lit habitats have mostly rods.

F. <u>Binocular vision</u>.

1) Having two eyes is useful in several ways:

    a. They allow flatworms to determine the direction of a light source.

    b. <u>Binocular vision</u>.

        i. The forward-facing eyes of predators and omnivores (like humans) each have slightly different but extensively overlapping visual fields.

        ii. This allows depth perception and more accurate judgement of the size and distance of an object from the eyes.

    c. The widely-spaced eyes of herbivores have little overlap in their visual fields.
      i. This allows a nearly 360° field of view for better detection of approaching predators.
      ii. Accurate depth perception is sacrificed.

5. **Chemical senses: olfaction and taste.**

  A. Chemical senses allow animals to:
    1) Find food.
    2) Avoid poisonous materials.
    3) Locate homes.
    4) Find mates.

  B. Terrestrial vertebrates have <u>two types of chemoreceptors</u>:
    1) For "smelling" airborne molecules (<u>olfaction</u>).
    2) For "tasting" chemicals dissolved in aqueous medium (<u>taste</u>).

  C. <u>Olfactory receptors</u> are nerve cells in tissues lining the back of the nasal cavity.
    1) Hair-like dendrites protrude which sample incoming air.
    2) Individual odors are sensed due to stimulation of particular receptor molecules in these hairs.

  D. <u>Taste buds</u> on the tongue are clusters of taste receptors and supporting cells.
    1) Microvilli protrude through small pores and dissolved chemicals bind to receptors on the microvilli.
    2) Four types of taste receptors exist: sweet, sour, salty, and bitter. The great variety of tastes is due to:
      a. Stimulation of several types of receptors at once.
      b. Interaction with the sense of smell.

6. **Pain**: a somewhat non-specific sense, caused by a number of different stimuli.

  A. Most pain is caused by tissue damage which releases an enzyme that converts blood components into <u>bradykinin</u>.

  B. Bradykinin receptors are present on the dendrites of specific sensory neurons that respond with action potentials interpreted by the brain as pain.

  C. Drugs that relieve pain block synapses in the pain pathways of the brain or spinal cord. The brain can modulate its perception of pain by producing <u>endorphins</u>.

****

REVIEW QUESTIONS:

1. Name and describe two ways that stimuli can produce receptor potentials in sensory neurons.

BRIEFLY DESCRIBE the function of the following structures related to hearing:

2. External ear.
3. Hammer, anvil, and stirrup bones.
4. Oval window.
5. Eustachian tube.
6. Cochlea.

----

7. Name three types of mechanoreceptors.
8. How are different pitches of sound determined in the ear?
9. Describe the functions of the semicircular canals in the ear.
10. Name, in order, the eye structures passed by a photon of light as it travels to the retina.
11. Describe the roles of the cornea and the lens in image focusing.
12. Contrast the functions of rods and cones in the retina.
13. What are the pros and cons of binocular vision vs. the widely spaced eyes of herbivores?
14. Describe the two chemical senses.
15. What do bradykinins and endorphins have to do with pain?

****

ANSWERS TO REVIEW QUESTIONS:

1. see 1.B.2)

2. see 3.D.1)

3. see 3.D.2)a.

4. see 3.D.2)b.

5. see 3.D.2)c.

6. see 3.E.

7. see 2.B.1)a.-c.

8. see 3 F.5)a.-c.

9. see 3.G.1)-3)

10. Cornea → aqueous humor →
iris (pupil) → lens →
vitreous humor → retina;
[see 4.C.1)-8)]

11. see 4.D.1)-5)

12. see 4.E.3)-4)

13. see 4.F.1)b.-c.

14. see 5.A.-D.

15. see 6.A.-C.

****

CHAPTER 38:   ACTION AND SUPPORT: THE MUSCLES AND SKELETON

CHAPTER OUTLINE.

This chapter begins by describing muscle tissue and the complex mechanisms underlying muscle contraction. It then describes the tissues and mechanics of the skeleton.

1.   **Muscle**.

   A.   Many animal cells are capable of movement.
      1)   At the cellular level, motion occurs due to the relative movements of two types of protein strands, <u>actin</u> and <u>myosin</u>.
      2)   Muscles have evolved from this pre-existing system.

   B.   <u>Types of muscles</u>:
      1)   <u>Skeletal</u> (<u>striated</u>) <u>muscle</u>.
      2)   <u>Cardiac muscle</u>.
      3)   <u>Smooth muscle</u>.

   C.   <u>Anatomy of skeletal (striated) muscle</u>.
      1)   Individual cells (<u>muscle fibers</u>) run the entire length of a muscle (up to 35 cm long) and are transversely striped in appearance.
      2)   Each muscle fiber contains many individual contractile subunits (<u>myofibrils</u>) extending the length of the fiber.
         a.   Each myofibril is surrounded by endoplasmic reticulum-derived membranes called <u>sarcoplasmic reticulum</u>.
         b.   Deep indentations of the cell membrane (called <u>transverse</u> or <u>T tubules</u>) extend into the muscle fibers, passing close to portions of the

344

sarcoplasmic reticulum containing large concentrations of calcium ions (Ca^+).

3) Within each myofibril is a precise arrangement of <u>actin</u> and <u>myosin</u> protein filaments, forming units called <u>sarcomeres</u>.

    a. Sarcomeres are attached end to end with junction points called <u>Z-lines</u>.

    b. Actin plus two types of accessory proteins are attached to the Z-lines and form <u>thin filaments</u>.

    c. Suspended between the thin filaments are <u>thick filaments</u> made of myosin protein.

    d. Small arms (called <u>cross-bridges</u>) extend from the myosin strands and touch the actin filaments.

    e. Each actin subunit has a binding site for a myosin cross-bridge, but in relaxed muscles the binding sites in actin are covered by the accessory proteins.

D. <u>Sliding filament model</u> of muscle contraction. As a muscle contracts:

1) The accessory proteins of the thin filaments are moved aside, exposing the actin's binding sites.

2) Myosin cross-bridges attach to the actin.

3) Using energy from ATP, the crossbridges repeatedly bend, release, and reattach further along (like climbing hand over hand).

4) The thin filaments are pulled past the thick filaments, shortening the sarcomere and contracting the muscle.

E. <u>Control of muscle contraction</u>.

1) Movement of accessory protein exposes or covers the binding sites in actin. The position of these proteins is controlled by the concentration of Ca^+ around the filaments and this is controlled by the nervous system.

2) Skeletal muscles are much like neurons since they:

    a. Have resting potentials.

    b. Can be stimulated by synaptic contact with neurons.

    c. Produce action potentials when stimulated above threshold by motor neurons from the spinal cord.

3) Motor neurons form synapses with muscle fibers at <u>neuromuscular junctions</u>.

    a. Action potentials from the neurons release a neurotransmitter (<u>acetylcholine</u>) across the synaptic gaps.

    b. Receptors in the muscle fiber membranes are excited to form a postsynaptic action potential.

    c. Every action potential in a motor neuron causes a large enough excitatory synaptic potential to evoke a muscle action potential.

    d. The muscle action potential passes down the T tubules deep into the cell.

    e. This potential causes the sarcoplasmic reticulum to release Ca^+ into the myofibrils.

    f. The Ca^+ binds to the accessory proteins on the thin filaments, changing their shapes and removing them from the binding sites of the actin. Myosin cross-bridges then attach and muscle contraction occurs.

    g. When the action potential fades, active transport proteins in the sarcoplasmic reticulum pump the Ca^+ back into the reticulum and the accessory proteins move back to cover the actin binding sites, ending the contraction.

4) The <u>degree</u> and <u>intensity</u> of muscle contraction depends on:

    a. The number of muscle fibers in a <u>motor unit</u> (the group of fibers in which a single neuron synapses).

    b. The frequency of action potentials in each fiber. The higher the frequency, the fuller will be the contraction due to <u>summation</u> (each action potential adds to the effect of the others).

    c. If rapid firing is prolonged, the muscle produces a sustained maximal contraction called <u>tetany</u>.

F. <u>Cardiac muscle</u>.

1) Found only in the heart.

2) Is striated due to the regular arrangement of the <u>sarcomeres</u>.

3) Action potentials spreading into a cell cause release of calcium from the sarcoplasmic reticulum, but calcium also enters from the extracellular fluid.

4) Cardiac muscle fibers can initiate their own contractions.

    a. This is especially well-developed in the cardiac muscle fibers of the <u>sinoatrial node</u> (the heart's pacemaker).

    b. Action potentials originating in the pacemaker spread rapidly throughout the heart due to <u>intercalated discs</u> (areas where membranes of adjacent muscle fibers are connected with numerous gap junctions).

    c. Intercalated discs allow electrical potentials to travel from cell to cell, synchronizing their contractions.

G. <u>Smooth muscle</u>.

1) Surrounds blood vessels and most hollow organs.

2) Lacks the regular arrangement of sarcomeres seen in skeletal and cardiac muscle.

3) Produces either sustained contractions (e.g., of arteries to elevate blood pressure when stressed) or slow, wavelike contractions (e.g., peristalsis of the digestive tract).

4) Often directly connected to each other for synchronized contraction.

5) Lacks sarcoplasmic reticulum: all calcium for contraction flows in from extracellular fluid during the action potential.

6) Activated by stretch, hormones, and/or nervous signals.

2. **Cartilage and bone**.

A. <u>Cartilage</u>.

1) Roles:

    a. Covers the ends of bones at joints.

    b. Supports the flexible portion of the nose and external ears.

    c. Connects the ribs to the sternum.

    d. Provides a framework for the larynx, trachea, and bronchi of the respiratory system.

    e. Found in knee joints and intervertebral discs between the vertebrae.

2) Living cells of cartilage are <u>chondrocytes</u>.

    a. These secrete flexible, elastic, non-living collagen that surrounds chondrocytes and forms the cartilage bulk.

        b.   No blood vessels penetrate into cartilage; cells survive via diffusion of nutrients and wastes, and have a very slow metabolic rate.

        c.   Damaged cartilage repairs itself slowly.

    3)   During embryonic development, the skeleton is first formed from cartilage and later replaced by bone.

B.   <u>Bone</u>.

    1)   Consists of an outer shell of <u>compact bone</u>, with <u>spongy bone</u> in the interior.

        a.   Compact bone is dense and strong and provides attachment sites for muscles.

        b.   Spongy bone is lightweight, rich in blood vessels, and highly porous.

        c.   Bone marrow, origin of blood cells, is a part of spongy bone.

    2)   The collagen fibers of bone are hardened by deposits of calcium phosphate.

    3)   Bone is well-supplied with blood capillaries.

    4)   The three types of bone cells:

        a.   <u>Osteoblasts</u>: bone-forming cells.

        b.   <u>Osteocytes</u>: mature bone cells.

        c.   <u>Osteoclasts</u>: bone-dissolving cells.

    5)   Early in development, when bone is replacing cartilage, osteoclasts dissolve the cartilage and osteoblasts replace it with bone.

        a.   Osteoblasts form a thin layer covering the outside of the bone.

        b.   Osteoblasts secrete a hardened matrix of bone and gradually become entrapped within it.

        c.   They then stop secreting matrix and become osteocytes.

    6)   Osteocytes connect to each other via processes extending through narrow channels in the bone, and are nourished by nearby capillaries.

    7)   Substances secreted from osteocytes control the continuous <u>remodeling of bone</u>.

        a.   Each year, 5-10% of your bones are dissolved away and replaced.

        b.   This allows your skeleton to alter its shape in response to physical demands.

           i.   The bones of an archer's right and left arms differ in thickness.

           ii.   An arm immobilized in a cast rapidly loses calcium.

           iii.   During aging, the remodeling process slows, and bones become more brittle.

        c.   Turnover of bone allows for maintenance of constant levels of calcium in the blood, a process is regulated by the hormones <u>calcitonin</u> and <u>parathormone</u>.

        d.   Bone remodeling results from the coordinated activity of:

           i.   Osteoclasts that dissolve bone by secreting acids and enzymes.

           ii.   Osteoblasts that rebuild bone.

           iii.   Hard bone is made up of <u>Haversian systems</u>, each consisting of osteocytes embedded in concentric layers of bone that surround a central canal containing a capillary.

           iv.   Osteoclasts and osteoblasts also play a crucial role in the repair of bone fractures.

3.   **The skeleton**: a supporting framework.

    A.   Among animals, <u>three types of skeletons</u> exist:

  1) Internal <u>endoskeletons</u>: chordates and echinoderms.

  2) <u>Hydrostatic skeletons</u>: fluid-filled sacs of worms, molluscs, and cnidarians rely on surrounding body wall muscles for shape.

  3) <u>Exoskeletons</u>: arthropods.

    a. Vary tremendously in thickness and rigidity.

    b. Thin and flexible at joints, allowing for complex movements.

B. <u>The vertebrate bony skeleton</u>.

  1) Functions:

    a. A rigid framework that supports the body and protects internal organs.

    b. Produces red and white blood cells and platelets in the red bone marrow found in porous bone areas of the sternum, ribs, upper arms and legs, and hips.

    c. Serves as a storage area for calcium and phosphorus.

    d. Participates in sound transduction (the tiny bones of the middle ear).

  2) The 206 bones of the human skeleton are of two types:

    a. <u>Axial skeleton</u>: bones of the head, vertebral column, and rib cage.

    b. <u>Appendicular skeleton</u>: bones of the extremities and their attachments to the axial skeleton (pectoral and pelvic girdles, arms, legs, hands, feet).

C. <u>Body movement: muscle-skeleton interactions</u>.

  1) The skeleton provides a framework that muscles can move.

  2) Antagonistic muscles alter the configuration of the skeleton.

    a. Muscles can only contract. To extend, the muscle must be pulled out again by the action of another muscle.

    b. <u>Antagonistic muscle pair</u>: one muscle contracts and causes the other to extend.

  3) <u>Movement around joints</u>.

    a. The vertebrate skeleton provides attachment points for skeletal muscles.

    b. Muscles move the skeleton around regions called <u>joints</u> (flexible attachment sites between adjacent bones).

    c. Skeletal muscles are attached to bones on either side of the joint by <u>tendons</u> (bands of tough fibrous connective tissue).

      i. The bone region that forms the joint is coated with a layer of cartilage.

      ii. Bones are joined to each other at joints by <u>ligaments</u> (bands of fibrous connective tissue).

    d. Skeletal muscles are arranged in <u>antagonistic pairs</u>.

      i. One is called a <u>flexor</u> and the other (on the opposite side of the joint) is an <u>extensor</u>.

      ii. When one contracts, it moves a bone and simultaneously stretches out the opposing muscle.

    e. In a joint, usually the contraction of antagonistic muscle pairs moves one bone back and forth while the other bone remains unmoved.

    f. <u>Hinge joints</u>:

      i. Are movable in only two directions.

      ii. Contain pairs of muscles in the same plane as the joint.

      iii. One end of each muscle, the <u>origin</u>, is fixed to a relatively immovable joint bone.

      iv. The other end of the muscle, the <u>insertion</u>, is attached to a movable joint bone.

v.   Contraction of the flexor muscle bends the joint, while contraction of the extensor muscle straightens the joint.

g.   <u>Ball and socket joints:</u>

i.   The rounded end of one bone fits into a depression in another bone, allowing movement in several planes.

ii.   Have at least two pairs of antagonistic muscles oriented perpendicular to each other.

****

REVIEW QUESTIONS:

1.-9.   **MATCHING TEST**: types of muscle cells.

Choices: A.   Cardiac muscle
B.   Smooth muscle
C.   Striated muscle

1.   Surrounds blood vessels and hollow organs.
2.   Skeletal muscle.
3.   Found only in the heart.
4.   Has a striped appearance under the microscope.
5.   Capable of synchronized contraction.
6.   Can initiate its own contractions.
7.   Peristalsis.
8.   Under the direct control of the nervous system.
9.   Activated by stretch and hormones.

10.-12. **BRIEFLY DESCRIBE** the following skeletal muscle structures:

10.   Sarcoplasmic reticulum.
11.   T-tubules.
12.   Sarcomeres.

----

13.   Briefly describe the sliding filament model of muscle contraction.
14.   Beginning with the neuromuscular junction, describe the events that occur leading to muscle contraction.
15.   What factors influence the degree and intensity of muscle contraction?
16.   What is meant by the "origin" and "insertion" of a muscle?
17.   Explain the difference between a "flexor" and an "extensor" muscle.

18.-25. **MATCHING TEST**: bone and cartilage cells.

Choices: A.   Chondrocytes   C.   Osteoblasts
B.   Osteocytes   D.   Osteoclasts

18. Bone-dissolving cells.
19. Mature bone cells.
20. Survive by diffusion of nutrients and wastes.
21. Bone-forming cells
22. Rebuild bone during bone remodeling.
23. Dissolve bone by secreting acids and enzymes.
24. Living cartilage cells.
25. Dissolve cartilage.

----

26. Briefly describe the three types of skeletons.
27. Define the axial and appendicular portions of the human endoskeleton.
28. What is the difference between a tendon and a ligament?

****

ANSWERS TO REVIEW QUESTIONS:

1.-9.	[see 1.C.,F.,G.]				15.	see 1.E.4)a.-b.		
	1.	B	6.	A				
	2.	C	7.	B	16.	see 3.C.3)f.iii.-iv.		
	3.	A	8.	C				
	4.	A,C	9.	B	17.	see 3.C.3)d.i.-ii.		
	5.	A,B						

		18.-25.	[see 2.A.-B.]		
10.	see 1.C.2)a.	18.	D	22.	C
		19.	B	23.	D
11.	see 1.C.2)b.	20.	A	24.	A
		21.	C	25.	D
12.	see 1.C.3)a.-e.				
		26.	see 3.A.1)-3)		
13.	see 1.D.1)-4)				
		27.	see 3.B.2)a.-b.		
14.	see 1.E.3)a.-g.				
		28.	see 3.C.3)c.		

****

CHAPTER 39: ANIMAL REPRODUCTION

CHAPTER OUTLINE.

This chapter considers sexual and asexual types of reproduction, and describes the anatomy and physiology of men and women as representative vertebrates. Hormonal influences are detailed as well as the menstrual cycle, ovulation with and without fertilization, and what happens during pregnancy.

1. **Introduction.**

   A. Reproduction evolved to pass one's genes to another generation, cheating death and achieving immortality through one's offspring.

   B. The 3 stages of an organism's life:
       1) Birth and growth to sexual maturity.
       2) Gathering the environmental resources needed to reproduce (food, strength, a physical territory).
       3) Reproduction.

2. **Reproductive strategies.**

   A. <u>Types of reproduction</u>:
       1) <u>Sexual</u>.
           a. Animals produce haploid gametes through <u>meiosis</u>.
           b. 2 gametes fuse (<u>fertilization</u>) to form a diploid offspring which is genetically different from either parent.
       2) <u>Asexual</u>.
           a. A single animal produces offspring through repeated mitosis of cells in some part of its body.
           b. The offspring are genetically identical to the parent.
           c. Is more efficient than sexual reproduction but does not generate new gene combinations necessary for evolution.

B. Asexual reproduction.
    1) Budding in sponges, Hydra and some anemones: producing a small new animal by direct growth from the body of an adult.
    2) Regeneration in sea stars occurs if a fragment contains part of the central disc.
    3) Fission in some corals, flatworms, and annelids: splitting to form 2 smaller individuals that then regenerate missing parts.
    4) Parthenogenesis in some female bees, insects, fish, amphibians, and reptiles: development of unfertilized diploid eggs into new female offspring.

C. Sexual reproduction: occurs when a haploid sperm fertilizes a haploid egg, producing a diploid offspring.
    1) Types of sexual individuals:
        a. Dioecious ("two houses"): males and females are separate individuals.
            i.   Females produce eggs (large, non-motile cells containing substantial food reserves in the cytoplasm).
            ii. Males produce sperm (small, motile cells with nearly no cytoplasm or food reserves).
        b. Monoecious ("one house").
            i.   Sperm and eggs are produced by the same individuals called hermaphrodites.
            ii. Often seen in animals that are relatively immobile and usually isolated from other members of their species.
            iii. Examples: tapeworms, many pond snails.
        c. For dioecious species and hermaphrodites that cannot self-fertilize, successful reproduction requires that sperm and eggs from different animals be brought together for fertilization.
    2) External fertilization: parents release sperm and eggs into water ("spawning") and sperm must swim to reach the eggs.
        a. Animals that spawn must:
            i.   Reproduce in water.
            ii. Synchronize their reproductive behavior both temporally (spawn at the same time) and spatially (spawn in the same place).
        b. Breeding usually occurs only during certain seasons of the year and animals rely on environmental cues to spawn in precise synchrony. Example: grunion fish spawn only on fall nights of the highest tides (during full moons) off the California coast.
        c. Some animals synchronize spawning by releasing pheromones into the water.
            i.   A pheromone is a chemical released by one animal that affects the behavior of other animals.
            ii. In many mussels and sea stars, females release eggs and pheromone into the water and nearby males react to the pheromone by releasing sperm.
        d. Many fish and amphibians have courtship rituals, bringing males and females close together as they release their gametes. Frogs practice amplexus (males mount females and stimulate them to lay eggs as they release sperm).
    3) Internal fertilization: sperm enter the body of the females where fertilization occurs.
        a. Advantages:

      i.   Sperm have a direct moist path to the eggs (important in terrestrial animals).

      ii.   It improves the likelihood that most eggs will be fertilized.

  b.   <u>Copulation</u> usually occurs.

      i.   Usually, the penis of the male is inserted into the body of the female and releases sperm.

      ii.   In some amphibians, arachnids, and squids, males package their sperm into gelatinous <u>spermatophores</u> ("sperm carriers") which either they or the female pick up and insert into the female's reproductive tract.

  c.   In most mammals, copulation occurs only at a certain season of the year, when the female signals her readiness to mate, usually coinciding with female ovulation.

      i.   In some animals, like rabbits, ovulation is triggered by copulation.

      ii.   In some snails, females store sperm for months until eggs are ready for fertilization.

3.   **Mammalian reproduction.**

  A.   <u>General Facts</u>.

     1)   Mammals are dioecious.

     2)   Many mammals reproduce only during certain seasons of the year.

     3)   Human reproduction is similar to other mammals except for loss of seasonality.

  B.   <u>The male reproductive tract</u>: paired gonads and accessory organs.

     1)   <u>Testes</u> (male gonads).

       a.   Produce both sperm and male sex hormones.

       b.   Located in the <u>scrotum</u> pouch and kept about $4°$ C cooler than body temperature.

       c.   Coiled <u>seminiferous</u> tubules, in which sperm are produced, nearly fill each testis.

       d.   <u>Interstitial</u> <u>cells</u> between the tubules make male sex hormone (<u>testosterone</u>).

       e.   <u>Spermatogonia</u> (diploid "germ" cells) lie within the wall of each seminiferous tubule. Spermatogonia may either:

         i.   Undergo mitosis to produce new spermatogonia.

         ii.   Undergo <u>spermatogenesis</u> (meiosis and differentiation) and become sperm.

       f.   Steps in <u>spermatogenesis</u>:

         i.   Growth and differentiation of spermatogonia into <u>primary</u> ($1°$) <u>spermatocytes</u>.

         ii.   Each diploid $1°$ spermatocyte undergoes meiosis I to produce 2 haploid <u>secondary</u> ($2°$) <u>spermatocytes</u>.

         iii.   Each $2°$ spermatocyte undergoes meiosis II to produce 2 <u>spermatids</u> (total of 4 spermatids per $1°$ spermatocyte).

         iv.   Differentiation of spermatids into <u>sperm</u> <u>cells</u>.

       g.   <u>Sertoli</u> <u>cells</u>.

         i.   Large cells found within the seminiferous tubules.

         ii.   Regulate spermatogenesis and nourish the developing sperm.

         iii.   Spermatogonia are embedded in invaginations of the Sertoli cells and migrate up towards the lumen (tubular cavity) of the

seminiferous tubules as meiosis occurs.
iv.   Mature sperm are liberated into the lumen.

h.   Human sperm.
    i.   Have little cytoplasm; the haploid nucleus nearly fills the head.
    ii.   The acrosome (a specialized lysosome) lies atop the nucleus and contains enzymes to dissolve protective layers surrounding the egg so that a sperm may enter and fertilize it.
    iii.   The sperm body is packed with mitochondria to provide energy to move the protein tail (a long flagellum).

i.   Spermatogenesis begins in puberty due to the interplay of 3 hormones:
    i.   LH (luteinizing hormone) and FSH (follicle stimulating hormone) from the anterior pituitary gland, and testosterone from the testes.
    ii.   LH stimulates the interstitial cells to produce testosterone, and testosterone plus FSH stimulate the Sertoli cells and spermatogonia, causing spermatogenesis.

j.   Testosterone also:
    i.   Stimulates development of secondary sexual characteristics.
    ii.   Maintains sexual drive.
    iii.   Is required for successful intercourse, with or without the presence of sperm.
    iv.   A good male birth control pill would suppress FSH (blocking spermatogenesis) but not LH (allowing testosterone production).

2)   Accessory structures.
a.   2 seminiferous tubules → single large epididymis tube → vas deferens tube. Sperm are stored in the latter 2 tubes.
b.   The vas deferens leaves the scrotum and joins with the urethra from the bladder, forming a single tube leading to the tip of the penis and serving as a common path for sperm and urine.
c.   Semen ejaculated from the penis consists of sperm and fluids from 3 glands:
    i.   The seminal vesicles.
    ii.   The prostate gland.
    iii.   The bulbourethral gland.
    iv.   These secretions activate swimming by the sperm, provide energy for swimming, and neutralize the acidic fluids of the vagina.

C.   Female reproductive tract: paired gonads and accessory structures that accept and conduct sperm to the egg, and nourish the embryo.
1)   Ovaries (female gonads): in the abdominal cavity.
a.   Oogonia (diploid ovarian cells): divide by mitosis to form diploid primary (1°) oocytes while the female is still a fetus in her mother's womb. By the third month of fetal development, no oogonia remain.
b.   Primary (1°) oocytes.
    i.   Between the third and ninth months of fetal development, all 1° oocytes begin meiosis I but all halt during prophase I before birth.
    ii.   At birth, the ovaries contain 2 million 1° oocytes but many die

each day until at puberty only 400,000 remain.

    iii.   After puberty begins, a few oocytes resume meiosis each month until menopause.

c.   <u>Follicles</u>: an oocyte plus a surrounding layer of smaller accessory cells that nourish the developing oocyte and secrete female sex hormones.

d.   <u>Steps in oogenesis</u> (egg production).

    i.   Each month between puberty and menopause, pituitary hormones stimulate development of about 12 follicles, although usually only one completely matures.

    ii.   The $1°$ oocyte completes meiosis I to produce a large <u>secondary</u> ($2°$) <u>oocyte</u> and a small <u>polar body</u>.

    iii.   Meanwhile, the small cells of the follicle multiply and secrete estrogen. The follicle grows, eventually erupting from the ovary and releasing the $2°$ oocyte.

    iv.   In the oviduct, the $2°$ oocyte undergoes meiosis II, forming an <u>egg</u> and a smaller <u>polar body</u> only if a sperm nucleus has entered the $2°$ oocyte.

    v.   The end result of meiosis in females is one egg cell and 3 polar bodies.

    vi.   Some follicle cells leave with the $2°$ oocyte. Those remaining in the ovary enlarge and become glandular, forming the <u>corpus luteum</u> that secretes estrogen and progesterone hormones. The corpus luteum degenerates if fertilization does not occur.

2)   <u>Accessory structures</u>.

    a.   <u>Oviducts</u> (<u>Fallopian tubes</u>).

        i.   Lie adjacent to each ovary.

        ii.   The open end of each oviduct is fringed with ciliated "fingers" of cells (called <u>fimbriae</u>) that beat to set up a current that carries the $2°$ oocytes into an oviduct where fertilization occurs.

        iii.   The zygote eventually enters the <u>uterus</u>, implants, and grows for 9 months.

    b.   The <u>uterus</u> (womb) has 2 layers of tissue.

        i.   <u>Myometrium</u> (outer muscular layer): contracts strongly during childbirth, delivering the infant out of the uterus.

        ii.   <u>Endometrium</u> (inner layer): has many blood vessels, forming the mother's part of the <u>placenta</u> that transfers O_2, CO_2, nutrients, and waste materials between mother and fetus.

        iii.   The bottom end of the uterus is nearly closed off by a ring of connective tissue, the <u>cervix</u>.

        iv.   Beyond the cervix is the <u>vagina</u> which opens to the outside. It serves as a receptacle for the penis during intercourse, and as the birth canal.

    c.   <u>Maintenance of the endometrium</u>:

        i.   Developing follicles secrete estrogen which stimulates the endometrium to grow a network of blood vessels and nutrient-producing glands.

        ii.   After ovulation, the corpus luteum produces estrogen and progesterone that promote continued growth of the endometrium.

        iii.   If the $2°$ oocyte is not fertilized, the corpus luteum disintegrates, hormone levels fall, and the endometrium breaks down. The myometrium then contracts, causing <u>menstrual cramps</u>, squeezing out the remains of the endometrium as a flow of blood called <u>menstruation</u>.

D.   <u>Menstrual cycle</u>: Hormonal interactions among the hypothalamus, pituitary gland, and ovaries to coordinate ovulation and development of the uterine endometrium.

1)   Neurosecretory cells of the hypothalamus constantly produce <u>GnRH</u> (gonadotropin releasing hormone) which stimulates the anterior pituitary gland to release <u>FSH</u> and <u>LH</u>.

2)   FSH and LH flow to the ovaries and initiate development of several follicles, the small cells of which secrete <u>estrogen</u>. These 3 hormones stimulate follicles to grow for about 2 weeks and for $1°$ oocyte to store food, proteins, and RNA molecules needed for early embryonic development.

3)   Maturing follicles secrete increasing amounts of estrogen that:
    a.   Promote continued development of the follicle and $1°$ oocyte.
    b.   Stimulate growth of the uterine endometrium.
    c.   Stimulate both hypothalamus and pituitary, resulting in a surge of LH and FSH at the 12th day of the cycle.

4)   The <u>surge</u> of LH has 3 consequences.
    a.   It triggers <u>resumption of meiosis I</u> in the $1°$ oocyte.
    b.   It causes <u>ovulation</u> of the follicle.
    c.   It transforms the remnants of the follicle in the ovary into the <u>corpus luteum</u>.

5)   The corpus luteum secretes both estrogen and progesterone, which together:
    a.   <u>Inhibit</u> the hypothalamus and pituitary, shutting down release of FSH and LH. This prevents development of any more follicles.
    b.   <u>Stimulate</u> further growth of the endometrium until it reaches 5 mm in thickness.

6)   If no pregnancy occurs, the corpus luteum self-destructs.
    a.   It can survive only while stimulated by LH from the pituitary (or by a similar hormone from the embryo).
    b.   But, it produces progesterone that shuts down the pituitary's release of LH.
    c.   By day 21, the corpus luteum dies if no pregnancy occurs.

7)   When the <u>corpus luteum dies</u>:
    a.   Estrogen and progesterone levels drop off, killing off the endometrium.
    b.   Menstrual flow begins on day 27 or 28 of the cycle.
    c.   Reduced progesterone levels allow the hypothalamus to induce the pituitary to release FSH and LH again, initiating a new menstrual cycle.

8)   The menstrual cycle includes both <u>positive</u> and <u>negative feedback</u>.
    a.   During the <u>first half</u> of the cycle:
        i.   FSH and LH stimulate estrogen production.
        ii.   High estrogen levels then <u>stimulate</u> a surge of FSH and LH release at midcycle (positive feedback).
    b.   During the <u>last half</u> of the cycle, estrogen and progesterone together <u>inhibit</u> release of FSH and LH (negative feedback).

E.   <u>Copulation</u>.
1)   <u>Erection</u> of the penis occurs.

     a.   The penis is usually flaccid due to constriction of its arterioles, allowing little blood flow.

     b.   When psychological and/or physical stimulation occurs, the arterioles dilate and:

        i.   Blood flows into vascular spaces within the penis.

        ii.   Veins that drain the penis close off.

        iii.   Pressure builds up, causing an erection.

2)   <u>Ejaculation</u> occurs.

     a.   Movements of male or female stimulate touch receptors on the erect penis.

     b.   Muscles circling the epididymis, vas deferens, and urethra contract, forcing semen out of the penis.

     c.   3-4 ml of semen (300-400 million sperm) are released per ejaculation.

     d.   Ejaculation coincides with orgasm.

3)   In the <u>female</u>.

     a.   Sexual excitement causes increased blood flow to vagina, vulva, and clitoris which becomes erect.

     b.   Stimulation may result in an organism, but this is not necessary for fertilization to occur.

F.   <u>Fertilization</u>.

1)   Both egg and sperm are short-lived cells, dying in about 1 or 2 days. So, fertilization succeeds only if copulation occurs within 2 days of ovulation.

2)   As the 2° oocyte leaves the ovary:

     a.   It is surrounded by small follicle cells that form the <u>corona</u> <u>radiata</u>, a barrier between sperm and the 2° oocyte.

     b.   A second barrier, the <u>zona</u> <u>pellucida</u> ("clear zone") membrane lies between the corona and the oocyte.

3)   In the oviduct, many sperm surround the corona, each releasing enzymes from its acrosome.

     a.   These enzymes weaken the corona and zona, allowing sperm to reach the oocyte.

     b.   Too few sperm won't produce enough enzyme to expose the oocyte.

     c.   One sperm finally contacts the surface of the 2° oocyte.

        i.   The cell membranes of oocyte and sperm fuse.

        ii.   The sperm head (nucleus) enters the oocyte's cytoplasm.

     d.   Two changes then occur in the 2° oocyte:

        i.   Vesicles release chemicals into the zona, strengthening it and not allowing more sperm to enter.

        ii.   The second meiotic division occurs, producing the larger <u>egg</u> and a smaller <u>polar</u> <u>body</u>.

     e.   <u>Fertilization</u> occurs as the haploid nuclei of sperm and egg fuse, forming the diploid <u>zygote</u> nucleus.

G.   <u>Pregnancy</u>.

1)   As it travels down the oviduct towards the uterus, the zygote begins to divide, becoming the <u>blastocyst</u> (a hollow ball of cells) after about 7 days.

     a.   The <u>inner</u> <u>cell</u> <u>mass</u> becomes the embryo.

     b.   The sticky outer ball will adhere to the uterine wall and burrow into the endometrium, a process called <u>implantation</u>.

      c.   The embryo then develops.

2)   About a week after an ovulation not resulting in pregnancy, the corpus luteum disintegrates, lowering estrogen and progesterone levels needed to maintain the endometrium, leading to menstruation. In the pregnant female, however:

      a.   Shortly after implantation occurs, the outer cells of the blastocyst start secreting an LH-like hormone called <u>chorionic gonadotropin</u> (<u>CG</u>) into the bloodstream.

      b.   CG travels to the ovary, preventing degeneration of the corpus luteum, and the endometrium continues to develop.

      c.   Some CG is excreted in the mother's urine; pregnancy tests are assays for CG in maternal blood or urine.

3)   The embryo initially gets nutrients directly from the endometrium for the first 7-14 days while the placenta develops.

      a.   The <u>placenta</u> is composed of interlocking tissues of the embryo and the endometrium.

      b.   Through the placenta, the embryo will receive nutrients and O_2 and disposes of wastes into the maternal circulation.

4)   As the fetus grows, changes occur in the mother's breasts.

      a.   Each breast contains milk glands arranged around the nipple, each with a duct leading to the nipple.

      b.   The breasts begin enlarging at puberty due to estrogen.

      c.   Their glandular structure does not fully develop until pregnancy occurs.

            i.   Large quantities of estrogen and progesterone stimulate the milk glands to grow, branch, and become capable of secreting milk.

            ii.   Milk secretion is promoted by prolactin hormone, which rises in concentration from the fifth week of pregnancy until birth.

            iii.   After birth, estrogen and progesterone levels fall and prolactin stimulates milk secretion.

      d.   Suckling stimulates nerve endings in the nipples.

            i.   This signals the hypothalamus to trigger more prolactin and oxytocin from the pituitary.

            ii.   Oxytocin causes muscles around the milk glands to contract, ejecting the milk into the ducts leading to the nipples.

      e.   Just following birth, the milk glands secrete a thin, yellowish fluid called <u>colostrum</u>.

            i.   Colostrum is high in protein and has antibodies to help the newborn fight some diseases.

            ii.   Colostrum is gradually replaced by mature milk that is higher in fat and milk sugar and lower in protein.

H.   <u>Delivery</u> ("birth").

1)   Immediate causes of delivery are a series of complex interactions among:

      a.   Prostaglandins released by the uterus and fetal membranes.

      b.   Stretching of the uterus, triggering contractions of smooth muscle.

      c.   Oxytocin released by the posterior pituitary gland.

2)   <u>Labor</u> begins with intense contractions of the uterus due to release of prostaglandins.

      a.   The baby's head pushes against and dilates the cervix.

      b.   This signals the hypothalamus to trigger oxytocin release.

      c.   The baby is expelled.

      d.   After a brief rest, the uterus contracts further, shrinking in size and expelling the placenta through the cervix (the <u>afterbirth</u>).

    3)   Further prostaglandin release in the <u>umbilical</u> <u>cord</u> causes muscular contractions around the fetal blood vessels there, shutting off blood flow.

4.   On limiting fertility.

    A.   <u>Introduction</u>.
        1)   During most of human evolution, infant mortality was high and natural selection favored people who produced many children.
        2)   Today, people have smaller families and longer life spans.
        3)   However, today's humans have sex drives appropriate to prehistoric times.
        4)   Every four days, a million new people are added to the earth.
        5)   Over the last 20 years, several effective techniques to prevent pregnancy have been developed.

    B.   <u>Permanent contraception</u> by <u>sterilization</u>.
        1)   This is the most effective method since the pathways through which sperm or eggs travel are surgically interrupted.
        2)   <u>Vasectomy</u>: in men, the vas deferens are severed.
        3)   <u>Tubal ligation</u>: in women, the oviducts are cut.
        4)   Sterilization is generally permanent though a surgeon can sometimes reconnect the tubes.

    C.   <u>Temporary contraception</u>.
        1)   <u>Three general categories</u>:
            a.   Preventing ovulation.
            b.   Preventing sperm and egg from meeting when ovulation does occur.
            c.   Preventing implantation of a fertilized egg in the uterus.
        2)   <u>Preventing ovulation</u>.
            a.   During a normal menstrual cycle, ovulation is triggered by a midcycle surge of LH.
            b.   One way to prevent ovulation is to suppress LH release by providing a continuous supply of estrogen and progesterone via <u>birth control pills</u>.
            c.   The pill must be taken daily for 21 days each menstrual cycle.
            d.   <u>Norplant</u>: a new form of long-term contraception.
                i.   It consists of six rods, inserted under the upper arm skin, containing a synthetic hormone levonorgestrel that prevents ovulation.
                ii.   The rods supply gradual steady diffusion of hormone into the bloodstream for five years, but can be removed at any time.
        3)   <u>Barrier methods</u>.
            a.   <u>Diaphragm</u>: a rubber cap that fits snugly over the cervix, preventing sperm from entering the uterus.
                i.   Very effective when used with a spermicide.
                ii.   No side effects.
            b.   <u>Contraceptive sponge</u>: a soft spermicide-impregnated plug inserted in the vagina up against the cervical opening to physically block out and kill sperm.

c. <u>Condoms</u>: worn over the penis to prevent sperm from entering the vagina.

d. Less effective procedures include:
   i. Spermicides alone.
   ii. Withdrawal of the penis just before ejaculation.
   iii. Douching the vagina before sperm enter.
   iv. Rhythm: no intercourse during the ovulatory period of the menstrual cycle.

4) <u>Preventing implantation</u>: preventing pregnancy by not allowing the blastocyst to implant in the uterus.

a. <u>IUD</u> (intrauterine device): a small copper or plastic device inserted into the uterus for long periods of time. Seems to work by irritating the uterine lining so that it cannot accept the embryo.

b. <u>Morning after pill</u>: contains a massive dose of estrogen to induce extremely early abortion.

5) <u>Abortion</u>: termination of a pregnancy by dilating the cervix and removing the embryo and placenta by suction.

a. Compound <u>RU-486</u> pill is used in France to terminate pregnancy during the first two months. It binds to progesterone receptors and blocks the action of progesterone essential for pregnancy maintenance.

b. Abortions are expensive, dangerous, and controversial since the fetus is killed.

6) <u>Future methods</u>.
a. A removable silicone plug for the oviduct.
b. A once-a-month contraceptive pill.
c. a contraceptive vaccine that lasts for over a year.

****

REVIEW QUESTIONS:

1.-5.   **MATCHING TEST**: types of asexual reproduction.

Choices: A.   Parthenogenesis
             B.   Budding
             C.   Fission
             D.   Regeneration

1. Development of unfertilized eggs.
2. Occurs in sea stars if a fragment contains part of the central disc.
3. Occurs in some reptiles.
4. Producing a small new animal by direct growth from the adult's body.
5. Splitting of the parent into 2 smaller individuals that regenerate any missing parts.

----

6. What is the difference between monoecious and dioecious individuals?
7. What three requirements are necessary for successful "spawning"?
8. What are the advantages of internal fertilization over external fertilization?

9.   Briefly describe the steps in spermatogenesis.
10. What are Sertoli cells and what do they do?
11. Name the four structures found in a human sperm.
12. Name the three hormones that control spermatogenesis and briefly describe their effects.
13. Describe the development of primary oocytes from the time of conception through puberty.
14. What are follicle cells and what do they do?
15. Briefly describe the steps in oogenesis.
16. Describe the three regions of the uterus (myometrium, endometrium, and cervix).
17. Explain how the endometrium is maintained during a typical menstrual cycle.
18. What are the three consequences of the surge of LH during the menstrual cycle.
19. What happens in the menstrual cycle if no pregnancy occurs?
20. What are the corona radiata and zona pellucida regions of the secondary oocyte? Briefly describe their functions in fertilization.
21. Explain the origin and role of chorionic gonadotropin hormone during pregnancy.
22. What is the placenta and what are its functions during pregnancy?
23. Which hormones are involved in labor and delivery of a baby and what do these hormones do?

****

ANSWERS TO REVIEW QUESTIONS:

1.-5.	[see 2.B.1)-4)]				14.	see 3.C.1)c.,d.iii.,vi;
	1.	A	4.	B		3.D.3), 5)-7)
	2.	D	5.	C		
	3.	A			15.	see 3.C.1)d.i.-v.

6.   see 2.C.1)a.-b.

7.   see 2.C.2)a.i.-ii.

8.   see 2.C.3)a.i.-ii.

9.   see 3.B.1)f.i.-iv.

10. see 3.B.1)g.i.-iv.

11. see 3.B.1)h.i.-iii.

12. see 3.B.1)i.i.-ii.

13. see 3.C.1)b.i.-iii.

16. see 3.C.2)b.i.-iii.

17. see 3.C.2)c.i.-iii.

18. see 3.D.4)a.-c.

19. see 3.D.6)-7)

20. see 3.F.2)-3)

21. see 3.G.2)a.-c.

22. see 3.G.3)a.-b.

23. see 3.H.1)-3)

****

CHAPTER 40:   ANIMAL DEVELOPMENT

CHAPTER OUTLINE.

This chapter focuses on embryonic development. Mechanisms of cell differentiation are presented, direct and indirect development are compared, and the stages of animal development are given.   The events occurring during human embryonic development are then outlined.

1.   **Introduction.**

    A.   Haploid sperm and egg cells fuse to create a diploid <u>zygote</u>, the beginning of a new generation. The zygote gives rise to the trillions of cells of the adult body by repeated mitosis and <u>differentiation</u> (becoming specialized cells such as brain, muscle, or skin).

    B.   Unanswered questions in the study of <u>development</u> (the process by which an organism proceeds from zygote through adulthood to death).
        1)   How do cells position themselves correctly?
        2)   What governs the onset of puberty?
        3)   Why do animals age?
        4)   Is death inevitable?

2.   **Differentiation.**

    A.   Cells do not differentiate by progressively losing genes not necessary for a particular cell type to function.
        1)   <u>Gurdon's experiments</u> with frogs (Figure 40-1) demonstrated:
            a.   Gene loss cannot be the mechanism of differentiation.
            b.   Differentiated cells contain all the genetic information needed for the development of the entire organism.
        2)   Thus, differentiation must occur because of differential <u>use</u> of genes. Genes that are transcribed to mRNA and translated into proteins must differ among cell types.

362

B.   Gene regulation during development.
1)   See Chapter 13 for some mechanisms controlling gene transcription.
2)   Generally, in eukaryotic cells, differential gene action occurs by:
   a.   Cellular materials (protein and/or steroid hormones) that travel to the nucleus and bind to the chromosomes.
   b.   These binders block transcription of certain genes or promote transcription of other genes that determine the features of the cell.
3)   Two major types of information control gene usage:
   a.   Gene-regulating substances positioned in certain regions of the egg cytoplasm, somehow building up during oogenesis.
      i.   The zygote divides in particular orientations.
      ii.   The developmental features of a daughter cell are determined by the portion of the egg cytoplasm (containing gene regulating substances) it receives.
   b.   Chemical messages (nutrients, neurotransmitters, hormones) received from other cells which can alter gene transcription and enzyme activity.

3.   Indirect and direct development.

A.   Animal eggs contain yolk (lipid and protein food reserves).
1)   A zygote has no mouth or digestive tract.
2)   Yolk provides nourishment for the early embryo.

B.   Indirect development.
1)   The juvenile animal differs significantly from the adult (e.g., caterpillars → butterflies).
2)   Seen in most invertebrates and a few vertebrates (mostly amphibians: tadpoles → frogs).
3)   These animals usually produce huge numbers of eggs, each with little yolk.
   a.   The embryo develops rapidly into a small, sexually immature feeding stage (larva) that usually feeds on different organisms than do the adults.
   b.   Eventually, the larva changes dramatically (metamorphosis) into a sexually mature adult whose primary function is reproduction.

C.   Direct development in most reptiles, birds, mammals, and land snails.
1)   The newborn animal is a sexually immature, miniature version of the adult, requiring considerable embryonic development.
2)   Strategies to meet the increased food requirements of these embryos:
   a.   Laying large eggs containing huge amounts of yolk (e.g., in reptiles, birds, and land snails).
   b.   Having little yolk in the eggs but nourishing the developing embryo in the mother's body (e.g., in mammals, some snakes, and a few fish).
   c.   Either way, providing food is demanding on the female and relatively few offspring are produced.

4.   Reptiles, birds, mammals, and membranes.

A.   Birds and mammals evolved from reptiles.

B.   As an adaptation to direct development in a terrestrial environment, reptile and bird

embryos produce four <u>extraembryonic membranes</u>:
1) <u>Chorion</u>: lines the shell and exchanges oxygen and carbon dioxide through the shell.
2) <u>Amnion</u>: encloses the embryo in a watery environment.
3) <u>Allantois</u>: surrounds wastes.
4) <u>Yolk sac</u>: contains stored food.

C. Although mammalian eggs contain almost no yolk, much of the reptilian genetic program for development persists, including the four extraembryonic membranes.

5. **Stages of animal development.**

A. <u>Cleavage</u>: distributing gene-regulating substances.
1) Cleavage of the zygote reduces cell size and distributes gene-regulating substances to the daughter cells (no cell growth occurs between divisions).
2) Eventually, a solid ball of cells (the <u>morula</u>) forms.
3) Then, a cavity opens within the morula which becomes a hollow ball of cells (the <u>blastula</u>).
4) The blastula is shaped like a sphere in zygotes with little yolk (i.e., sea urchins and amphibians) or a flat disc in zygotes with much yolk (i.e., reptiles and birds).
5) Gene-regulating substances in the egg cytoplasm must be correctly distributed to the cells during cleavage or the embryo will not develop normally (see Figure 40-6).

B. <u>Gastrulation</u>: a process of cell movement giving rise to a 3-layered embryo called a <u>gastrula</u>.
1) A dimple (the <u>blastopore</u>) forms on one side of the blastula and cells migrate through this dimple to the inside (like punching in an underinflated beachball).
2) The enlarging dimple and the cells that migrate in (<u>endoderm</u> or inner skin cells) become the digestive tract.
3) The cells remaining outside become the skin and nervous system and are called <u>ectoderm</u> or outer skin cells.
4) Some cells also migrate between endoderm and ectoderm, forming a third layer called <u>mesoderm</u> or middle skin cells. Mesoderm cells produce muscle, skeleton, and the circulatory system.
5) <u>Influences from the cellular environment</u>.
   a. During gastrulation, the developmental fate of most cells is determined by chemical messages received from other cells, a process called <u>induction</u>.
   b. The "gray crescent" area of the amphibian zygote becomes the blastopore area of the blastula and the dorsal lip of the blastopore induces nearby cells to differentiate (see Figure 40-8).

C. <u>Organogenesis</u>: adult structures develop by rearrangement of ectoderm, mesoderm, and endoderm cell layers.
1) Usually occurs by induction. Example: in frog embryos, the optic cup induces formation of the eye lens from epidermal tissue.
2) For some cells, death is genetically programmed to occur at a precise time during development.
   a. Some cells die unless they receive a <u>survival signal</u> (e.g., embryonic

vertebrate motor neurons will die unless they successfully innervate a skeletal muscle which releases a survival chemical for the motor neuron).

  b. Some cells live unless they receive a <u>death signal</u> (e.g., in tadpoles, thyroid hormone stimulates cells in the tail to make enzymes that digest the tail away).

D. <u>Sexual maturation</u>.
  1) Animals continue to change throughout their lives.
  2) Animals become sexually mature at an age determined by both genes and environment (e.g., song birds become sexually mature in the spring, stimulated by increasingly long days).

E. <u>Aging</u>. No animal is static:
  1) Stomach cells are replaced every few weeks.
  2) Some nerve cells die each day and aren't replaced.
  3) Other cells function less efficiently or divide more slowly as they age.
  4) Is death (both cellular and organismal) a programmed part of life?
    a. A cell lineage has a built-in maximum life span that varies from species to species.
    b. Longevity depends on the cell's ability to repair DNA damage.
    c. All normal cells die eventually, but cancer cells survive and reproduce indefinitely in cell culture.
  5) Scientists do not understand the mechanism that regulates a cell's life span.

6. **Human development**: strongly reflects our evolutionary heritage (see Figure 40-10).

A. <u>The first 2 months</u>.
  1) Early human embryonic development.
    a. Fertilization occurs in the oviduct, as do the first few cleavage divisions.
    b. Just before implantation in the uterine wall, the embryo is a <u>blastocyst</u> consisting of a thin-walled hollow ball with a thicker <u>inner cell mass</u> on one side.
    c. The thin-walled ball becomes the chorion which becomes the embryonic part of the placenta.
    d. The inner cell mass grows and splits forming 2 fluid-filled sacs separated by a double layer of cells (the <u>embryonic disc</u>).
      i. One sac forms the <u>amniotic cavity</u>.
      ii. The other sac becomes the <u>yolk sac</u> although it contains no yolk in humans.
  2) <u>Gastrulation</u> begins about day 14.
    a. The <u>primitive streak</u> (= blastopore) slit appears down the center of the ectoderm and ectoderm cells migrate through it into the interior of the embryo, forming the mesoderm.
    b. The mesoderm first forms the notochord.
  3) During the third week of development:
    a. The ventral surface (endoderm cells) of the embryo curls to form the <u>gut</u>.
    b. The notochord induces the dorsal surface (ectoderm cells) of the embryo to curl, forming the <u>neural tube</u> (future brain and spinal cord).
  4) The <u>placenta</u> (chorion and endometrium) replaces the yolk as a food supply, and allantois as a place for gas exchange and waste elimination.

5)   By the end of the sixth week, the embryo has developed chordate features (notochord, gill arches, tail) that disappear as gestation continues. In addition, the embryo has:
   - a.   Primitive eyes.
   - b.   A beating heart.
   - c.   Traces of fingers on it's tiny hands.
   - d.   A large, developing brain.

6)   By the eighth week, nearly all organs have formed (including testes or ovaries) and the embryo becomes a <u>fetus</u> because it begins to look distinctly human.

7)   Rapid differentiation and growth occurs in the embryo during the first 2 months and this is a time of considerable sensitivity to drugs and environmental insults (viruses, alcohol, tobacco toxins, etc.) since many rapid cell divisions occur.

B.   <u>The placenta</u>.
   1)   Formation.
      - a.   Early in pregnancy, embryonic cells burrow into the thickened uterine lining.
      - b.   The embryo's outer cells form the chorion that penetrates the uterine lining with fingerlike projections called <u>chorionic villi</u>.
   2)   Major functions.
      - a.   Secretes estrogen and progesterone hormones.
        - i.   Estrogen stimulates growth of uterus and mammary glands.
        - ii.   Progesterone stimulates mammary glands and inhibits premature uterine contractions.
      - b.   Regulates selective exchange of materials between blood of mother and blood of fetus without allowing them to mix.
        - i.   Chorionic villi have dense network of fetal capillaries and are bathed in polls of maternal blood.
        - ii.   Diffusion occurs between the bloodstreams:
          - a)   Oxygen and nutrients from mother to fetus.
          - b)   Carbon dioxide and urea from fetus to mother.
        - iii.   Capillary and chorionic villi membranes are barriers preventing passage of some disease organisms, large proteins, and most cells.
        - iv.   Many harmful substances can penetrate the placental barrier.

C.   <u>The last 7 months</u>.
   1)   The brain continues to develop rapidly.
   2)   As the brain and spinal cord grow, the fetus develops noticeable behaviors (response to touch, thumb sucking, etc.).
   3)   Body organs enlarge and become functional.

D.   <u>Birth</u>.
   1)   The baby becomes positioned head downward in the uterus (crown of skull resting on the cervix).
   2)   Somehow, the fetus initiates the birth process:
      - a.   The weight of the baby and uterine contractions force the cervix to dilate.
      - b.   The baby is usually delivered head first.

****

REVIEW QUESTIONS:

1. How do we know that cells do not differentiate by progressively losing genes?
2. What are the two major types of information that cells use to control gene activity?
3. Compare indirect development and direct development.

4.-13. **MATCHING TEST**: stages of animal development.

Choices:
A. Organogenesis
B. Gastrulation
C. Cleavage
D. Maturation and aging

4. Process of cell movement, producing a 3-layered embryo.
5. Distribution of cytoplasmic gene-regulating substances.
6. Animals become sexually mature.
7. A blastopore dimple forms.
8. Adult structures develop by rearrangement of tissue layers.
9. Produces a solid ball of cells (the morula).
10. Ectoderm, mesoderm, and ectoderm tissues form.
11. Death occurs.
12. Induction first occurs.
13. A blastula (hollow ball of cells) develops.

---

14. Explain why, for some cells, death is a genetically programmed part of life.

15.-19. **MATCHING TEST**: embryonic membranes in reptiles.

Choices:
A. Allantois
B. Amnion
C. Chorion
D. Yolk sac

15. Surrounds wastes.
16. Surrounds the embryo and the other membranes.
17. Encloses the embryo within a watery environment.
18. Exchanges gases with the air.
19. Surrounds the embryo's food supply.

20.-26. **Fill in the blanks:**

20. Fertilization occurs in the mammalian _____.
21. Just before implantation, the mammalian embryo is a _____.

22. The inner cell mass of a mammalian embryo grows and splits into 2 sacs separated by a double layer of cells called the _____ _____.
23. The mammalian embryo's blastopore is called a _____ _____.
24. During the third week of development of a mammalian embryo, the _____ induces the dorsal surface of the embryo to curl, forming the _____ _____.
25. The chorion and endometrium together form the _____ in mammals.
26. By the eighth week of development, when the embryo begins to look distinctly human, it is called a _____.

****

ANSWERS TO REVIEW QUESTIONS:

1. see 2.A.1)-2)

2. see 2.B.3)a.-b.

3. see 3.B.-C.

4.-13. [see 5.A.-E.]

4.	B	9.	C
5.	C	10.	B
6.	D	11.	D
7.	B	12.	B
8.	A	13.	C

14. see 5.E.4)a.-c.

15.-19. [see 4.B.1)-4)]

15.	A	18.	A
16.	C	19.	D
17.	B		

20. Oviduct; [see 6.A.1)a.]

21. Blastocyst; [see 6.A.1)b.]

22. Embryonic disc; [see 6.A.1)d.]

23. Primitive streak; [see 6.A.2)a.]

24. Notochord, neural tube; [see 6.A.3)b.]

25. Placenta; [see 6.A.4)]

26. Fetus; [see 6.A.6)]

****

CHAPTER 41: ANIMAL BEHAVIOR I: INDIVIDUAL BEHAVIOR.

CHAPTER OUTLINE.

This chapter considers various aspects of individual behavior. Instinctive behaviors include taxes, kineses, reflexes, and fixed action patterns. Learned behaviors include imprinting, habituation, conditioning (classical and operant), trial-and-error learning, and insight.

1. **Introduction.**

    A.   Activities resulting in successful reproduction demand a lifetime of behavior finely tuned to the physical and social environment. Proper behavior involves the proper meshing of:
    1)   Anatomy.
    2)   Physiology.
    3)   Genetic programming (instinct).
    4)   Learning.

    B.   Ethologists study behavior as it occurs under natural conditions.
    1)   Behavior has evolved as an adaptive trait through natural selection.
    2)   Behavior is a genetically programmed response to specific stimuli in the natural environment.
    3)   This chapter mainly uses examples from ethology.

    C.   Comparative physiologists treat behavior as a lab science.
    1)   Evolutionary aspects and adaptive values of behavior are minimized in their studies.
    2)   They use a few types of lab animals (rats and pigeons) under highly controlled unnatural conditions to study the mechanisms of learning.

    D.   Behavioral neurobiologists study how the brain directs behavior.
    1)   They incorporate elements of psychology, neuroanatomy, brain biochemistry, and properties of individual neurons.

2)   Some use invertebrates to study how simple nervous systems direct behavior.

2.   **All behavior has some genetic basis**.

A.   Much behavior consists of responses to environmental stimuli.

B.   Sense organs, developed under directions from the genes, receive and filter the stimuli. Some organs are extensions of the genetically coded nervous system.

C.   Each species perceives the world differently and behavior is governed by these perceptions. Examples:
1)   Bats hear very high pitched sounds.
2)   Bees see ultraviolet light.
3)   Snakes can detect body heat from nearby mice.
4)   Ticks smell butyric acid in mammalian skin.

D.   Heredity sets narrow boundaries for innate (instinctive) behavior and wider boundaries for learned behavior.

3.   **Innate (instinctive) behavior**.

A.   Innate behaviors are acts performed in reasonably complete form the first time an animal at the right age and motivational state encounters a particular stimulus.
1)   Examples:
     a.   Fruit flies move towards light and away from gravity.
     b.   Human infants turn towards food stimuli and attempt to suckle.
2)   Innate behaviors are entirely programmed by the genes and passed from parents to offspring.
3)   Innate behaviors are performed without learning or prior experience and tend to be highly stereotyped.
4)   The 4 categories of innate behaviors are kineses, taxes, reflexes, and fixed action patterns.

B.   Kineses: behaviors in which an organism changes its speed of random movement in response to an environmental stimulus.
1)   The animal blunders into a favorable environment by chance and stays there, or blunders into a hostile condition and speeds away.
2)   Example: pillbugs seek moist environments by moving faster (but in no particular direction) in dry air surroundings until they encounter a damp area where they slow down and stop.

C.   Taxes: a directed movement of the entire body towards or away from a stimulus.
1)   Most behaviors of simple (and unicellular) organisms are taxes.
2)   Examples:
     a.   Moths fly towards light.
     b.   Euglena moves towards dim light (a positive taxis) but moves away from chloroplast damaging bright light (a negative taxis).
     c.   Grayling butterflies fly toward the sun only when pursued, temporarily blinding their predators.
     d.   Female mosquitoes fly towards the warmth, humidity, and CO_2 exuded

by their prey.

      e.   A fish swims with its dorsal surface away from gravity and towards light.

D.   <u>Reflexes</u>: movement of a body <u>part</u> in response to a stimulus.
     1)   Usually stereotyped and rapid since the brain is not involved.
     2)   Examples:
          a.   Eye-blinking.
          b.   Knee-jerking.
          c.   Withdrawing the hand from a hot stove.

E.   <u>Fixed action patterns</u>: instinctive and stereotyped, often complex series of movements performed correctly the first time the appropriate stimulus (called the <u>releaser</u>) is presented.
     1)   Are performed <u>without</u> <u>prior</u> <u>experience</u>: they are instinctive and innate. Examples:
          a.   Nut-burying in squirrels is innate, with the nuts serving as releasers of the behavior.
          b.   Cuckoo chicks, immediately after hatching, shove the nest-owner's eggs out of the nest.
     2)   Encountering the releaser in the wrong circumstances will elicit <u>inappropriate</u> <u>behavior</u>. Examples:
          a.   <u>Bird</u> <u>banding</u>: the shiny metal bands placed on the legs of baby birds elicited the parents to clean up the nest by throwing out the bands (along with the baby birds).
          b.   Male redwinged blackbirds will <u>attempt</u> <u>copulation</u> with the stuffed tail of a female provided the feathers are oriented properly.
          c.   <u>Exaggerated</u> <u>releasers</u> ("supernormal stimuli") are far more effective than their normal counterparts (e.g., some birds prefer volley-ball "eggs" to their own because of size).
     3)   <u>Behavioral hybrids</u> can be produced by mating closely related species that differ in a particular behavior, indicating a genetic basis for the behavior (e.g., the way African lovebirds carry nesting material to their nests).

F.   <u>Importance of innate behaviors</u>. They are <u>adaptive</u> since:
     1)   Survival may depend on the behavior being properly performed the first time (e.g., camouflage behavior to avoid predators).
     2)   Animals with simple nervous systems can't learn such behaviors.
     3)   Social interactions necessary for survival may depend on proper behavior (e.g., roles in insect societies; mating rituals).

4.   **Learning.**

A.   Survival is enhanced if an animal can modify its innate behaviors based on experience. Generally, the more complex an animal's nervous system, the more its behavior can be modified by experience (the easier it is to "learn"). Five categories of learning exist (B.-F. below).

B.   <u>Imprinting</u>: a strong association learned during a particular stage (the "sensitive period") in an animal's life. Examples:
     1)   The <u>following</u> <u>response</u>: birds learn to follow the animal or object (usually the

mother) they most frequently encounter during a sensitive period (13-14 hours after hatching in mallard ducks).

2) During sexual imprinting, a young bird learns the traits of its future mate.

3) Also, parents may imprint on their offspring, learning which young are their own.

C. Habituation: a decline in response to a harmless, repeated stimulus, resulting in less time and energy wasted on inconsequential stimuli.

1) Unicellular organisms habituate to stimuli like touch.

2) Humans habituate to a variety of stimuli (night sounds, odors, etc.).

D. Classical and operant conditioning: complex learning, usually demonstrated in the lab.

1) Classical conditioning: an animal learns to perform a response (normally caused by one type of stimulus) to a new stimulus. Examples:

a. Pavlov's studies with dogs where ringing a bell elicited salivation.

b. Dogs that salivate at the sound of a can opener.

c. Flatworms that learn to associate a flash of light with an electrical shock.

2) Operant conditioning: an animal learns to perform a behavior (pushing or pecking something) to receive a reward or avoid a punishment. Examples:

a. "Skinner box" experiments where a rat or pigeon trains itself (push a lever to get a pellet of food).

b. Training pigeons for sea search and rescue missions.

E. Trial and error learning: animals acquire new and appropriate responses to stimuli through experience.

1) Animals with complex nervous systems learn this way through play and exploratory behavior.

2) Trial and error learning can modify the releaser for innate behavior (e.g., a toad will learn to avoid bees as a feeding releaser if it gets stung on the tongue once or twice).

F. Insight: manipulating concepts in the mind to arrive at adaptive behavior (like mental trial and error learning).

1) Requires the ability to remember past experiences and to apply those lessons in creative ways to new situations.

2) Example: a hungry chimp will stack boxes to reach a banana suspended from the ceiling without training.

5. **Innate and learned behaviors**: in adult animals, every behavior is an intimate mixture of instinct and learning.

A. Innate learning is rigidly programmed by the genes. Examples:

1) Imprinting in young birds.

2) Language learning in young children.

3) Learning songs in young birds.

4) Bees learning the location of their hive during only the first flight each day.

B. Learning may influence innate behavior.

1) As herring gull chicks learn about their parents appearance, they cease pecking at models not closely resembling their parents.

2) Habituation learning fine-tunes an organism's innate responses to environmental

stimuli. Example: crouching by young experienced birds (to avoid being seen) is released only by the very specific shape of predatory birds while inexperienced birds crouch when anything flies overhead. Such habituation is adaptive.

3) Trial and error learning results in more appropriate response to releasers. Example: fly-catching toads quickly learn not to capture bees.

****

REVIEW QUESTIONS:

1.-6. **MATCHING TEST**: fields of behavior.

Choices: A. Ethology   C. Behavioral
          B. Comparative physiology   neurobiology

1. Treats behavior as a lab science.
2. Study of behavior under natural conditions.
3. Studies mechanisms of behavior in lab animals under controlled conditions.
4. Studies properties of individual cells that affect behavior.
5. Studies behavior as an adaptive trait, evolved through natural selection.
6. Studies psychological and biochemical aspects of behavior.

----

7. Give several examples of different species perceiving the world differently, and its effect on their behavior.

8.-19. **MATCHING TEST**: innate (instinctive) behavior.

Choices: A. Kineses   D. Fixed action patterns
          B. Taxes   E. All of these
          C. Reflexes   F. None of these

8. Movement of a body part in response to a stimulus.
9. Entirely programmed by the genes.
10. Moths fly towards light.
11. Performed correctly without learning or prior experience.
12. Nest-cleaning in birds.
13. An organism changes its speed of random movement in response to a stimulus.
14. Eye-blinking.
15. Directed movement of the entire body towards or away from a stimulus.
16. Nut-burying in squirrels.
17. Pillbugs seeking a moist environment.
18. A fish swims with its dorsal surface towards light.
19. Complex series a movements in response to the appropriate releaser stimulus.

----

20. Describe several pieces of evidence indicating that fixed action patterns are genetically programmed, instinctive behavior.
21. Why are innate behaviors important?
22. How does learning differ from instinct?

23.-36. MATCHING TEST: categories of learning.

	Choices:	A.	Classical conditioning	E.	Operant conditioning
		B.	Habituation	F.	Trial and error learning
		C.	Insight	G.	All of these
		D.	Imprinting	H.	None of these

23. Decline in response to a harmless, repeated stimulus.
24. Modification of behavior by experience.
25. A strong association learned during a sensitive period in life.
26. Training pigeons for sea search and rescue missions.
27. Primarily instinctive.
28. Humans ignoring night sounds while asleep.
29. An animal learns to associate an old behavior with a new stimulus.
30. A toad will avoid bees if it is stung on the tongue.
31. The "following response" of young birds.
32. Manipulating concepts in the mind to arrive at adaptive behavior.
33. An animal learns to perform a behavior to receive a reward.
34. Dogs salivating at the sound of a can opener.
35. Acquiring new and appropriate responses to stimuli by experience.
36. A hungry chimp will stack boxes to reach bananas suspended from the ceiling.

****

ANSWERS TO REVIEW QUESTIONS:

1.-6.	[see 1.B.-D.]				20.	see 3.E.1)-3)		
	1.	B	4.	C				
	2.	A	5.	A	21.	see 3.F.1)-3)		
	3.	B	6.	C				
					22.	see 4.A.; 5.A.-B.		
7.	see 2.C.1)-4)							
					23.-36.	[see 4.B.-F.]		
8.-19.	[see 3.B.-E.]					23.	B	30.	F
	8.	C	14.	C		24.	G	31.	D
	9.	E	15.	B		25.	D	32.	C
	10.	B	16.	D		26.	E	33.	E
	11.	E	17.	A		27.	H	34.	A
	12.	D	18.	B		28.	B	35.	F
	13.	A	19.	D		29.	A	36.	C

****

CHAPTER 42: ANIMAL BEHAVIOR II: PRINCIPLES OF SOCIAL BEHAVIOR

CHAPTER OUTLINE.

This chapter deals with mechanisms of communication, competitive behaviors within species (for food, territory, and mates), cooperative behavior in insect and vertebrate societies, and the study of human behavior.

1. **Communication**: the basis of all social behavior.

    A.   <u>Definition</u>: the production of a signal by one organism that causes another to change its behavior in a way beneficial to one or both.

        1)   Most communication occurs between members of the same species, resolving conflicts that result from competitive interactions while minimizing harmful encounters.

        2)   Communication mechanisms are quite diverse and utilize all the senses.

    B.   <u>Visual communication</u>: in animals with well developed eyes.

        1)   <u>Active signals</u>: a specific movement conveys a message.

        2)   <u>Passive signals</u>: the size, shape, or color of an animal conveys information, especially about its reproductive state.

        3)   <u>Advantages</u> of visual communications:

            a.   They are instantaneous.

            b.   They may be graded, conveying the intensity of a motivational state.

            c.   They are quiet and unlikely to alert predators.

        4)   <u>Disadvantages</u>:

            a.   The signaller becomes conspicuous to nearby predators.

            b.   They are limited to close-range communication.

            c.   They aren't effective in the dark (except for fireflies).

C. Sound communication.
1) Is almost instantaneous.
2) Can convey a variety of messages.
3) Can be transmitted through darkness, vegetation, and water.
4) Can carry farther than the eye can see.
5) Can be graded in intensity to convey motivational states.

D. Chemical (Scent) communication.
1) Pheromones are chemicals that influence the behavior of others.
2) Can carry messages over long distances but take very little energy to produce.
3) May not be detected by other species.
4) Persists over time after the animal has departed (e.g., dog and wolf urine).
5) Fewer different messages are communicated with chemicals than with sight or sound (less variety and gradation possible).
6) Releaser pheromones cause an immediate overt behavior in the animal detecting them. They may "say":
   a. This area is mine.
   b. I am ready to mate now.
   c. Food is this way.
7) Primer pheromones cause a physiological change in the animal detecting them, entering through the nose or mouth.
   a. Most affect the reproductive state of the receiver.
   b. Are crucial to the maintenance of complex insect societies (termites, ants, bees).
   c. "Queen substance" made by queen honeybees and eaten by her female hive mates prevents them from becoming sexually mature.
   d. Urine of mature male mice influences female reproductive hormones:
      i. It will initiate estrus.
      ii. It will cause a female newly pregnant by another male to abort and become sexually receptive to the urinating male.
   e. Lab-synthesized sex pheromones of Japanese beetles or gypsy moths are used to lure these insects into traps or are sprayed around to disrupt normal mating. This type of pest control:
      i. Is specific to one species (harmless to others).
      ii. Can't cause resistance to evolve.

E. Touch communication.
1) Establishes social bonds among group members.
   a. Kissing, nuzzling, patting, petting, and grooming in primates.
   b. Mutual licking, sniffing, and gentle nipping around the mouth in wolves and dogs.
2) Tightens bonds between parents and offspring.
3) Precedes sexual activity.

2. **Competition for resources.**

A. Aggression.
1) An obvious result of competition for resources such as food, space, and mates.
2) Usually involves harmless symbolic displays or rituals for resolving conflicts, allowing resolution without inflicting wounds.

      3)   <u>Visual aggressive displays</u>:
             a.   Exhibit "weapons" (fangs, claws, teeth).
             b.   Often include behaviors to make the animal appear larger (standing upright, fur fluffing, erecting the feathers, ears, fins, etc.).
             c.   Are often accompanied by intimidating sounds whose intensity can be a factor in deciding the winner.
      4)   <u>Ritualized combat</u> may occur: harmless clashing of "weapons", shoving (not slashing).

  B.   <u>Dominance hierarchies</u>: one way to resolve competition without wasteful aggression.
      1)   Each animal establishes a rank that determines its social status.
      2)   Example: "pecking order" in chickens in which hens defer to more dominant birds when feeding. Conflict is minimized because each bird "knows its place".
      3)   Dominance in bighorn sheep depends on horn size.
      4)   In wolf packs, each sex has a dominant (alpha) individual to whom all others are subordinate.
      5)   Dominant individuals obtain most access to the resources needed for reproduction (food, space, mates).

  C.   <u>Territoriality</u>: defense of an area where important resources are located.
      1)   Territorial animals restrict some or all of their activities (feeding, mating, raising young, storing food) to the defended area and advertise their presence there.
      2)   Territorial behavior is most often seen in adult males who defend the area against members of the same species.
      3)   Territoriality is an example of convergent evolution (seen in animals as diverse as worms and mammals), suggesting that it conveys some important evolutionary advantages:
             a.   "Good fences make good neighbors" (reduces needless aggression) because boundaries are recognized and respected.
             b.   Helps to keep population numbers within the limits set by the available resources (inversely related to the size of territories).
      4)   Territories are "advertised" through:
             a.   Sight and behavior of the defender.
             b.   Smell: scent-marking the boundaries using pheromones.
             c.   Sound: e.g., male sea lions, male crickets, male songbirds.

3. Reproduction.

  A.   Successful reproduction requires that the animals identify one another as:
      1)   Members of the same species, often through vocal communication.
      2)   Members of the opposite sex, often through vocal communication and visual displays.
      3)   Being sexually receptive.

  B.   Pheromones play an important role in reproductive behavior. Sexually receptive females often release powerful sex pheromones (e.g., female dogs in heat).

  C.   In solitary animals, most encounters between individuals are competitive and aggressive. While courting, sexual encounters often include submissive signals or the presentation of gifts to ward off aggression (e.g.: a male of some carnivorous insect species offers a female dead insects that she eats while he mates and runs).

4. **Cooperation**.

    A.   <u>Disadvantages of group living</u>. Increased:
        1)   Competition within the group for limited resources.
        2)   Risk of infection.
        3)   Risk that offspring will be killed by other members of the group.
        4)   Risk of being spotted by predators.

    B.   <u>Advantages of social behavior</u>. Increased:
        1)   Ability to detect, repel, and confuse predators.
        2)   Hunting efficiency or ability to find food.
        3)   Division of labor within the group.
        4)   Conservation of energy.
        5)   Likelihood of finding mates.

    C.   Social interactions such as aggression and cooperation have evolved because they help the individual and its offspring survive.
        1)   Cooperative interactions between solitary individuals are restricted to brief aggressive encounters and mating.
        2)   Some animals cooperate based on changing needs (e.g., solitary coyotes will hunt in packs only when food is scarce).
        3)   Some animals naturally form loose social groupings (herds, schools, flocks) to better deter predators and find food.
        4)   A few highly integrated cooperative societies are found primarily among insects and mammals.

    D.   <u>Insect societies</u>.
        1)   The individual is a cog in an intricate machine and can't function alone.
        2)   Individuals are born into one of several castes, genetically programmed to perform a specific function. Example: the 3 major pre-ordained roles in honeybees are:
            a.   <u>Queen</u> (one per hive) who functions to:
                i.   Produce eggs (up to 1000/day).
                ii.   Regulate the lives of the workers.
            b.   <u>Male drones</u>: mates for the queen.
                i.   Mate with the queen during her first week of life (she mates up to 15 times, storing enough sperm to fertilize 3 million eggs).
                ii.   After mating, the drones are driven from the hive or killed.
            c.   Sterile <u>female workers</u> whose tasks are determined by their ages and conditions in the colony:
                i.   Young workers are <u>waitresses</u>, carrying food to other bees.
                ii.   Later special wax glands become active and they become <u>builders</u> of hexagonal cells where the queen deposits eggs. They also act as <u>maids</u> (cleaning the hive) and <u>guards</u> (driving intruders from the hive).
                iii.   Older workers become <u>foragers</u>, gathering nectar and pollen for food for the hive. They inform others of the location of nectar by the "waggle dance".

       d.   Pheromones play a major role in regulating the lives of social insects, especially the "queen substance" in bee colonies.

E.   <u>Vertebrate societies</u>.
- 1)   With the exception of humans, vertebrates have less complex societies than insects.
- 2)   Each individual is unique and in vertebrates this is enhanced because they exhibit more flexible learned behavior. They are less predictable and less robotic than is necessary to maintain complex insect societies.
- 3)   <u>Bullhead catfish</u>: a simple vertebrate in which complex social interactions are based almost entirely on pheromones sensed by smell.
  - a.   Bullheads clearly recognize one another as individuals by scent and have a long memory.
  - b.   Status or change of status of an individual is communicated by scent.
  - c.   Under crowded conditions, an antiaggression pheromone is produced, minimizing conflict.
- 4)   <u>Prairie dogs</u>.
  - a.   Live in "towns" of up to 1000 individuals.
  - b.   The primary social unit is a "coterie" of about 10 individuals who share the same burrow, recognize each other, and together defend a small territory around their burrow.
  - c.   When a predator is spotted, the watchdog produces a yipping bark and warns others of the danger.
- 5)   <u>Hamadryas baboons</u> of Africa exhibit social organization at three levels:
  - a.   The <u>one-male unit</u>: a male and his harem (2-5 females and their young).
    - i.   He dominates them with slaps, bites or aggressive stares.
    - ii.   <u>Bachelor units</u>: gathering of excess males caused by the harem system.
  - b.   <u>Bands</u>: several one-male units congregate each evening and travel to water holes and sleeping cliffs together.
  - c.   <u>Troops</u>: several bands (hundreds of individuals) may congregate at large sleeping cliffs.

5.   **Human ethology**.

A.   Scientists disagree as to the existence and extent of genetic influences on human behavior (is human behavior entirely a product of learning or do many human tendencies have a genetic basis?).

B.   <u>Studies of young children</u>.
- 1)   Instinctive behavior of infants (genetic in origin) include:
  - a.   Rhythmic movement of the head in search of the mother's breast.
  - b.   Suckling.
  - c.   Grasping with hands and feet.
  - d.   Walking movements when the body is supported.
  - e.   Smiling at a human face or reasonable facsimile.
- 2)   Blind and deaf children produce normal smiles, laughter, and expressions of frustration and anger, a strong indication of inherent genetic tendencies.
- 3)   The ability of young children (ages 1-8) to acquire language rapidly and easily is an example of developmentally programmed learning.

C. Exaggerating human releasers.
1) A behavior has an instinctive component if the stimulus releasing it can be exaggerated beyond reality and elicit as even stronger response.
2) Examples:
a. Eyespots cause young infants to smile.
b. Baby features elicit protective feelings in adults and youngsters. So, products with large domed heads, chubby checks, small noses, short arms and legs, and small rounded bellies release "mothering" behaviors.
c. Involuntary enlargement of the eye pupil occurs when we view something pleasant or loved.
i. We react to this signal in others.
ii. In the middle ages, women artificially enlarged their pupils using the drug belladonna ("beautiful woman").

D. Comparative cultural studies. Simple acts performed by people from isolated and diverse cultures are similar, implying an instinctive basis. Examples:
1) Facial expressions of pleasure, rage, and disdain.
2) Movements like the "eye flash" and a hand upraised in greeting.

E. Human pheromones may elicit unconscious responses: primer pheromones may influence female reproductive physiology.
1) Menstrual cycles of roommates and close friends become significantly more synchronous.
2) Women who spend more time with men have shorter menstrual cycles than females that spend less time with men.

F. Studies of twins.
1) Fraternal twins are genetically dissimilar but are exactly the same age and share a very similar environment.
2) Identical twins are genetically identical and are exactly the same age and share a very similar environment.
3) Studies of identical twins separated shortly after birth indicate that many human behavioral traits are heritable (identical co-twins are more similar than fraternal co-twins). Examples:
a. Activity level.
b. Alcoholism.
c. Sociability.
d. Anxiety.
e. Intelligence.
f. Dominance.
g. Political attitude.
h. Personality.

****

REVIEW QUESTIONS:

1. Define communication.
2. List three advantages and three disadvantages of visual communication.
3. List two advantages of sound communication over visual communication.

4.-12. **MATCHING TEST:** modes of communication.

Choices:
    A. Visual
    B. Sound
    C. Chemical
    D. Touch

4. Dog urine.
5. Easily alerts predators.
6. Uses pheromones.
7. May be active or passive.
8. Persists after the animal has departed.
9. Can establish social bonds among group members.
10. Limited to close-range communications.
11. Ignored by other species.
12. Grooming in primates.

13.-23. **MATCHING TEST:** mechanisms of competition.

Choices:
    A. Aggression
    B. Dominance hierarchies
    C. Territoriality
    D. All of these
    E. None of these

13. Includes behavior that makes the animal appear larger.
14. Defense of an area where important resources are located.
15. Instinctive behavior.
16. Establishing a rank that determines social status.
17. Harmless symbolic displays or rituals for resolving conflicts without fighting.
18. Adult males defend an area against members of the same species.
19. Scent-marking boundaries with pheromones.
20. Exhibiting fangs, claws, or teeth.
21. Pecking orders in chickens.
22. "Good fences make good neighbors".
23. The sheep with the biggest horns gets most access to necessary resources.

---

24. Name four advantages and four disadvantages of social behavior/group living.

25. Briefly describe the various roles played by honeybees in a hive society.
26. Discuss the following five lines of evidence indicating that many human behaviors have a genetic basis:

        A.   Studies of young children.
        B.   Exaggerated human releasers.
        C.   Comparative cultural studies.
        D.   Effects of human pheromones.
        E.   Studies of twins.

****

ANSWERS TO REVIEW QUESTIONS:

1.	see 1.A.			13.-23.	[see 2.A.-C.]		
				13.	A	19.	C
2.	see 1.B.3)-4)			14.	C	20.	A
				15.	D	21.	B
3.	see 1.C.1)-6)			16.	B	22.	C
				17.	A	23.	B
4.-12.	[see 1.B.-E.]			18.	C		
	4.	C	9.	D			
	5.	B	10.	A,D	24.	see 4.A.-B.	
	6.	C	11.	C			
	7.	A	12.	D	25.	see 4.D.2)a.-c.	
	8.	C					
				26.	see 5.B.-F.		

****

UNIT V EXAM:

<u>Chapters 30 - 42</u>. All questions are dichotomous. Circle the correct choice in each case.

Chapter 30.

1. Circulatory systems with capillaries are (closed / open).
2. (Closed / Open) circulatory systems are characteristic of arthropods.
3. All vertebrates have (closed / open) circulatory systems.
4. Detoxification of substances occurs in the (kidneys / liver).
5. The larger pumping chamber of the heart is the (atrium / ventricle).
6. Pulmonary circulation utilizes the (left / right) side of the heart.
7. The pacemaker is the (atrioventricular / sinoatrial) node.
8. The (parasympathetic / sympathetic) nervous system slows down the heart rate.
9. (Arteries / Veins) have valves.
10. Blood pressure is greater in the (arteries / veins).

Chapter 31.

1. In animals, the production of energy by cellular respiration requires (oxygen / sunlight).
2. A waste product of cellular respiration is (oxygen / carbon dioxide).
3. Respiration be simple diffusion is sufficient in (large / small) animals living in (moist / dry) environments.
4. Diffusion is supplemented by a circulatory system in (flatworms / earthworms).
5. The movement of fluids in unison through large areas is (bulk flow / diffusion).
6. Arachnids have (book lungs / lungs).
7. Insects have (book lungs / tracheae).
8. The lungs of (mammals / birds) are more efficient.
9. In human lungs, gas is exchanged in sacs called (alveoli / bronchi).
10. Each hemoglobin molecule can bind (one / four) oxygen molecules.

Chapter 32.

1. (Carbohydrates / Fats) are most commonly used to store energy in animals.
2. (Glycogen / Starch) is stored in our liver as an energy source.
3. Vitamins (can / cannot) be synthesized by humans.
4. (Fat-soluble / Water-soluble) vitamins can be stored by the body.
5. Digestive enzymes are most likely to be found in the (lysosome / ribosome).
6. The muscular organ of the earthworm that pulverizes food is the (crop / gizzard).
7. The chemical breakdown of food and absorption of the simple molecules occurs in the (large / small) intestine.
8. Bile is produced in the (gallbladder / liver).
9. The pancreas secretes (hydrochloric acid / sodium bicarbonate).
10. The large intestine is (longer / shorter) than the small intestine.

Chapter 33.

1. The (flame cell / nephridium) is the simplest excretory system.
2. Urea is produced by (body cells / kidneys).
3. Purified or filtered blood is found in the renal (arteries / veins).
4. Urine is passed from the kidney to the bladder by the (ureter / urethra).
5. The inner layer of the kidney is the (cortex / medulla).
6. Reabsorption takes place in (Bowman's capsule / tubules).
7. Movement of substances like salts and nutrients from the filtrate out of the tubule is accomplished by (active transport / simple diffusion).
8. The (distal / proximal) tubule is located closest to Bowman's capsule.
9. Vasopressin (decreases / increases) the permeability of the distal convoluted tubule and the collecting duct to water.
10. If the blood volume is reduced or the concentration of salts in the blood is increased, then (less / more) vasopressin is released.

Chapter 34.

1. The (external / internal) defense mechanism is the more effective defense mechanism of the body.
2. The (immune / inflammatory) response is nonspecific and occurs first.
3. Histamine causes blood flow to (decrease / increase).
4. The B cells mature in the (marrow / thymus).
5. Antibodies are (polysaccharides / proteins).
6. (Memory / Plasma) cells do not release antibodies.
7. (B / T) cell immunity is called humoral immunity.
8. The (first / second) immune response is faster.
9. Allergies are associated with the release of (antihistamines / histamines).
10. (B / T) cells are involved in the destruction of cancer cells.

Chapter <u>35</u>.

1.   Endocrine glands usually (do not have / have) ducts.
2.   The action of hormones is dependent upon (gland / target) cells.
3.   (Endocrine / Exocrine) glands produce chemicals that exert their effects outside the body of the animal producing them.
4.   Neurohormones are hormones that (affect / are produced by) nerve cells.
5.   (Peptide / Steroid) hormones may enter cells more easily.
6.   Animals regulate hormone release through (negative / positive) feedback.
7.   An increase in antidiuretic hormone (vasopressin) will (decrease / increase) blood pressure.
8.   The (anterior / posterior) pituitary is more like part of the brain than an endocrine gland.
9.   The (anterior / posterior) pituitary produces the most hormones.
10.   The breakdown of glycogen into glucose is favored by the presence of (glucagon / insulin).
11.   Insulin (decreases / increases) blood glucose.
12.   Adrenalin and noradrenalin stimulate the (parasympathetic / sympathetic) nervous system.
13.   Adrenalin causes blood to flow (away from / toward) the stomach.
14.   The adrenal medulla secretes a (female / male) hormone.
15.   Aspirin (counteracts / enhances) the effects of prostaglandins.

Chapter <u>36</u>.

1.   (Axons / Dendrites) carry an impulse away from the nerve cell body and are the long extensions of a nerve cell.
2.   (Axons / Dendrites) initiate an impulse.
3.   Nerves pass on impulses with (diminished / undiminished) intensity.
4.   The resting potential inside a nerve cell is always (negative / positive) within the cell.
5.   When the threshold level is reached, the (potassium / sodium) channel opens.
6.   The size of the action potential is (dependent on / independent of) the strength of the stimulus.
7.   Myelin covers some (axons / dendrites).
8.   Impulses transmitted by nerve cells covered with myelin travel (faster / slower) than those without myelin.
9.   Receptors for neurotransmitters are located on the (postsynaptic / presynaptic) neuron.
10.   A centralized nervous system is characteristic of (radially / bilaterally) symmetrical forms.
11.   The (cerebellum / medulla) controls coordination.
12.   The (corpus callosum / thalamus) connects the two sides of the cerebrum.
13.   The (left / right) side of the brain is associated with creativity.
14.   Short-term memory is (chemical / electrical).
15.   Long-term memory is (chemical / morphological).

Chapter <u>37</u>.

1.   Structures that change when acted upon by stimuli are (receptors / acceptors).
2.   Structures that convert signals from one form to another are (reducers / transducers).
3.   Thermoreceptors respond to (infrared radiation / blue light).
4.   High frequency vibrations of air or water is (sound / taste).
5.   The cochlea is part of the human (middle / inner) ear.
6.   The sense of vision involves (chemical / physical) changes.

7. The fovea contains virtually no (cones / rods).
8. The retinas of nocturnal animals are made up of (cones / rods).
9. Pain is a special kind of (chemical / physical) sense.
10. (Cones / Rods) are responsible for color vision.

Chapter 38.

1. The thin filaments of muscle contain the protein (actin / myosin).
2. The chemical (acetylcholine / serotonin) carries the nerve signal across a synapse.
3. (Skeletal / Cardiac) muscle fibers can initiate their own contractions.
4. (Smooth / Cardiac) muscle lacks a regular arrangement of sarcomeres.
5. In relaxed muscle, the cross-bridges are (extended / retracted).
6. When a muscle is contracted, calcium is found in (sarcomere / reticulum) regions.
7. Most motor neurons innervate (more than one / only one) muscle fiber.
8. Chondrocytes are found in (bone / cartilage).
9. Bone-forming cells are (osteoclasts / osteoblasts).
10. Bones are joined to each other by (ligaments / tendons).

Chapter 39.

1. (Asexual / Sexual) reproduction requires more energy.
2. The greatest advantage of sexual reproduction is (stability / variability).
3. Parthenogenic bees are (females / males).
4. Worker bees are (diploid / haploid).
5. Organisms that have separate sexes are called (dioecious / monoecious).
6. The seminiferous tubules produce (hormones / sperm).
7. (Oogenesis / Spermatogenesis) does not begin until puberty.
8. During spermatogenesis, meiosis produces (primary / secondary) spermatocytes.
9. Sperm are filled with (mitochondria / ribosomes) to provide energy for movement.
10. When a human female is born, her ovaries contain many (primary / secondary) oocytes.
11. The second meiotic division of a human oocyte occurs in the (ovary / oviduct).
12. Human fertilization occurs in the (oviducts / uterus).
13. (Estrogen / Progesterone) stimulates the production of pituitary hormones.
14. The (clitoris / vagina) is the female counterpart of the penis.
15. (Chorionic gonadotropin / Luteinizing hormone) prevents the disintegration of the corpus luteum during pregnancy.

Chapter 40.

1. The normal specialization of cells that occurs during growth is called (development / differentiation).
2. The (egg / sperm) contributes the most to cytoplasmic regulation of the zygote genome.
3. Larval forms are characteristic of (direct / indirect) development.
4. Large numbers of eggs, each with little yolk, usually are produced by animals that demonstrate (direct / indirect) development.
5. In embryonic development, the (blastula / gastrula) comes first.

6. Human eggs have (little / much) yolk material.
7. The (chorion / amnion) surrounds the human embryo with a watery environment.
8. The (chorion / allantois) collects wastes from the reptilian embryo.
9. The chorion and endometrium are called the (placenta / uterus) in a pregnant human female.
10. Human embryo development more closely resembles that of the (amphibian / reptilian) embryo.

Chapter 41.

1. (All / Some) behavior has some genetic basis.
2. A rattlesnake responds to (heat / sound).
3. Flexible behaviors are (innate / learned).
4. Direction of movement is influenced by the direction of the stimulus in (kinesis / taxis).
5. (Kineses / taxes) are characterized by random movement.
6. The brain (is / is not) involved in reflex actions.
7. The trigger for a fixed action pattern is known as a (releaser / stimulus).
8. The (more complex / simpler) the animal, the more it relies on instinct.
9. In insect societies, almost all behavior is (innate / learned).
10. (Habituation / Imprinting) is characterized by a crucial time during which it can become part of an animal's behavior.

Chapter 42.

1. (Interspecific / Intraspecific) communication is more common and important.
2. If an animal assumed a particular posture to communicate with another animal, this behavior is considered (active / passive).
3. Both sound and visual signals (can / cannot) be of graded intensity.
4. Competition is greatest between members of (different / the same) species.
5. Most aggressive encounters between members of the same species are (real / symbolic).
6. Territories are more commonly laid out by (females / males).
7. Territories are usually defended against invasion by members of (different / the same) species.
8. Established territories promote (conflict / harmony) among members of the same species.
9. The majority of encounters between (social / solitary) animals are competitive and aggressive.
10. Vertebrates have (less / more) flexible behavior patterns than invertebrates.
11. An animal that exposes its neck is displaying (aggressive / submissive) behavior.
12. Sociobiologists believe that human behavior is (innate / learned).
13. Suckling by babies is (instinctive / learned) behavior.
14. Menstrual cycles of roommates are usually (independent / synchronized).
15. Women isolated from men tend to have (longer / shorter) menstrual cycles.

****

ANSWER KEY:

Chapter 30:

1. closed
2. open
3. closed
4. liver
5. ventricle
6. right
7. sinoatrial
8. parasympathetic
9. veins
10. arteries

Chapter 31:

1. oxygen
2. carbon dioxide
3. small, moist
4. earthworms
5. bulk flow
6. book lungs
7. tracheae
8. birds
9. alveoli
10. four

Chapter 32:

1. fats
2. glycogen
3. cannot
4. fat-soluble
5. lysosome
6. gizzard
7. small
8. gallbladder
9. sodium bicarbonate
10. shorter

Chapter 33:

1. flame cell
2. body cells
3. veins
4. ureter
5. medulla
6. tubules
7. active transport
8. proximal
9. increases
10. more

Chapter 34:

1. external
2. inflammatory
3. increase
4. marrow
5. proteins
6. memory
7. B
8. second
9. histamines
10. T

Chapter 35:

1. do not have
2. target
3. exocrine
4. are produced by
5. steroid
6. negative
7. increase
8. posterior
9. anterior
10. glucagon
11. decreases
12. sympathetic
13. away from
14. male
15. counteracts

Chapter 36:

1. axons
2. dendrites
3. undiminished
4. negative
5. sodium
6. independent of
7. axons
8. faster
9. postsynaptic
10. bilaterally
11. cerebellum
12. corpus callosum
13. right
14. electrical
15. morphological

Chapter 37:

1. receptors
2. transducers
3. infrared radiation
4. sound
5. inner
6. chemical
7. rods
8. rods
9. chemical
10. cones

Chapter 38:

1. actin
2. acetylcholine
3. cardiac
4. smooth
5. retracted
6. sarcomere
7. more than one
8. cartilage
9. osteoblast
10. ligaments

Chapter 39:

1. sexual
2. variability
3. males
4. diploid
5. dioecious
6. sperm
7. spermatogenesis
8. secondary
9. mitochondria
10. primary
11. oviduct
12. oviducts
13. estrogen
14. clitoris
15. chorionic
    gonadotropin

Chapter 40:

1. differentiation
2. egg
3. indirect
4. indirect
5. blastula
6. little
7. amnion
8. allantois
9. placenta
10. reptilian

Chapter 41:

1. all
2. heat
3. learned
4. taxis
5. kineses
6. is not
7. releaser
8. simpler
9. innate
10. imprinting

Chapter 42:

1. intraspecific
2. active
3. can
4. the same
5. symbolic
6. males
7. the same
8. harmony
9. solitary
10. more
11. submissive
12. innate
13. instinctive
14. synchronized
15. longer

****

UNIT VI: ECOLOGY

CHAPTER 43:   POPULATION GROWTH AND REGULATION

CHAPTER OUTLINE.

This chapter considers the variables that control population size, including birth and death rates, biotic potential, and environmental resistance. Spatial distribution patterns for individuals, and survivalship curves are discussed as are the growth characteristics of human populations in the world and in the U.S.

1. **Introduction to Ecology.**

   A. **Ecosystem**: a complex, interrelated network of living organisms and their non-living surroundings.
      1) Sizes vary from puddles to oceans.
      2) All the various organisms in an ecosystem constitute a <u>community</u>.
      3) A community consists of many <u>populations</u>, each consisting of all the members of a single species.

   B. **Ecology**: the science dealing with the interrelationships among populations of living things and their environment. The environment is divided into two components:
      1) <u>Abiotic</u> <u>portion</u>: the non-living components of soil, water, and weather in an ecosystem.
      2) <u>Biotic</u> <u>portion</u>: all life forms within an ecosystem.

2. **Population growth.**

A. In undisturbed ecosystems, populations tend to remain relatively stable in size over time. Three factors determine how much the size of a population changes: births, deaths, and migration.
   1) Organisms enter a population through <u>birth</u> or <u>immigration</u>.
   2) Organisms leave a population through <u>death</u> or <u>emigration</u>.
   3) A population remains stable if about as many individuals enter as leave.
      a. Population growth occurs when births + immigration > deaths + emigration.
      b. Populations decline when the reverse occurs.
      c. Thus, <u>change in population size</u> = (births - deaths) + (immigrants - emigrants).
         Expressed in "per thousand individuals/year".
      d. In many populations, changes occur mainly through birth and death rates, since few individuals enter or leave relative to births and deaths that occur.

B. Ignoring migration, the <u>ultimate size</u> of a population is a result of a balance between 2 major opposing forces:
   1) <u>Biotic potential</u>: maximum rate at which the population can increase, assuming ideal conditions allowing a maximum birth rate and minimum death rate.
   2) <u>Environmental resistance</u>: limits set by the biotic and abiotic environments present (how much food and space, competition, predation, and parasitism). These factors usually decrease the birth rate and increase the death rate.
   3) Interactions between environmental resistance and biotic potential result in a balance between population size and available resources.

C. <u>Biotic potential</u>: the capacity of organisms to replace themselves manyfold by reproduction, causing exponential growth of the population.
   1) Ignoring migration, the <u>rate of growth</u> ($\underline{r}$) of a population is:
      $$r = b - d, \text{ where}$$
      b = birth rate (number/unit time/ave. population)
      d = death rate (number/unit time/ave. population)
   2) If r = +, exponential growth will occur since r is multiplied by the original population size (N).
      a. Example:
         N = 10,000 individuals
         Each year, 1500 are born but 500 die.
         r = 0.15 - 0.05 = 0.10 (the population increases by 10% each generation).
         So, rN = (0.10) (10,000) = 1000 new individuals are added this year and the population becomes 11,000.
         If r remains at 0.10, next year rN = 1100 new individuals are added and the population becomes 12,100.
         So, each year N increases and the population increases faster.
      b. Since population size grows at an ever-accelerating pace, the increase is exponential. A graph of this type of growth pattern produces a J-shaped line.
      c. This will occur if, on the average, each individual produces > 1 surviving offspring during its lifetime.
   3) Several factors influence biotic potential:

a.   The age at which the organisms first reproduce: delayed childbearing significantly slows population growth.

b.   The frequency at which reproduction occurs.

c.   The average number of offspring produced each time.

d.   The length of the reproductive lifespan of the organisms.

e.   The death rate of individuals under ideal conditions: higher death rates significantly slow population growth.

4)   In nature, exponential growth curves are observed only under special conditions (abundant food, space, and rainfall) and only for limited periods of time, causing "boom and bust" cycles in short-lived organisms: rapid population growth followed by massive die-offs.

5)   In longer-lived species, populations tend to become relatively stable in size with minor fluctuations in response to environmental variables like weather and food availability, although exponential growth may occur under special circumstances (e.g., elimination of predators or invading a new favorable habitat).

6)   All exponential growth curves must eventually either flatten or crash.

D.   <u>Environmental resistance</u>: limits to growth.

1)   Exponential growth carries the seeds of its own destruction.

a.   As the population increases:

     i.   Competition for resources intensifies.

     ii.   Predators may increase in number.

     iii.   Parasites and disease spread more readily.

b.   Population size eventually stabilizes at or below the maximum size which the environment can sustain, and the rate of growth drops to about 0.

c.   This type of population growth, typical of long-lived species colonizing a new area, is represented graphically by a "sigmoid" or S-curve.

2)   <u>Carrying capacity</u>: the maximum number of organisms that an area can support on a sustained basis.

a.   It is determined primarily by 2 types of resources:

     i.   <u>Nonrenewable</u> <u>resources</u>: space.

     ii.   <u>Renewable</u> <u>resources</u>: nutrients, water, and light energy. Excess demands may damage the ecosystem (e.g., over-grazing on dry western grasslands allows sagebrush to flourish).

b.   In nature, populations are maintained at or <u>below</u> the carrying capacity of an area due to environmental resistance factors of 2 types: density-independent and density-dependent.

3)   <u>Density-independent limits</u> to growth.

a.   Limits population size and growth regardless of the density of the population.

b.   Examples:

     i.   Weather factors such as the time of the first hard freeze.

     ii.   Man-made factors such as use of pesticides, pollution, and habitat destruction for housing developments and farms.

c.   Natural limits can be overcome by developing insulating fur, hibernating, and migrating long distances, or by plants becoming dormant in the winter.

4)   <u>Density-dependent limits</u> to growth.

a.   Become increasingly effective as population density increases, thus exerting a negative feedback effect on population size.

b.   Examples:
   - i.   Community interactions such as <u>predation</u> and <u>parasitism</u>.
   - ii.   <u>Competition</u> within species and between different species.

c.   <u>Predation and prey populations</u>.
   - i.   <u>Predators</u> kill and eat other organisms (the "prey").
   - ii.   Predation controls prey populations in a density-dependent manner (more effective in denser prey populations).
   - iii.   Many predators eat a variety of prey species, depending on relative abundance and ease of capture.
   - iv.   Predators may increase in number as prey population grows. Consequently:
     - a)   Some predators regulate number of offspring they produce according to abundance of prey.
     - b)   Sometimes, increase in predators causes a crash of the prey population followed by decline in predator numbers. Result: out-of-phase population cycles for both predators and prey.
   - v.   Influence of predators on prey populations varies:
     - a)   If predators feed primarily on prey weakened because of over-population, predation will maintain prey populations near their optimal population densities.
     - b)   Some predators maintain their prey at well below carrying capacity (i.e., cactus moth predator and prickly pear prey).

d.   <u>Parasitism and host populations</u>.
   - i.   <u>Parasites</u> live on or in their prey (called "hosts"), weakening but not killing them outright.
   - ii.   Parasites may indirectly increase death rates of hosts weakened by overcrowding.
   - iii.   Infestations by parasites makes hosts more vulnerable to predators.
   - iv.   Parasitism is density-dependent since less mobile parasites spread more readily between hosts forced closer together by high density.

e.   Parasites and predators destroy the least fit of the prey, leaving the better-adapted prey to reproduce.
   - i.   But when a predator or parasite enters an area in which it did not evolve, local prey species are not adapted to it and are usually completely decimated.
   - ii.   European conquest of the world was partly due to the disease organisms they carried with them into new areas where the natives had no resistance to the pathogens (smallpox ravaged natives of Hawaii, Argentina, and Australia).

f.   Both <u>intraspecific</u> (within the same species) and <u>interspecific</u> (between different species) competition for finite resources limits population size. Organisms have evolved several ways of dealing with intraspecific competition:
   - i.   Scramble <u>competition</u> in insect and plant species: a free-for-all with resources as the prize. Example: random seed dispersal throughout an area.

ii. <u>Contest competition</u> in most animals and a few plants: social or chemical interactions used to limit access to important resources. Examples: territoriality and dominance hierarchies.

g. <u>Overcrowding</u> may cause:

i. <u>Emigration</u> out of the area, but most animals die as a result since the new areas are often not hospitable to colonization.

ii. <u>Social stress</u> in small mammals, ultimately reducing reproduction rates and resistance to disease in lab studies.

3. **Patterns in populations**.

A. **Distribution of populations**: 3 major types of spatial distributions exist.

1) <u>Aggregated distributions</u> (the most common pattern).

a. Members of the population live in family or social groups (herds, packs, prides, flocks, schools, etc.).

b. Advantages:

i. Many eyes on the lookout for food or predators.

ii. To confuse predators by darting off in all directions.

iii. Aids in mating.

c. Some cluster because resources are localized (e.g., water in the dry savanna of Africa).

2) <u>Uniform distributions</u>.

a. Members maintain a relatively constant distance between individuals, often within distinct "territories".

b. Caused by need to defend scarce resources such as breeding sites in shorebirds or water supplies in creosote bushes.

3) <u>Random distributions</u> (the least common pattern).

a. Members form no social groups and resources are plentiful throughout the area.

b. No vertebrate species maintains random distribution throughout the year due to the need to breed.

B. **Survivalship in populations**.

1) <u>Survivalship curves</u>: patterns over time of population size caused by characteristic occurrences of deaths or survivalship. Three types exist.

2) <u>Convex survivalship curves</u>: low infant mortality with most individuals surviving to old age.

a. Seen in humans and many other large animals.

b. Few offspring are produced and these are protected by the parents.

3) <u>Constant survivalship curves</u>.

a. Individuals have an equal chance of dying at any time during their lifespans.

b. Seen in the American robin, the gull, and lab populations of asexually reproducing organisms (hydras and bacteria).

4) <u>Concave survivalship curves</u>.

a. Large numbers of offspring are produced and these are left to survive on their own.

b. Mortality is very high among the offspring, but those that reach adulthood usually survive to old age.

   c.   Seen in most invertebrates, most plants, and many fish.   In some
        populations of blacktailed deer, 75% of the population dies within the first
        10% of its lifespan.

4.   **The human population.**

   A.   **Human population growth.**
        1)   The human growth curve is exponential.
             a.   We have <u>overcome</u> environmental resistance (rather than reaching a
                  balance with it) by a series of revolutions:
                  i.   <u>Cultural</u> <u>revolution</u>: fire → tools and weapons → built shelters →
                        protective clothing → increased range of habitable areas.
                  ii.   <u>Agricultural</u> <u>revolution</u>, beginning about 8000 B.C.:   raising
                        domesticated crops and animals → increased dependable food
                        supply → increased carrying capacity → increased longevity and
                        reproductive span.
                  iii. <u>Industrial-medical</u> <u>revolution</u>, beginning in the mid-1700s in
                        England: medical advances → decreased death rates and
                        increased life span due to improved sanitation, use of antibiotics
                        and vaccines, and more sophisticated medical procedures.
             b.   Birth rates have declined in some developed countries due to:
                  i.   Better education.
                  ii.   Increased use of contraceptives.
                  iii. Shift to an urban life style.
                  iv.   More career options for women.
             c.   Birth rates have not declined in less developed countries.
                  i.   Children serve:
                        a)   As a form of "social security" to support parents in their
                             old age.
                        b)   As a source of labor in agricultural societies.
                        c)   As a source of pride and prestige to impoverished
                             parents.
                  ii.   Lack of both education and contraceptive impedes progress in
                        curbing population growth.
                  iii. So, birth rates are high while death rates have been reduced by
                        medical advances. Exponential growth continues to occur here.
                        Nearly 80% of all humans will live in these poor countries by
                        2000 A.D.

   B.   **Age structure and population growth.**
        1)   When the number of children (ages 0-14) exceeds the number of reproducing
             individuals (ages 15-45):
             a.   The population is growing.
             b.   The <u>age structure diagram resembles a pyramid</u>.
             c.   Seen in many less-developed countries.
             d.   Ironically, population growth in these countries is helping to perpetuate
                  the poverty and ignorance that in turn tend to sustain high birth rates.
        2)   Stable populations achieve replacement level fertility.
             a.   The number of children equals the number of reproducing adults.

b.   The <u>age structure diagram resembles a rocket or obelisk</u>.

c.   Seen in most of the developed countries.

3)   In shrinking populations:

a.   There are fewer children than reproducing adults.

b.   The <u>age structure diagram is constricted at the base</u>.

4)   <u>Age</u> <u>structure</u> <u>diagrams</u> for populations that are (a) growing, (b) stable, and (c) shrinking. The right side of each diagram represents females (who tend to live longer than males) and the left side males. The lower portion of the diagrams represent children, the middle regions are reproducing adults, and the upper portions are postreproductive individuals.

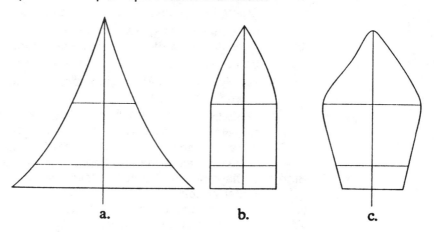

a.   b.   c.

C.   **Population growth in the U.S.**

1)   The U.S. population is growing exponentially, at a growth rate of 1% annually.

2)   <u>Replacement</u> <u>level</u> <u>fertility</u> (RLF) in the U.S. is 2.1 children for every woman. However, actual RLF is 2.0. Why does the population continue to grow?

a.   Immigration (about 500,000 legally per year, and more illegally): accounts for 30-40% of our annual growth rate.

b.   The past "baby boom": parents of the late 1940s-late 1960s had larger than replacement families. Due to this, in 1980, over a million more women entered their reproductive years than did in 1970.

3)   Even if fertility drops immediately to about 1.6 births per woman, the population will continue to grow for the next 50 years before beginning a gradual decline.

D.   **World population and carrying capacity.**

1)   World population may stabilize in 2110 A.D. at 10.5 billion people, twice the current number. Will this exceed the earth's carrying capacity?

2)   <u>At the present time</u>, we are depleting the world's ecosystems. Evidence:

a.   Each year, an area of once-productive land the size of Maine becomes desert due to overgrazing and deforestation. There will be 20% more desert area by the year 2000 A.D.

b.   20% of the world's farmland is being depleted through erosion, mismanagement, and urban sprawl.

c.   Forested areas are being reduced by an area the size of Cuba each year.

d.   The fish harvest per capita has fallen by 13% since 1970.

e.   Destruction of tropical rain forests may cause extinction of 1,000,000 species of plants and animals by 2000 A.D. due to loss of their habitats.

3) Mass starvations, like the 1985 and 1987 famines in Ethiopia, occur because resources are unequally distributed among and within countries. When human populations exceed the carrying capacity of the soil where they live, and lack the wealth to import and distribute adequate food, disaster strikes.

4) Merely redistributing the earth's wealth is not an adequate answer, and technological advances will not continue to increase the earth's carrying capacity into the indefinite future.

5) The human population will stop its exponential growth, either through:
   a. Voluntary <u>reduction</u> <u>of</u> <u>the</u> <u>birth</u> <u>rate</u> worldwide, or
   b. Various forces of environmental resistance that will <u>increase</u> <u>the</u> <u>death</u> <u>rate</u>.
   c. The choice is ours!

****

REVIEW QUESTIONS:

1. **Define** the following:
   A. Ecology.
   B. Ecosystem.
2. What is the formula for change in population size?
3. Define the two opposing forces that determine the ultimate size of a population.
4. In a population with 10,000 individuals, 3000 are born but 2000 die each year.
   A. What is the new population size at the end of the first year?
   B. At the end of the second year?
   C. What kind of change is occurring in population size over time?
5. **Define** the following terms:
   A. Biotic potential.
   B. Environmental resistance.
6. Explain why long-lived species colonizing a new area usually show optimum growth curves that are shaped like the letter S.
7. Give three examples each of density-independent and density-dependent limits to population growth.

8.-15. **MATCHING TEST**: spatial distribution patterns.

Choices:
A. Uniform distribution
B. Aggregated distribution
C. Random distribution

8. Members form no social groups and resources are plentiful.
9. Distinct territories set up.
10. Members live in social groups such as herds, flocks, etc.
11. Trees in tropical rain forests.
12. Response to localized resources.
13. Members maintain a relatively constant distance between each other.
14. Not seen in vertebrates due to the need to breed.

15.   Response to defense of scarce resources.

16.-25.   **MATCHING TEST**: patterns of survivalship through time.

              Choices: A.   Constant survivalship curve
                              B.   Concave survivalship curve
                              C.   Convex survivalship curve

16.   Most individuals die quite young.
17.   Few offspring produced and these are protected by the parents.
18.   Individuals have an equal chance of dying at any age.
19.   Most offspring survive until old age.
20.   Large numbers of offspring produced and these are not protected by the parents.
21.   Seen in humans.
22.   Seen in most invertebrates.
23.   Seen in bacteria.
24.   75% of the population dies within the first 10% of the lifespan.
25.   25% of the population dies within the first 90% of the lifespan.

----

26.   Describe how the cultural, agricultural, and industrial-medical revolutions have contributed to
         world population growth of humans.
27.   What factors contribute to a decline in birth rates in more-developed countries and to higher birth
         rates in less-developed countries?
28.   What happens to human populations when the number of children (ages 0-14) exceeds the
         number of reproducing adults (ages 15-45)?
29.   Describe population growth in the U.S.
30.   When the human population stabilizes (at 10.5 billion in 2110 A.D.), do you think it will exceed the
         earth's carrying capacity? Defend your answer.

****

****

ANSWERS TO REVIEW QUESTIONS:

1.   see 1.A.-B.

2.   see 2.A.3)a.-d.

3.   see 2.B.1)-3)

4.   [see 2.C.]

Rate of growth (rN) =
birth rate(r)-death rate (d)
  r = 0.30 - 0.20 = 0.10

A.   After the first year,
        rN = (0.10) (10,000) =
        1000 new individuals.
        So, population = 11,000
        after the first year.

B.   After the second year,
        rN = (0.10) (11,000) =
        1100 new individuals.
        So, population = 12,100
        after the second year.

C.   The population increases
        exponentially (faster
        each year).

5.   see 2.B.1)-2)

6.   see 2.D.1)a.-c.

7.   see 2.D.3)-4)

8.-15. [see 3.A.1)-3)]

8.	C	12.	B
9.	A	13.	A
10.	B	14.	C
11.	C	15.	A

16.-25. [see 3.B.1)-4)]

16.	B	21.	C
17.	C	22.	B
18.	A	23.	A
19.	C	24.	B
20.	B	25.	C

26.   see 4.A.1)a.i.-iii.

27.   see 4.A.1)b.-c.

28.   see 4.B.1)a.-d.

29.   see 4.C.1)-3)

30.   see 4.D.1)-5)

****

CHAPTER 44: COMMUNITY INTERACTIONS

CHAPTER OUTLINE.

This chapter deals with community interactions of three types: competition for ecological niches, predation between species (and coevolutionary changes in both predator and prey), and symbiotic relationships (parasitism, commensalism, and mutualism). Also, the basic principles of community succession are discussed.

1. **Introduction.**

   A. An ecological <u>community</u> consists of all the interacting populations within an ecosystem.
      1) Usually, populations within a community have coevolved.
      2) <u>Coevolution</u>: different species acting as agents of natural selection on each other. Examples:
         a. Predators limit their prey populations without eliminating them.
         b. Animals that are preyed upon have evolved elaborate defenses that help them survive.

   B. Three categories of community interactions.
      1) <u>Competition</u>: populations compete for limited resources.
      2) <u>Predation</u>: one organism kills and eats another.
      3) <u>Symbiosis</u>: 2 species live together in close association over an extended time.

   C. <u>Succession</u>: periods of gradual change as communities replace each other. Long-established communities eventually reach a complex and self-sustaining state with interacting populations remaining relatively stable.

2. **The ecological niche.**

A. Ecological <u>niche</u>: all aspects of a species' way of life, including:
1) Its physical home or habitat in an ecosystem.
2) Its occupation or role, including its food, behavior, and predators.
3) All its physical environmental factors necessary for survival.
4) Its role in the community.

B. Each species occupies a unique ecological niche, although some aspects of the niche may be shared with other species.

3. **Competition between species** (Interspecific competition).

A. Occurs when two or more species attempt to use the same limited resources.
1) The intensity of competition depends on how similar the requirements of the two species are.
2) The degree of competition is proportional to the amount of niche overlap between competing species.

B. <u>Competitive exclusion principle</u> of Gause (1934): Two different species cannot inhabit the same ecological niche.
1) If two species with similar niches compete for limited resources, one will outcompete the other and the less well adapted will die out.
2) R. MacArthur, investigating five species of North American warblers that hunt for insects and nest in the same type of spruce tree, found that each species hunted in different areas of the tree, hunted differently, and nested at different times.
   a. By petitioning the resource, the warblers minimized niche overlap and reduced competition.
   b. This shows that when species with similar requirements coexist, they often occupy smaller niches than if no competition was present, a phenomenon called <u>resource partitioning</u>.
   c. So, resource partitioning results from coevolution of species with extensive niche overlap.
      i. <u>Coevolution</u> occurs when several species practice natural selection on each other.
      ii. Natural selection favors individuals with fewer competitors.
      iii. Over time, competing species develop physical and behavioral adaptations that minimize competitive interactions.

C. Interspecific Competition and Population Size.
1) Closely related species compete directly for limited resources.
2) This may restrict the size and distribution of competing populations. Examples: barnacles.

4. **Predators and their Prey.**

A. <u>Predators</u> are organisms that kill and eat other organisms.
1) Examples:
   a. Lions eat zebras.
   b. Antelopes eat sagebrush, killing the part of the plants that they eat.
   c. Sundew plants "eat" insects they trap.

2) Predators are either larger than their prey or hunt collectively.
3) Predators are less abundant than their prey.

B. <u>Coevolution</u> of predators and prey.

1) As prey become more difficult to catch, predators must become more adept at hunting, and vice versa.

2) Coevolution has endowed the cheetah with speed and camouflage spots, and its zebra prey with speed and camouflage stripes. Likewise, hawks evolved keen eyesight while prairie dogs evolved warning calls.

3) Coevolution of <u>herbivores</u> and their <u>plant prey</u>.
   a. Plants have evolved a variety of adaptations to deter predators. Examples:
      i. Milkweed plants evolved the ability to make toxic and distasteful chemicals.
      ii. Animals either learn not to eat foods that make them sick, or evolve immunity: efficient ways to detoxify or store the harmful plant chemicals.
   b. <u>Result</u>: nearly every toxic plant species is eaten by at least one species of insect. Examples:
      i. Monarch butterfly caterpillars eat toxic milkweed plants, tolerating their poison and storing it in their tissues as a defense against bird predators that are sickened by the toxin.
      ii. As grasses evolved tough silicone substances in their blades to reduce predation, horses evolved long teeth with thick enamel that resist wear.

4) Bats and moths have coevolved:
   a. Bats hunt at night and locate prey by echolocation (emitting high frequency and high intensity sounds pulses , locating nearby objects by analyzing the returning echoes).
   b. Some moths evolved simple ears sensitive to bat echolocation frequencies. When moth hears a bat, it flies erratically or drops to the ground.
   c. Bats countered this by switching frequencies of sonar away from the moth's sensitive range.
   d. Moths countered by jamming bat's sonar by producing ultrasonic clicks.
   e. Bats countered this by turning off their sonar and following the moths' clicks.

5) <u>Protective coloration</u>.
   a. The best hiding place is in plain sight. Some predators and prey have evolved colors and patterns that make them inconspicuous even in plain sight.
   b. <u>Camouflaged</u> animals:
      i. May resemble their general surroundings (e.g., tree bark).
      ii. May resemble specific but uninteresting objects like leaves, twigs, or bird droppings.
      iii. Prey behavior is important (must remain motionless rather than fleeing).
   c. <u>Warning coloration</u>: distasteful or poisonous animals are brightly colored and conspicuous, warning predators that they are harmful.

              i.   Predators usually learn after one bad experience.

              ii.   A common warning pattern results in easier learning by predators, protecting all the species showing the common pattern (e.g., stripes on bees, hornets, and yellowjackets).

       d.   <u>Mimicry</u>: tasty animals evolve resemblance to distasteful species.

              i.   Example: tasty viceroy butterflies resemble poisonous monarchs.

              ii.   With mimicry, one tasty animal species (the <u>mimic</u>) evolves to resemble a poisonous or distasteful species (the <u>model</u>).

              iii.   A few predators have evolved <u>aggressive</u> <u>mimicry</u>, resembling a harmless animal to fool their prey.

       e.   <u>Startle coloration</u> is seen in some prey species.

              i.   Color patterns resembling "eyespots" on the wings of some moths are revealed to startle a predator about to attack.

              ii.   The moth then quickly covers the "eyespots" and flees before the predator recovers.

    6)   <u>Chemical Warfare</u>. Both predators and prey evolve toxic chemicals for attack or defense.

       a.   Many plants produce defensive toxins.

       b.   Spider and snake venom paralyze prey and deter predators.

       c.   Some molluscs (squid and octopi) and sea slugs emit clouds of "ink" when attacked and this "smoke screen" confuses predators while they escape.

       d.   The bombardier beetle, when bitten by an ant, releases abdominal secretions: enzymes catalyze an explosive chemical reaction that shoots a toxic, boiling hot spray onto the attacking ant.

5.   **Symbiosis** ("living together").

  A.   <u>Definition</u>: close interactions of organisms of different species for an extended time. Although one species always benefits, the second species may be:

    1)   Harmed (<u>parasitism</u>).

    2)   Not affected (<u>commensalism</u>).

    3)   Benefitted (<u>mutualism</u>).

  B.   **Parasitism.**

    1)   Parasites eat others in subtle ways and may not be immediately fatal.

    2)   Parasites usually are smaller and more numerous than their prey (called <u>hosts</u>).

    3)   Parasites live on or in their hosts.

    4)   Examples: tapeworms, fleas, disease microbes.

    5)   <u>Trypanosoma</u>, a parasitic protozoan causes sickness or death:

       a.   African antelope, which coevolved with this parasite, are relatively unaffected by it.

       b.   Cattle bred in infested areas for generations suffer but are not killed.

       c.   Newly imported cattle generally die due to infection if untreated.

       d.   Thus, powerful forces of coevolution occur between parasitic microorganisms and their hosts.

  C.   **Commensalism.** Examples:

    1)   Large, herd-forming mammals (bison) and certain birds that eat insects disturbed

403

by the bison. Bison are not affected but the birds benefit.

2) Birds build nests in trees obtaining shelter and protection, but the trees are not affected.

3) Other examples: orchids attach to trees without harming them, and barnacles are harmless hitchhikers on marine animals and benefit from a free ride through food-rich waters.

D. **Mutualism**: both species benefit. Examples:
1) Flowering plants and their pollinators.
2) Protists and bacteria live in the digestive tracts of cows and termites, finding food and shelter while producing certain vitamins for the host.
3) Nitrogen-fixing bacteria live in the roots of legume plants, obtaining food and shelter while converting nitrogen gas into a form plants can use.
4) Ants and acacia plants.
5) Is rare in vertebrates.

6. **Succession**: a change in a community and its non-living environment over time.

A. Succession is a kind of <u>community relay</u> in which assemblages of plants and animals replace one another in a sequence that is somewhat predictable. Examples:
1) Freshwater lakes → marshes → dry land.
2) Shifting sand dunes → stabilized by creeping plants → forest.

B. <u>Basic principles of succession</u>.
1) Succession begins by a few hardy invader species called <u>pioneers</u> and ends with a diverse and stable <u>climax community</u>.
2) As succession proceeds, the organisms gradually alter the non-living environment to favor competitor species that displace the existing populations.
3) The climax community no longer alters the environment, and thus is not replaced. It persists unless external forces (e.g., human activities) alter it.

C. General <u>trends in ecosystem structure</u> during succession:
1) <u>Soil</u> increases in depth and organic content.
2) Overall <u>productivity</u> (amount of organic material produced per time) increases.
3) <u>Number</u> of different species increases, as do interspecific interactions.
4) <u>Longer-lived</u> species come to dominate, and the rate at which populations replace each other slows down. The communities become more stable and resistant to change.
5) In the <u>climax community</u>, total weight of the organisms reaches a maximum, and the ecosystem is no longer altered by the species there.

D. **Forms of succession**.
1) <u>Primary</u>: begins with no trace of a previous community (bare rock, sand, or a clear glacial pool). It is the formation of an ecosystem "from scratch" and takes a long time (thousands of years).
2) <u>Secondary</u>: Occurs after an existing ecosystem is disturbed (forest fire, abandoned farm field) and proceeds rapidly (hundreds of years).

E. <u>Primary succession</u> in a temperate climate. Sequence of plant species: bare rock

→ pioneer lichens (algae and fungi) → mosses → woody shrubs (juniper) → trees (pine and spruce) → climax shade tolerant trees (firs, birch).

F.   <u>Secondary succession</u> on an abandoned farm in the southwestern U.S.: pioneer fast-growing annual weeds (crab-grass, ragweed) → perennial plants (goldenrod, grasses) and woody shrubs (blueberry) → pines and fast-growing deciduous trees (tulip, sweetgum) → climax shade-resistant, slower growing hardwoods (oak, hickory).

G.   The <u>climax community</u>: the end result of succession.
   1)   The exact nature of the climax community is determined by numerous geological and climactic variables, and the nature of human activities.
   2)   Many ecosystems persist in a <u>sub-climax</u> stage, never reaching climax due to periodic fires and human agriculture.
      a.   Grains are specialized grasses characteristic of early successional stages, and farmers spend much energy preventing competitors (weeds and shrubs) from taking over.
      b.   The suburban lawn is a sub-climax ecosystem, with mowing and herbicides keeping competitors out.

****

REVIEW QUESTIONS:

1.   **Define** the following terms:
   A.   Community.
   B.   Coevolution.
   C.   Succession.

2.   What is an "ecological niche"?
3.   What is meant by the "competitive exclusion principle" of Gause?

4.-9.   **True or False**: If the statement is false, correct it.

4.   Different species can maintain identical niches as long as they do not compete with each other.
5.   Niche overlap has little to do with competition between species.
6.   The greater the niche overlap, the less intense will be competition between the species.
7.   Coevolution assures that different species can have extremely overlapping niches without serious consequences to the species involved.
8.   Natural selection favors individuals whose niches overlap the most with the niches of other species.
9.   Closely related species will compete directly for limited resources.

----

10.   Define predation and give several characteristics of predators relative to their prey.
11.   Describe coevolutionary changes in herbivores and their plant prey, and give several examples.
12.   Describe the coevolution of bats and moths.

13.-24.   **MATCHING TEST**: protective coloration.

Choices: A.   Camouflage
B.   Warning coloration
C.   Mimicry
D.   Startle coloration

13.   Resemblance of tasty animals to distasteful animals.
14.   Resemblance to the general surroundings.
15.   Yellow and black stripes on stinging bees.
16.   Resemblance to a specific uninteresting object (like a leaf or a bird dropping).
17.   Resemblance to a poisonous "model".
18.   Bright, conspicuous colors or patterns.
19.   "Eyespot" patterns on the wings of moths.
20.   Must remain motionless to be effective.
21.   Prey momentarily frighten predators with a flash of color or pattern.
22.   Advertising distastefulness or toxicity.
23.   A predator resembles a harmless animal to fool the prey.
24.   White stripes down the backs of skunks.

---

25.   Define symbiosis.

26.-36.   **MATCHING TEST**: symbiosis.

Choices: A.   Mutualism
B.   Commensalism
C.   Parasitism

26.   Birds eating insects disturbed by buffalo.
27.   Flowering plants and their pollinators.
28.   One species is harmed by the other species.
29.   Fleas and dogs.
30.   One species is not affected by the other species.
31.   Protists in the intestines of termites.
32.   Tapeworms and their hosts.
33.   Both species benefit from their association.
34.   Ants and acacia plants.
35.   Birds building nests in trees.
36.   Nitrogen-fixing bacteria and legume plants.

---

37.   Succession begins with a few hardy species called _____ and ends with a diverse and stable
_____ _____.
38.   List the five general trends in ecosystem structures seen during succession.
39.   Name two important differences between primary and secondary succession.
40.   List the sequence of plant types expected in primary succession occurring in a temperate climate.

41. List the sequence of plant types expected in secondary succession occurring on an abandoned farm in the southwestern U.S.
42. Relate a suburban lawn to succession. Why isn't it a stable community?

****

ANSWERS TO REVIEW QUESTIONS:

1. see 1.A.-C.

2. see 2.A.1)-4)

3. see 3.B.

4.-9. [see 2.B.1)-2)]

4.	False	7.	False
5.	False	8.	False
6.	False	9.	True

10. see 4.A.1)-3)

11. see 4.B.3)a.-b.

12. see 4.B.4)a.-e.

13.-24. [see 4.B.4)a.-e.]

13.	C	19.	D
14.	A	20.	A
15.	B	21.	D
16.	A	22.	B
17.	C	23.	C
18.	B	24.	B

25. see 5.A.

26.-36. [see 5.A.-D.]

26.	B	32.	C
27.	A	33.	A
28.	C	34.	A
29.	C	35.	B
30.	B	36.	A
31.	A		

37. pioneer, climax community; [see 6.B.1)]

38. see 6.C.1)-5)

39. see 6.D.1)-2)

40. see 6.E.

41. see 6.F.

42. see 6.G.2)

****

CHAPTER 45: THE STRUCTURE AND FUNCTION OF ECOSYSTEMS

CHAPTER OUTLINE.

This chapter focuses on energy flow through ecosystems and considers trophic levels and food chains. Nutrient cycles and the ways that human interventions have disrupted them are discussed.

1. **Introduction.**

    A. Energy is:
        1) Trapped by photosynthesis.
        2) Released by cellular respiration.
        3) Used to construct the complex molecules of life.

    B. All the activity of life is provided by the energy from sunlight.
        1) Each time energy is used, some of it is lost as heat.
        2) Solar energy constantly bombards the earth and is continuously lost as heat. Energy is renewable.

    C. Energy flows through ecosystems, but nutrients cycle and recycle.
        1) Nutrients remain on earth.
        2) They may change in form and distribution but they do not leave the world ecosystem.

2. **The flow of energy.**

    A. Primary productivity.
        1) The sun fuses hydrogen into helium, releasing energy, some of which reaches

earth.

2) Only about 1% of the sun's energy reaches earth and 3% of that is trapped by green plants to power life (the rest is lost in the atmosphere, clouds, and land surface).

3) The energy that powers ecosystems enters through <u>photosynthesis</u> (solar energy combines CO_2 and H_2O into sugar and releases O_2).

    a. Photosynthetic organisms are called <u>autotrophs</u> ("self feeders") or <u>producers</u> (they produce food).

    b. Organisms that rely on food produced by autotrophs are called <u>heterotrophs</u> ("other feeders") or <u>consumers</u>.

4) The energy that producers make available for consumers is called <u>net primary productivity</u>.

    a. An ecosystem's productivity is influenced by many environmental variables such as:

        i. Amount of nutrients available.

        ii. Amount of sunlight.

        iii. Availability of water.

        iv. Temperature.

    b. The amount of life an ecosystem can support is determined by the net primary productivity.

    c. It is measured in units of energy (calories), or as dry weight of organic material/unit/ year.

    d. When essential resources are abundant (in estuaries or tropical rain forests), productivity is high.

B. <u>Trophic levels.</u>

1) Energy flows through communities from the photosynthetic producers to several categories of consumers. Each category is a <u>trophic</u> (feeding) <u>level</u>.

    a. <u>Producers</u>: The first (lowest) trophic level, deriving energy from sunlight. They are the most abundant organisms in an ecosystem.

    b. <u>Herbivores</u> (feed exclusively on producers): the second trophic level.

        i. These are the <u>primary</u> consumers.

        ii. Examples: grasshoppers, giraffes.

    c. <u>Carnivores</u> feeding on herbivores: the third trophic level.

        i. These are the <u>secondary</u> consumers.

        ii. Examples: spider, wolf.

    d. <u>Carnivores</u> feeding on other carnivores: the fourth trophic level. These are the <u>tertiary</u> consumers.

2) <u>Food chains</u>: A theoretical, linear feeding relationship in an ecosystem (producer eaten by herbivore eaten by carnivore).

3) <u>Food webs</u>: describe actual feeding relationships within a community, including the fact that many animals are <u>omnivores</u> (acting at different times as primary, secondary, and tertiary consumers). Thus, food webs are more complex than food chains.

4) <u>Detritus feeders and decomposers</u>: liberate nutrients for reuse.

    a. <u>Decomposers</u>: primarily fungi and bacteria that digest food outside their bodies.

    b. <u>Detritus feeders</u>: small animals and protists that live on the refuse of others.

        i. Include earthworms, arthropods, nematode worms, protists, and vultures.

  ii. Eat dead organic matter and excrete it in a decomposed state, reducing it to simple molecules.

  iii. Thus, once living complex substances are reduced to simple molecules needed to build more living organisms.

 c. Essential to life on Earth.

C. <u>Energy flow through trophic levels</u>.

 1) Energy use is <u>never completely efficient</u> in living organisms (some energy is lost as heat and as new tissue is made).

 2) Transfer of energy from one trophic level to the next is also very inefficient.

  a. Only some of the energy captured by the first trophic level is available to organisms in the second level since the producers use some of the energy themselves.

  b. Likewise, only some of the energy consumed by the second level's individuals is available to the third level, etc.

  c. The net energy transfer between trophic levels is roughly <u>10% efficient</u>. So, primary consumers can capture only 10% of the energy consumed by the producers. And, secondary consumers capture only 10% of the energy captured by the primary producers.

  d. In other words, for every 100 calories of solar energy captured by grass, 10 calories are converted into herbivores, and 1 calorie into carnivores. This is visualized as an <u>ecological pyramid</u>.

 3) Because of the inefficient transfer of energy through an ecosystem:

  a. The predominant organisms are plants (producers).

  b. The most abundant animals are the herbivores (primary consumers).

  c. Carnivores are always relatively rare.

  d. For humans, the lower the trophic level they occupy, the more food energy will be available. More people can be fed on grain than on meat.

 4) <u>Biological magnification</u>: an unfortunate side effect of energy transfer, in that concentrations of certain persistent toxic insecticide chemicals concentrate in the bodies of carnivores, including humans.

3. Cycling of nutrients.

A. The same pool of nutrients has been supporting life for over 3 billion years.

B. <u>Nutrients</u> are all the chemical building blocks of life.

 1) <u>Macronutrients</u> are required in large quantities (H_2O, CO_2, hydrogen, nitrogen, oxygen, phosphorus, calcium).

 2) <u>Micronutrients</u> are required in only trace amounts (zinc, molybdenum, iron, selenium, iodine).

 3) <u>Nutrient cycles</u> describe the pathways of these substances as they move from living to nonliving parts of ecosystems and back again to living tissue. The <u>reservoir</u> (major source) of important nutrients is generally in the nonliving environment.

  a. The reservoir of carbon and nitrogen is the atmosphere (atmospheric cycles).

  b. The reservoir of phosphorus is rocks (the sedimentary cycle).

C. Carbon cycle (an atmospheric cycle).
1) Major reservoirs of CO_2: atmosphere (0.033% of gases there) and oceans (dissolved in seawater).
2. Producers trap CO_2 during photosynthesis and it thus enters the living community. Producers also release some of it into the atmosphere through cellular respiration while incorporating some into their tissues.
3) Carbon in plant bodies is passed to herbivores which respire some of it and incorporate some into their tissues.
4) Predators, detritus feeders, and decomposers ultimately return most carbon to the atmosphere as CO_2.
5) Some carbon cycles move very slowly:
    a. Example: molluscs.
        i. Extract CO_2 from water and form calcium carbonate in their shells.
        ii. These shells accumulate in undersea deposits of limestone.
        iii. Limestone dissolves slowly in water, freeing the carbon again.
    b. Example: fossil fuels (remains of ancient plants and animals).
        i. The carbon in these organisms is present in coal, oil, and natural gas.
        ii. Humans release the carbon by burning these fossil fuels.

D. Nitrogen cycle (an atmospheric cycle).
1) The atmosphere is 79% N_2 gas, but plants and animals cannot use this gas directly.
2) Ammonia (NH_3) is made by certain bacteria via nitrogen fixation (combining nitrogen and hydrogen) or by decomposing amino acids and urea found in dead bodies and wastes.
3) Other bacteria convert ammonia into nitrates. Nitrates are also made by electrical storms.
4) Humans use fertilizers containing ammonia and nitrates on their farms, gardens, and lawns.
5) Plants incorporate the nitrogen into amino acids, proteins, nucleic acids, and vitamins. The plants are eventually consumed by primary consumers, detritus feeders, and decomposers.
6) Some nitrogen is liberated in wastes and dead bodies which decomposer bacteria convert back into nitrates and ammonia.
7) N_2 is continuously returned to the atmosphere by denitrifying bacteria in mud, bogs, and estuaries.

E. Phosphorus Cycle (a sedimentary cycle).
1) Phosphorus is found in biological molecules such as ATP, nucleic acids, phospholipids, and vertebrate teeth and bones.
2) Reservoir of phosphorus is crystalline rock in the form of phosphate (PO_4^{3-}).
3) Rainwater dissolves phosphate which is absorbed by producers (plants, protists, and blue-green bacteria).
4) Producers pass it through food webs, where excess phosphorus is excreted.
5) Ultimately, decomposers return phosphorus in dead bodies to the soil and water as phosphate.
6) Some phosphate in fresh water is carried to the oceans where some is absorbed by marine producers and eventually incorporated into the bodies of invertebrates

and fish.

7) Some of these are consumed by seabirds who excrete large quantities of phosphorus onto the land as guano.

8) Phosphorus-rich rock is also mined and used to make fertilizers. Soil that erodes from fertilized fields carries large quantities of phosphates into lakes, streams, and oceans, stimulating growth of producers.

F. The Water Cycle (the hydrologic cycle).
1) Differs from other cycles since the water remains unchanged.
2) Major reservoir is the ocean (97% of available water).
3) The hydrologic cycle is driven by solar energy (evaporates water) and gravity (water falls to earth as precipitation).
4) Water falling on land may:
   a. Evaporate from soil, lakes, and streams.
   b. Run off the land back to the oceans.
   c. Small amounts may enter underground reservoirs.
   d. Enter the biotic portion of ecosystems:
      i. Absorbed by plant roots and evaporated from plant leaves.
      ii. Combined with CO_2 during photosynthesis.
      iii. Heterotrophs get water from food or by drinking.
5) Distribution of life and composition of biological communities depends on and largely is determined by patterns of precipitation and evaporation.

4. Human intervention in energy flow and nutrient cycling.

A. Many environmental problems of modern society result from human interference in ecosystem functioning.
1) Primitive people were sustained solely by solar energy and produced wastes readily assimilated back into nutrient cycles.
2) Modern humans subvert these natural processes through:
   a. Mining substances such as lead, arsenic, cadmium, mercury, oil, and uranium that are foreign to natural ecosystems and harmful to life.
   b. Factories that make synthetic substances harmful to life.
   c. Reliance on energy from fossil fuels (rather than sunlight) for heat, light, transportation, industry, and agriculture.
3) This augmentation of natural energy flow has disrupted certain normal nutrient cycling.
   a. Burning of coal and oil produce large amounts of byproducts found naturally in small amounts in ecosystems.
   b. Vast quantities of these nutrients far exceed the capacity of ecosystems to process them, causing problems such as acid rain and the greenhouse effect.

B. Acid Rain: Overloading the nitrogen and sulfur cycles.
1) Each year, U.S. discharges 30 million tons of sulfur dioxide into the atmosphere, mostly from power plants burning coal or oil.
2) Human industry accounts for 90% of the sulfur dioxide in the atmosphere.
3) 25 million tons of nitrogen oxides are released by the U.S. each year. Of this total:

a.   40% comes from transportation sources.

b.   30% comes from power plants.

c.   Almost 30% comes from industry.

4)   Excess of these compounds causes <u>acid rain</u>:

    a.   Combined with water vapor in the atmosphere nitrogen oxides become nitric acid and sulfur dioxide becomes sulfuric acid.

    b.   Days later and many miles away, the corrosive acid falls either dissolved in rain or as dry microscopic particles ("<u>acid deposition</u>").

    c.   The acids damage trees and crops, renders lakes lifeless, and damages buildings and statues.

5)   Sources of sulfur dioxide are concentrated in the upper Ohio valley, Indiana, and Illinois, where old power plants burn high-sulfur coal with few emission controls.

    a.   Winds carry the acids to New England where rocks do <u>not</u> contain calcium carbonate that can buffer and neutralize acidity in the atmosphere.

    b.   Thus, lakes and forests in New England are quite vulnerable.

    c.   About 75% of the U.S. is vulnerable to acid rain due to lack of buffering capacity in the rocks and soil.

6)   <u>Damage to aquatic ecosystems</u>.

    a.   Acid rain destroys much of the food web sustaining fish, which then die.

    b.   Invertebrates die first, then amphibians, and finally fish.

    c.   Loss of insect larvae and crustaceans contributes to drastic declines in black duck populations in the mountains of New York state.

7)   <u>Damage to forests and farms</u>.

    a.   Acid rain interferes with growth and yield of many farm crops:

        i.   It leaches out essential minerals (calcium and potassium).

        ii.   It kills decomposer microorganisms, preventing return of nutrients to the soil.

        iii.   Plants become weak and more susceptible to infection and insect attack.

    b.   Effects have been seen in the Mountains of Vermont, Mount Mitchell in North Carolina, and the Black Forest of Germany.

8)   <u>Increased exposure to toxins</u>.

    a.   Acid deposition increases exposure to toxic metals (aluminum, lead, nickel, mercury, cadmium, and copper).

    b.   These are relatively inert in natural ecosystems, but are far more soluble in acidified water.

    c.   Drinking water in some households has become dangerously contaminated from lead dissolved by acid water from lead pipes.

    d.   Fish in acidified water often have dangerous levels of mercury in their bodies.

9)   <u>Damage to structures</u>. Acid deposition is corroding bridges, eating away at buildings and monuments, and destroying ancient works of art.

C.   <u>The Greenhouse Effect</u>: Short-circuiting the carbon cycle.

    1)   Between 345-280 million years ago (Carboniferous period), bodies of plants and animals were buried in sediments, escaping decomposition.

        a.   Heat and pressure converted their bodies to fossil fuels (coal, oil, natural gas).

        b.   Humans now burn them as fuels, releasing CO_2 into the atmosphere at a rapid pace.

2) Another source of additional CO_2 is global deforestation (cutting millions of acres of forests each year).

    a. Deforestation is occurring principally in the tropics.

    b. The carbon stored in these trees returns to the atmosphere as they decompose or are burned.

3) The CO_2 content of the atmosphere (0.035%) has increased by 25% since 1850, and probably will double within the next century.

4) CO_2 in the earth's atmosphere traps heat.

    a. Atmospheric CO_2 acts like the glass in a greenhouse, allowing sunlight to enter but holding the energy in once it is converted to heat.

    b. "Greenhouse gases" besides CO_2 are methane, chloroflourocarbons, and nitrous oxide.

    c. This greenhouse effect could cause a rise of about 4° C by 2050.

5) Consequences of global warming.

    a. As the icecap and glaciers melt, sea levels will rise, flooding coastal wetlands that are the breeding grounds for many animals species whose populations will be reduced.

    b. There will be a shift in global distribution of temperature and rainfall.

        i. Small temperature changes will alter the paths of major air and water currents.

        ii. Agricultural disruptions could be disastrous for human populations.

    c. The impact of warming on forests could be catastrophic, altering their extent and species compositions.

        i. Although temperatures further north would become hospitable for displaced tree species, the ability of forests to move northward is limited by slow seed dispersal, patterns of rainfall, and composition of the soil.

        ii. Some tree species may die out entirely.

        iii. The great forests of Mississippi and Georgia might be replaced by grasslands.

D. What can we do?

1) Shift to energy-efficient technologies.

2) Reverse forest destruction.

3) Implement a tree-planting campaign.

4) Increase fuel prices to encourage fuel efficiency of cars, reduce driving, and develop/use efficient mass transit.

5) Develop other sources of electricity.

6) Purchase efficient appliances.

7) Reduce use of electricity for heating and cooling by increased insulation and weatherproofing of homes.

8) Recycling of glass, aluminum, and paper.

9) Stabilize human population sizes.

****

REVIEW QUESTIONS:

1. **Explain** the following statements:
    A.   Energy is renewable but nutrients are non-renewable.
    B.   Energy flows through ecosystems but nutrients cycle and recycle.

2.-4.   **Define** the following terms related to trapping and using solar energy:

2.   Producers.
3.   Consumers.
4.   Net primary productivity.

5.-19.   **MATCHING TEST**: trophic levels.

	Choices:	A.	Herbivores
		B.	Carnivores eating carnivores
		C.	Producers
		D.	Carnivores eating herbivores
		E.	All of the above

5.   Rarest organisms in an ecosystems.
6.   Third trophic level.
7.   Feed exclusively on producers.
8.   First trophic level.
9.   Secondary consumers.
10.   Derive energy directly from sunlight.
11.   Fourth trophic level.
12.   Primary consumers.
13.   Part of a food chain.
14.   Most abundant organisms in an ecosystem.
15.   Second trophic level.
16.   Tertiary consumers.
17.   Use energy inefficiently.
18.   Wolves.
19.   Giraffes.

---

20.   Why is energy transfer from one trophic level to the next very inefficient?
21.   Why is it true that more people can be fed on grain than can be fed on meat?
22.   What is "biological magnification"?
23.   Why are food webs more complex than food chains?
24.   What are detritus feeders and decomposers, and why are they important?
25.   Define "nutrients" and distinguish between macronutrients and micronutrients.
26.   Describe the carbon cycle.

27. Describe the nitrogen cycle.
28. Describe the phosphorous cycle.
29. Describe the water cycle.
30. Name two ways that augmenting sunlight energy with fossil fuel energy has disrupted natural ecosystems.
31. What causes "acid rain" and what problems does it cause in ecosystems around the world?
32. What is the "greenhouse effect" and what causes it? What effects will it have on the earth's environment?
33. Name six ways that humans can eliminate the causes of acid rain and the greenhouse effect.

****

ANSWERS TO REVIEW QUESTIONS:

1.	see 1.A.-C.				24.	see 2.B.4)a.-c.
2.-4.	[see 2.A.1)-4)]				25.	see 3.B.1)-2)
5.-19.	[see 2.B.1)-5)]				26.	see 3.C.1)-5)
	5.	B	13.	E		
	6.	D	14.	C	27.	see 3.D.1)-7)
	7.	A	15.	A		
	8.	C	16.	B	28.	see 3.E.1)-8)
	9.	D	17.	E		
	10.	C	18.	D	29.	see 3.F.1)-5)
	11.	B	19.	A		
	12.	A			30.	see 4.A.3)a.-b.
20.	see 2.C.1)-2)				31.	see 4.B.1)-9)
21.	see 2.C.3)a.-d.				32.	see 4.C.1)-5)
22.	see 2.C.4)				33.	see 4.D.1)-9)
23.	see 2.D.					

****

CHAPTER 46: THE EARTH'S DIVERSE ECOSYSTEMS

CHAPTER OUTLINE.

This chapter presents the basic patterns of ecosystem diversity seen in nature. Characteristics of life on land and in 10 typical terrestrial biomes are described. Characteristics of life in the oceans is then presented, followed by facts about three aquatic ecosystems. Finally, the characteristics of human ecosystems are enumerated and evaluated.

1. **Patterns within diversity.**

    A.   Communities are very diverse, yet have similar <u>patterns of diversity</u>.

    1)   Around the world, although the plant species may differ, very similar groups of plants are found wherever a particular climate exists.

    2)   Each type of community is dominated by organisms that are specifically adapted for a particular environment.

    B.   In regions of the earth with similar environmental conditions, we find similar types of organisms organized into similar types of communities.

    1)   The distribution of communities on earth is determined by the environment, differing in the abundance of <u>4</u> <u>basic</u> <u>resources</u>: nutrients, energy, water, and appropriate temperatures for metabolic reactions.

    2)   The availability of sunlight, water, and appropriate temperatures determine the <u>climate</u> of a given region.

2. **Climate.**

    A.   Both weather and climate affect life on land.

    1)   <u>Weather</u>: short term fluctuations in temperature, humidity, cloud cover, wind, and precipitation that vary over periods of hours or days.

2)   <u>Climate</u>: weather patterns that prevail from year to year in a particular region.
3)   Weather affects individuals, while climate influences and limits the overall distribution of entire species.

B.   <u>Solar radiation and climate</u>.
   1)   Solar energy drives the wind, ocean currents, and global water cycle.
   2)   Sunlight is modified by the atmosphere:
      a.   The stratospheric ozone layer absorbs much ultraviolet radiation that can damage biological molecules.
      b.   Dust, water vapor, and clouds scatter light, reflecting some energy back into space.
      c.   Gases in the air absorb infrared (heat) wavelengths, trapping warmth in the atmosphere (natural greenhouse effect).
      d.   Thus, only half of solar radiation reaching the atmosphere actually strikes the earth's surface.
         i.   Small fractions are reflected back into space or used for photosynthesis.
         ii.   The rest is absorbed as heat.
      e.   Solar energy absorbed as heat by the atmosphere and the earth's surface (2/3 of the total) maintains the earth's relative warmth and controls its climate.

C.   <u>Factors influencing climate</u>.
   1)   <u>Latitude</u>:
      a.   Near the equator, sunlight hits the earth's surface nearly at a right angle, causing constant warmth and little seasonal variation.
      b.   Further north or south, the surface is oblique to the sun's rays, resulting in lower temperatures.
      c.   Because the earth tilts on its axis as it revolves around the sun, higher latitudes experience pronounced seasonal changes in temperature.
   2)   <u>Air currents</u>:
      a.   In the tropics, heated air rises, laden with water evaporated by solar heat.
      b.   As the water-saturated air rises, it cools a bit:
         i.   Water, condensing from the cooler rising air, falls as rain.
         ii.   This creates a rainy band around the equator called the tropics, the warmest and wettest region on earth.
      c.   The cooler, but still relatively warm air then flows away from the equator, cooling further as it moves.
      d.   At around 30° north and south latitudes, the cooled air begins to sink:
         i.   As it sinks, it is warmed again by compression and heat from the earth.
         ii.   As it reaches the surface, it is warm and very dry.
         iii.   The deserts are found at these latitudes.
      e.   This air then flows back towards the equator.
      f.   Further north and south, this general circulation pattern is repeated, dropping moisture at about 60° north and south latitudes and creating extremely dry conditions at the poles.
   3)   <u>Ocean currents</u>.
      a.   Ocean currents are driven by:
         i.   Earth's rotation.

        ii.   The wind.

        iii.   Direct heating of water by the sun.

    b.   Continents interrupt the currents, creating circular patterns called gyres that rotate clockwise in the northern hemisphere and counterclockwise in the south.

    c.   Since water both heats and cools more slowly than land, ocean currents reduce temperature extremes on nearby land.

  4)   Continents and mountains.

    a.   Continents (heat and cool quickly) and oceans (heat and cool slowly) alter the flow of wind and water and contribute to the irregular distribution of ecosystems.

    b.   Variations in elevation add further complications:

        i.   The atmosphere at higher elevations is thinner and retains less heat.

        ii.   For each 100 meters of elevation, there is a decrease of $0.6°$ C.

    c.   Mountains modify rainfall patterns.

        i.   Water-laden air cools as it rises up a mountain. Thus, water condenses as rain or snow on the windward (near) side of the mountain.

        ii.   The cool dry air is warmed again as it descends the far side of the mountain, absorbing water from the land and creating a local dry area called a rain shadow.

3.   **The Requirements of Life.**

  A.   The four fundamental resources required for life:

    1)   Nutrients to construct living tissues.

    2)   Energy to construct tissues.

    3)   Liquid water for chemical reaction solvents.

    4)   Appropriate temperatures to carry out life processes.

  B.   Life on land.

    1)   Land organisms are restricted in their distribution largely by the availability of water and appropriate temperatures.

    2)   Water and favorable temperatures are limited and unevenly distributed in place and time.

    3)   Land creatures must be adapted to obtain water when available and conserve it when scarce.

  C.   Terrestrial biomes.

    1)   Biomes: Large land areas with similar environmental conditions and characteristic types of plants.

    2)   Are dominated and defined by their plant life.

    a.   Plants are relatively immobile, so they must be precisely adapted to the existing climate.

    b.   Similar types of plants are found wherever a particular climate exists (e.g, different types of cacti occur in different deserts).

    3)   Tropical rain forests: Large areas of Africa and South America near the equator.

    a.   Temperature between $25°$-$30°$ C with little variation, and rainfall of 250-

400 cm yearly (evenly warm and moist).
- b. Dominated by huge broad-leaf evergreen trees 30-50 m tall.
- c. Cover 6% of the earth's surface but contain about 5 million species (2/3 of the world's total).
- d. Typically stratified with several layers of vegetation.
- e. Much animal life is <u>arboreal</u> (living in trees) since there is little edible plant material close to the ground.
- f. Virtually all the nutrients in a rain forest are tied up in the vegetation. The soil is quite infertile and unsuitable for farming, especially since the heavy rainfalls will quickly wash away any remaining nutrients.
- g. <u>Human impacts</u>.
  - i. 50 million acres of rain forests are felled for lumber and agriculture each year (100 acres per minute).
  - ii. All unprotected rain forests could be felled within the next 30 years with a loss of half the world's species.
  - iii. Removal of these forests is contributing to CO_2 buildup in the atmosphere (greenhouse effect).
  - iv. In West Africa, forest burning is causing acid rain.
  - v. Currently, only 1 acre is replanted for every 10 destroyed.
  - vi. Ultimate solution: sustainable use--harvesting products without permanent damage to trees or the ecosystem.

4) <u>Tropical deciduous forests</u> in India and Southeast Asia.
- a. Slightly farther away from the equator than the rain forests.
- b. Rainfall is not as constant, creating pronounced wet and dry seasons.
- c. Plants shed their leaves during the dry season to decrease evaporative loss of water.

5) <u>Savannas</u> in Central Africa.
- a. Located along the edges of the tropical deciduous forests.
- b. Grasses are the predominant vegetation, with scattered trees and thorny scrub forests.
- c. Have a rainy season when all the year's rain falls (<30 cm). During the dry season, no rain falls for months.
- d. Only a few specialized tree species (thorny acacia and baobab) can survive the dry seasons.
- e. Contains the most diverse array of large mammals on earth.
  - i. Herbivores: antelope, buffalo, giraffe.
  - ii. Carnivores: lion, leopard, hyena, wild dogs.
- f. Humans in East Africa use savannas for grazing cattle, building fences that disrupt the migration patterns of native mammals. Also, poaching has brought several species to the brink of extinction.

6) <u>Deserts</u> (<25 cm water yearly).
- a. Not enough rainfall occurs to maintain even drought-resistant grasses.
- b. Found on every continent at about 20°-30° north and south latitudes and in the rain shadows of major mountain ranges.
- c. If present, plants are spread out very evenly, though intermittently. The plants are perennials with large, shallow root systems and waxy waterproof coatings on thick-bodied stems (that store water) and spiny leaves.
- d. Desert animals are highly adapted to survive on little water.
  - i. They are active only at night when the desert cools down.

      ii.   Smaller animals never drink, getting water from their food and from cellular respiration.

      iii.   Large animals are dependent on water holes.

  e.   <u>Human impacts</u>.

      i.   Human activities are causing the spread of deserts (<u>desertification</u>):

         a)   Overpopulation, exceeding the carrying capacity of the land.

         b)   Drilling deep wells.

         c)   Increasing livestock herds that destroy natural vegetation.

         d)   Cultivating fragile land without allowing the soil to recover.

      ii.   Massive famines are the tragic result.

7)   <u>Chaparral</u>: in Southern California and the Mediterranean.

  a.   In coastal regions bordering on deserts.

  b.   Annual rainfall is similar to deserts, but proximity to the sea increases the length of the rainy season in the winter and frequent fogs reduce evaporation.

  c.   Plants consist of small trees or large bushes with thick waxy or fuzzy evergreen leaves that conserve water.

8)   <u>Grasslands</u>: the <u>prairies</u> of temperate North America.

  a.   Found east of the rain shadow deserts, in areas where rainfall gradually increases; grasslands stretch nearly halfway across the continent.

  b.   Have a continuous cover of grass and virtually no trees except along rivers.

  c.   Summer droughts and fires prevent trees from outcompeting and replacing the grasses.

  d.   The more eastern grasslands contain the most fertile soil on earth due to the growing and decomposing of grasses for thousands of years. The former prairies have become the "breadbasket" of North America.

  e.   The more western shortgrass prairie is now grazed by cattle and sheep. Overgrazing is turning this area into a cool sagebrush desert.

9)   <u>Temperate deciduous forest</u>: begins at the eastern edge of the North American grasslands.

  a.   Higher precipitation than in the prairies (75-150 cm), especially in the summer. Thus, trees can grow, shading out the grasses.

  b.   Has cold winters with hard frosts, so trees shed their leaves to conserve available liquid water.

  c.   Insects and other arthropods are conspicuous. These feed on bacteria, earthworms, fungi, and small plants on the forest floor.

  d.   A variety of vertebrates (mice, shrews, squirrels, raccoons, birds) are found, and larger mammals were formally abundant but were driven out by humans.

  e.   <u>Human impacts</u>. In the U.S., the forests have been reduced due to clearing for lumber and agriculture.

10)   <u>Temperate rain forests</u>: along the Pacific coast of North America from Washington state to Alaska.

  a.   Rain is abundant year-round (400+ cm/year) with no hard freezes due to the warm Pacific Ocean.

        b.   Trees are evergreen conifers, with mosses and ferns as ground cover.

11)   <u>Taiga (Northern coniferous forests):</u>   north of the grasslands and temperate forests of North America, mainly in southern Canada.

        a.   Longer, colder winters and shorter growing seasons than to the south.

        b.   Trees (black spruce, birch) are evergreen conifers with small waxy needles that retard water loss. These trees grow slowly year-round.

        c.   Diversity of plant life is low due to the harsh climate.

        d.   Large mammals (grizzly bears, moose, wolf) and small mammals (wolverines, foxes, deer, snowshoe hare) roam the taiga.

        e.   <u>Human impacts</u>. Huge expanses have been clearcut for lumber, and some species have become endangered (e.g., the spotted owl).

12)   <u>Tundra</u>: nearest the polar ice caps, bordering the Arctic Ocean.

        a.   A vast treeless region.

        b.   Severe weather conditions (-40° C in winter, 50-100 km/hour winds, 25 cm precipitation/ year): a freezing desert.

        c.   Growing season is only a few weeks long during "summer".

        d.   <u>Permafrost</u> is present: a permanently frozen layer of soil about 0.5 m below the surface. So, during the summer thaw, melted snow and ice cannot soak into the ground, and the tundra becomes a huge marsh.

        e.   Vegetation includes small perennial flowers, several inch tall dwarf willows, and a large lichen called "reindeer moss".

        f.   Mosquitoes thrive during the marshy summers, and are eaten by birds that have migrated north for the summer.

        g.   Tundra vegetation supports caribou and lemmings. Wolves, snowy owls, arctic foxes, and grizzly bears eat the lemmings.

        h.   Tundra is the most fragile of all biomes due to its short growing season.

D.   <u>Interactions between temperature and rainfall.</u>

    1)   Temperature strongly influences the effectiveness of rainfall in providing soil moisture for plants, and standing water for animals to drink.

    2)   Places that receive similar rainfall can have strikingly different vegetation due to differences in temperature.

    3)   Examples of differences in biomes, all of which receive about 30 cm (12.5 inches) of rain annually.

        a.   Sonoran Desert (near Tucson, AZ) has average temperature of 20° C and vegetation includes saguaro cactus and low drought-resistant brush.

        b.   In eastern Montana, shortgrass prairie exists due to lower average temperature (7° C).

        c.   In central Alaska, a taiga forest exists due to a colder temperature (-4° C) creating permafrost that results in a "swamp forest" during the summer thaw.

E.   <u>Diagram</u> relating the major terrestrial biomes to latitude, altitude, and moisture.

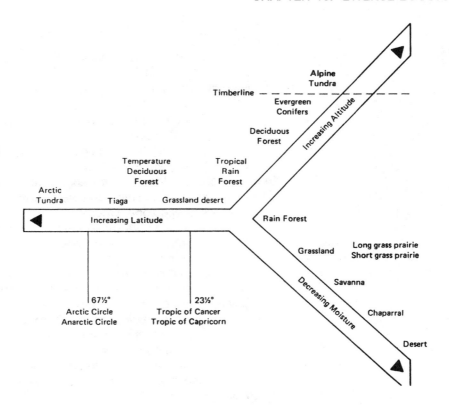

4.   **Life in water.**

    A.   <u>General facts</u>.
        1)   Saltwater aquatic ecosystems cover nearly 71% of the earth's surface as opposed to 1% for freshwater ecosystems.
        2)   <u>Common features</u> of aquatic ecosystems:
            a.   They are more moderate in temperature changes than land ecosystems.
            b.   Water absorbs much of the light energy that sustains life:
                i.   The intensity of light decreases rapidly with depth.
                ii.   In clear water, at depths of 200 meters or more, there is little light to power photosynthesis.
            c.   Nutrients tend to concentrate near the bottom. Separation of energy from nutrients limits aquatic life.
        3)   The major factors that determine the quantity and type of life in aquatic ecosystems are <u>energy</u> and <u>nutrients</u>.
        4)   Freshwater ecosystems include rivers, streams, ponds, lakes, and marshes.
        5)   Marine (saltwater) ecosystems include estuaries, tidepools, open ocean, and coral reefs.

    B.   <u>Freshwater lakes</u>.
        1)   Moderate to large lakes in temperate climates have distinct life zones and temperature stratification.
        2)   <u>Life zones</u>.
            a.   Distribution of life depends on access to light, nutrients, and a substrate

for attachment to the bottom.

b. Littoral zone (near the shore):
  i. Water is shallow and plants find abundant light, anchorage, and adequate nutrients.
  ii. Littoral zone communities are the most diverse, having:
     a) anchored plants.
     b) microscopic photosynthetic phytoplankton ("drifting plants") protists.
     c) zooplankton ("drifting animals") protists.
     d) the greatest diversity of animals in the lake.

c. Further from shore, plants cannot anchor to the bottom and still collect light for photosynthesis. This open water area is divided into:
  i. Upper limnetic zone, where:
     a) enough light penetrates for photosynthesis.
     b) primary producers are phytoplankton and blue-green bacteria.
     c) producers are eaten by protozoans and small crustacea that are themselves eaten by fish.
  ii. Lower profundal zone, where:
     a) light is insufficient for photosynthesis.
     b) nourishment is from detritis supplied from above and by incoming sediment.
     c) inhabitants include decomposer organisms and detritis feeders like snails and some insect larvae and fish.

3) Temperature stratification.
  a. In summer, lakes in temperate regions become persistently stratified, with a layer of warmer, less-dense water "floating" atop a later of denser cold (4° C) water.
     i. The upper region with photosynthetic organisms becomes depleted of nutrients as summer progresses.
     ii. Heterotrophic organisms near the bottom use up oxygen.
     iii. By late summer, limnetic populations are limited by lack of nutrients and profundal populations are limited by lack of oxygen.
  b. In fall, surface temperature drops and the stratification breaks down, allowing mixing of the nutrient-rich lower level and the oxygen-rich upper level.
     i. Similar mixing occurs in the spring as the surface ice melts.
     ii. Such fall and spring turnovers help lakes support their natural communities, and cause spurts of growth among phytoplankton and zooplankton.

4) Types of freshwater lakes:
  a. Oligotrophic ("poorly fed") lakes: very poor in nutrients.
     i. Often formed by glaciers in bare rock.
     ii. Fed by mountain streams carrying little sediment.
     iii. Are clear, allowing deep penetration by light so that the limnetic zone may extend all the way to the bottom.
     iv. Ideal for oxygen-loving fish (trout).
  b. Eutrophic ("well fed") lakes: receive larger inputs of sediments, organic materials, and inorganic nutrients.
     i. Are murkier from sediment and dense phytoplankton populations.

424

        ii.   The lighted limnetic zone is shallow, but dense "blooms" of algae occur seasonally.

        iii.   Dead algal bodies fall into the profundal zone which becomes oxygen-depleted as decomposers metabolize the detritis.

    c.   Lakes are transient ecosystems.

        i.   They gradually fill with sediment, undergoing succession to dry land.

        ii.   Oligitrophic lakes become eutrophic, a process called eutrophication.

5)   Human impacts.

    a.   Human activities may greatly accelerate the process of eutrophication:

        i.   Nutrients are carried into lakes from farms, feedlots, and sewage.

        ii.   Water from sewage treatment plants is often rich in phosphates and nitrates.

    b.   These added nutrients support excessive growth of phytoplankton that form a "scum" on the lake surface, depriving submerged plants of sunlight.

    c.   Plant bodies are decomposed by bacteria, depleting dissolved oxygen so that fish and invertebrates then die and are broken down by bacteria, further depleting oxygen.

    d.   Although full of life, the lake appears dead:

        i.   Most trophic levels are gone.

        ii.   The community is dominated by bacteria and microscopic algae.

C.   Marine ecosystems.

    1)   Zones.

        a.   Photic zone: the upper layer of ocean water where light is strong enough to support photosynthesis.

        b.   Benthic zone: lower zone where the only energy comes from excrement and bodies of organisms that sink or swim down.

    2)   Nutrients.

        a.   Most ocean nutrients are at or near the bottom where no photosynthesis can occur. When photic zone individuals die, their nutrient-rich bodies sink into the benthic zone.

        b.   Nutrient sources for photic zone:

            i.   The land, from which rivers remove and carry nutrients to the oceans.

            ii.   Upwelling: cold, nutrient-laden water from ocean depths is brought to the surface usually along coastlines by winds displacing surface water.

    3)   Major concentrations of life occur where abundant light is combined with nutrient sources:

        a.   Regions of upwellings.

        b.   Shallow coastal waters.

        c.   Coral Reefs.

    4)   Coastal waters.

        a.   The most abundant ocean life is found here, including the only area where large plants can grow anchored to the ocean bottom, since:

            i.   The water is shallow.

            ii.   A steady flow of nutrients is washed off the land.

      b.   Coastal waters include:
           i.   <u>Intertidal zone</u>: area alternately covered and uncovered by tidal water.
           ii.   <u>Near shore zone</u>: bays, salt marshes, estuaries, shallow sub-tidal areas.
      c.   Associated with large anchored plants are protists and animals from nearly every phylum.
      d.   Near shore areas are the breeding grounds for crabs, shrimp, and many types of fish.
      e.   <u>Human impacts</u>:
           i.   Wetlands once covered 6% of the earth's surface, but half have been eliminated by human activities (agriculture, industry, urban sprawl)
           ii.   Conflict between preservation and development of wetlands is becoming increasingly intense.

5)   <u>Coral reefs.</u>
      a.   Sheltered, shallow water environments created by animals and plants in warm (22-28° C) tropical waters with the proper bottom depth (<40 meters), wave action, and nutrients.
      b.   A mutualistic association exists between reef-building corals and dinoflagellate unicellular algae:
           i.   They build reefs from their skeletons of calcium carbonate.
           ii.   The algae benefit from the high CO_2 concentration in the coral tissues.
           iii.   The algae assist the coral by removing the CO_2 and aiding in the deposition of the skeleton.
      c.   These reefs serve as:
           i.   An anchoring place for other algae and bottom-dwelling animals.
           ii.   A place of shelter and food for the most diverse array of invertebrates and fish species in the ocean.
      d.   <u>Human impacts</u>. Coral reefs are extremely sensitive to:
           i.   Silt eroding from nearby land, since:
                 a)   Light is diminished.
                 b)   Photosynthesis is reduced.
                 c)   The reef will be buried in mud.
           ii.   Overfishing:
                 a)   Harvesting fish faster than they can replace themselves.
                 b)   Removal of fish predators disrupts ecological balance of the community, causing population explosion of coral-eating sea urchins.
           iii.   Both protection and sustainable use are crucial for survival of coral reefs.

6)   <u>The open ocean.</u>
      a.   Most life is limited to the photic zone where life forms are <u>pelagic</u> (free-swimming or floating for their entire lives).
      b.   Food base is <u>plankton</u>, a collection of pelagic:
           i.   <u>phytoplankton</u>: mainly diatoms and dinoflagellates, the ultimate food source in the open ocean.
           ii.   <u>zooplankton</u>: mostly minute crustaceans that feed on phytoplankton or each other.

      c.   Flotation devices for open ocean organisms:
- i.   Oil droplets within cells.
- ii.   Swim bladders in fish regulate buoyancy.
- iii.   Active swimming.
- iv.   Some crustaceans migrate to the surface at night to feed, and sink below during the day to avoid predation.

      d.   Below the photic zone, nutrients come mainly from excrement and dead bodies raining down from above.

7)   Vent communities.

      a.   Discovered in 1977 at the Galapagos Rift (area where crustal plates on the ocean floor were separating).

      b.   Surrounding vents that spew out superheated water, black with sulfur and minerals, are rich communities of fish, crabs, mussels, sea anemones, and tube worms.

      c.   Sulfur bacteria are the primary producers, harvesting energy from hydrogen sulfide using a process called chemosynthesis.
- i.   The bacteria proliferate in the warm water surrounding the vents.
- ii.   These bacterial bodies provide the food base for the vent community.

5.   **Humans and ecosystems**. Characteristics of ecosystems dominated by humans:

A.   Human ecosystems are simple, having few species and fewer community interactions than undisturbed ecosystems. Hence, they are fragile.

B.   Human ecosystems have become dependent on non-renewable energy from fossil fuels, and non-renewable energy from sunlight. Hence, they are energy-intensive and not permanent, and are changing the atmosphere.

C.   Human ecosystems tend to lose nutrients which natural ecosystems tend to recycle. Hence, they are not self-sustaining and renewable.

D.   Human ecosystems tend to pollute water and lose it rapidly, while natural ecosystems tend to store water and purify it through biological processes. Hence, they pollute the environment and waste water.

E.   Human ecosystems such as farms tend to be extremely unstable. Hence, they are vulnerable to destruction by unfavorable biological and environmental conditions.

F.   Human ecosystems have constantly growing populations, while natural populations are relatively stable in size. Hence, humans must expand their ecosystems, destroying natural ones and their species in the process.

****

REVIEW QUESTIONS:

1. Name the four basic resources whose relative characteristics and quantities determine the distribution of communities on earth.
2. What two resources control the distribution of life on land?
3. Distinguish between weather and climate.
4. How is sunlight modified by the atmosphere?
5. How does latitude influence climate?
6. Describe worldwide air current patterns and explain how they influence climate.
7. How do continents and mountains influence the distribution of ecosystems?
8. What is a "biome"?

9.-27.   **MATCHING TEST**: terrestrial biomes.

               Choices: A.   Chaparral
                                       B.   Savanna

- A. Chaparral
- B. Savanna
- C. Temperate deciduous forest
- D. Tropical rain forest
- E. Taiga (northern coniferous forest)
- F. Tundra
- G. Tropical deciduous forest
- H. Desert
- I. Temperate rain forest
- J. Grassland (prairie)

9. In southern Canada, with long, cold winters, evergreen trees, and large mammals.
10. In Central Africa, with the most diverse array of large mammals on earth.
11. In large areas of Africa and South America near the equator.
12. Along the Pacific coast of northern North America, with abundant rain but no hard winter freezes.
13. Treeless with severe winter conditions.
14. Dominated by tall, broad-leaf evergreen trees.
15. Found on every continent between 20°-30° north and south latitudes, with little water.
16. Most animal life is arboreal.
17. Nearest polar ice caps.
18. In India and southeast Asia; warm with pronounced wet and dry seasons.
19. Contains the most fertile soil on earth.
20. Most nutrients are tied up in the vegetation.
21. Have trees, hard winter frosts, and abundant insect and vertebrate life.
22. A "freezing desert" with permafrost.
23. In coastal regions bordering on deserts, with small evergreen trees or bushes having waxy leaves.
24. Has reindeer moss, dwarf willows, caribou, and arctic foxes.
25. 2/3 of the world's species live there.
26. Found east of rain-shadow deserts, with no trees except along rivers.
27. The most fragile of all biomes.

----

28.   What two resources control the distribution of life in the oceans?
29.   Describe the littoral zone of a lake.
30.   Explain the differences between the limnetic and profundal zones of a lake.
31.   Compare and contrast oligotrophic and eutrophic lakes.
32.   Name two ways that new nutrients enter the oceans.
33.   Where is the most abundant ocean life and why is it there?
34.   Where in the oceans are the most diverse assemblages of invertebrates and fish found and why?
35.   What types of organisms are the basis for all life in the open oceans?
36.   What are ocean "rifts" and why are the areas around rifts rich in sea animals?
37.   List seven characteristics of human ecosystems. In what ways do these characteristics differ from natural ecosystems?

****

ANSWERS TO REVIEW QUESTIONS:

1.   see 3.A.1)-4)

2.   see 3.B.1)-3)

3.   see 2.A.1)-3)

4.   see 2.B.2)a.-e.

5.   see 2.C.1)a.-c.

6.   see 2.C.2)a.-f.

7.   see 2.C.4)a.-c.

8.   see 3.C.1)-2)

9.-27.   [see 3.C.3)-12)

9.	E		19.	J
10.	B		20.	D
11.	D		21.	C
12.	I		22.	F
13.	F		23.	A
14.	D		24.	F
15.	H		25.	D
16.	D		26.	J
17.	F		27.	F
18.	G			

28.   see 4.A.3)

29.   see 4.B.2)b.

30.   see 4.B.2)c.i.-ii.

31   see 4.B.4)a.-c.

32.   see 4.C.2)b.

33.   see 4.C.4)a.

34.   see 4.C.5)a.-c.

35.   see 4.C.6)b.

36.   see 4 C.7)a.-c.

37.   see 5.A.-F.

****

UNIT VI EXAM:

Chapters 43 - 46. All questions are dichotomous. Circle the correct choice in each case.

Chapter 43.

1. Physical aspects of the environment, such as weather, are part of the (abiotic / biotic) environment.
2. Random distribution is more characteristic of (plants / animals).
3. The more common type of distribution found in nature is (aggregated / random) distribution.
4. Animals are more likely to be (aggregated / randomly distributed) during the mating season.
5. Movement out of an area is known as (emigration / immigration).
6. If a population is growing, then (deaths / births) are more common.
7. Environmental resistance will promote (exponential / S-shaped) growth curves.
8. Biotic factors in the environment are usually density (dependent / independent).
9. Weather conditions are density (dependent / independent).
10. Birth rates are higher in the (more / less) developed countries.

Chapter 44.

1. The things that an organism needs to survive define its (habitat / ecological niche).
2. (Predation / symbiosis) occurs when one organism kills and eats another.
3. Coevolution causes the niches of different species to overlap (more / less) greatly.
4. (Predators / parasites) are smaller than their prey and do not kill immediately.
5. An insect that stores toxic chemicals produced by its host plant is an example of (coevolution / mimicry).
6. An insect larva that resembles a bird dropping is an example of (warning coloration / camouflage).
7. The yellow and black stripes of yellowjacket bees is an example of (startle / warning) coloration.
8. Birds building nests in trees is an example of (commensalism / mutualism).
9. The end result of succession is a (pioneer / climax) community.
10. Reforestation of an abandoned farm is an example of (primary / secondary) succession.

Chapter 45.

1. The transfer of energy from one trophic level to another is (efficient / inefficient).
2. Food (chains / webs) are the best representation of what happens to energy in nature.
3. Detritus feeders are important in (energy / nutrient) cycling.
4. (Energy / Nutrients) can be recycled.
5. Eutrophication favors (aerobic / anaerobic) forms.
6. Heavy metals are more soluble in (acid / alkaline) conditions.
7. An increase in carbon dioxide in the air (increases / decreases) temperatures on Earth.
8. Because of pollution by humans, the level of ultraviolet radiation reaching the earth's surface will (fall / rise).
9. The (chlorine / fluorine) in the fluorocarbons destroys ozone.
10. Succession in an abandoned field is usually highly (predictable / unpredictable).
11. Succession following a fire is an example of (primary / secondary) succession.
12. Diversity (decreases / increases) during succession.
13. Productivity (decreases / increases) during succession.
14. Succession (increases / decreases) stability in a community.
15. (Primary / Secondary) succession occurs faster.

Chapter 46.

1. Lakes with high algae populations are usually (clear / cloudy).
2. Temperature extremes are greater (on land / in the sea).
3. The (benthic / photic) zone is near the sea surface.
4. Nutrient concentrations in the ocean are greater in (deep / shallow) water.
5. The clear blue water in some tropical seas is a reflection of (excessive / lack of) nutrients in the water.
6. Nutrients are (less / more) likely to be limiting on land than in aquatic environments.
7. Tropical rain forests are characterized by (maximum / minimum) supplies of nutrients in the soil.
8. (Grasslands / Savannas) are more likely to have scattered trees.
9. Deserts are found on the (eastern / western) side of mountains.
10. In the prairies, fire favors (grasses / trees).
11. The western part of the prairie has (long / short) grass.
12. (Fire / Overgrazing) changes a prairie into a sagebrush desert.
13. The top of a mountain above the tree line is characterized by vegetation known as (taiga / tundra).
14. Evergreen trees are characteristic of (taiga / tundra).
15. Human ecosystems have (simple / complex) community structures and tend to (recycle / lose) nutrients.

****

ANSWER KEY:

Chapter 43:

1. abiotic
2. plants
3. aggregated
4. aggregated
5. emigration
6. births
7. S-shaped
8. dependent
9. independent
10. less

Chapter 44:

1. ecological niche
2. predation
3. less
4. parasites
5. coevolution

6. camouflage
7. warning
8. commensalism
9. climax
10. secondary

Chapter 45:

1. inefficient
2. webs
3. nutrient
4. nutrients
5. anaerobic
6. acid
7. increases
8. rise
9. chlorine
10. predictable
11. secondary
12. increases

13. increases
14. increases
15. secondary

Chapter 46:

1. cloudy
2. on land
3. photic
4. shallow
5. lack of
6. less
7. minimum
8. savannas
9. eastern
10. grasses
11. short
12. overgrazing
13. tundra
14. taiga
15. simple, lose

****